T0190071

G. Drettakis
N. Max (eds.)

Rendering Techniques '98

Proceedings of the Eurographics Workshop
in Vienna, Austria,
June 29–July 1, 1998

Eurographics

SpringerWienNewYork

Dr. George Drettakis
INRIA, Grenoble, France

Prof. Dr. Nelson Max
Lawrence Livermore National Laboratory,
University of California, U.S.A

© 1998 Springer-Verlag/Wien

Typesetting: Camera ready by authors

Graphic design: Ecke Bonk

Printed on acid-free and chlorine-free bleached paper

SPIN: 10684767

With 231 partly coloured Figures

ISSN 0946-2767
ISBN 3-211-83213-0 Springer-Verlag Wien New York

Preface

This book contains the proceedings of the 9th Eurographics Rendering Workshop, which took place on June 29 and 30, and July 1, 1998, at the Vienna University of Technology, in Vienna, Austria. This annual workshop has become the main forum for new research results in rendering (second only to the Siggraph conference, which has space for only a few papers on this topic). Although it is held in Europe, participants come from all over the world.

This year we received a total of 80 submissions, which were reviewed by the 28 program committee members and the external reviewers, both listed following the contents pages. Several members agreed to join the program committee at the last minute, to handle the record number of submissions. Inevitably, many good papers had to be rejected due to lack of space in this volume and at the workshop sessions. This volume contains the 27 papers selected, plus two invited papers, from Tomoyuki Nishita (Fukuyama University) and Anthony Apodaca (Pixar). Almost all are accompanied by color illustrations, in the Appendix.

As in past years, many of the papers concern local and global illumination. These include papers on representing and approximating reflection functions, representing light sources, computing visibility, soft shadows, radiance for non-diffuse reflection, stochastic methods, and light scattering in natural scenes including sky, water, clouds, fog, smoke, and snow.

A new area that is rapidly becoming important is image-based rendering, and papers on this subject include model acquisition and display, ray tracing in layered depth images, parametrization, compression and decompression of light fields, and clever uses of texturing and compositing hardware to achieve depth-image warping and 3D surface textures.

There are also two interesting papers on how to use knowledge about human perception to concentrate the work of accurate rendering only where it will be noticed, and conversely, one on distorting realism in story-telling animation to draw the viewers attention to features important to the story.

While most papers present new algorithms and techniques, several emphasize the integration of such techniques into usable systems. This book should be a source of ideas for advancing the efficient generation of realistic images, on both the research and application fronts.

We wish to thank the organising chariman Robert Tobler and his colleagues at the Vienna University of Technology for their help in the production of this book, and in the local arrangements of the workshop, and also the staff at Springer Verlag for their help in producing this book on a rapid schedule. Finally we wish to thank the many authors of the submissions, and the many reviewers who read them and suggested improvements.

George Drettakis
Nelson Max

June, 1998

Contents

International Programme Committee

External Reviewers

Author Index

Light Scattering Models for The Realistic Rendering of Natural Scenes

Tomoyuki Nishita

Department of Electronic and Electrical Engineering
Fukuyama University
Fukuyama, 729-0292 Japan
nis@eml.hiroshima-u.ac.jp

Abstract: Image synthesis of realistic 3-dimensional models is one of the most widely researched fields these days. Display of natural scenes such as mountains, trees, the earth, and the sea, have been attempted. This paper discusses shading models for volumetric objects in natural scenes.

A precise shading model is required to display realistic images. This paper describes rendering methods for physically-based images taking into account light scattering due to particles in the atmosphere and water, which include sky color, sky light, clouds, fog effects, smoke, snow, the earth viewed from space, water color, and optical effects within water such as caustics and shafts of light. For these effects, a single scattering and multiple scattering models are discussed.

Keywords : Light Scattering, Shading model, Natural scenes, Atmospheric scattering, multiple scattering, Photo-realism

1 INTRODUCTION

The recent progress and spread of the application of computer graphics is remarkable, especially regarding techniques for realistic image synthesis concerning shapes, illumination, natural phenomena, etc. Research on the rendering of natural scenes, such as clouds, ocean waves, trees, terrain, grass, and fire, has become increasingly wide spread. This paper reports shading models for natural phenomena, such as the scattering effect due to particles.

In general the following functions for shading models are required: (a) precise calculation of illumination (including shadows), (b) light-source geometry (including luminous intensity distribution, types of light sources such as linear, area sources), (c) reflection/refraction from diffuse and/or non-diffuse surfaces, (d) the rendering of volumetric objects taking into account light scattering/absorption (e.g., clouds, sky, water). In previous work, (a), (b) and (c) have been researched. (d) has became one of the main current research topics. Most of the objects in natural scenes are considered as solid/liquid objects. For precise shading models, some of them should be considered as volumetric objects which include particles such as air molecules.

When designing a building, for instance, it is very important to create photorealistic images as accurately as possible before construction. In this case, additional components for the background image such as sky color, clouds, fog effects, snow, and water color are also indispensable to increase reality. These natural phenomena are caused by light scattering due to particles in the atmosphere or water.

The main topics in this paper are atmospheric scattering and water color. These are applicable to displays of buildings under various weather conditions and to flight/diving simulators.

2 Light Scattering for Realistic Rendering

Light rays are scattered and absorbed by particles while passing through the atmosphere/water before arriving at the eye of the viewer; as a result, for example, distant scenery appears dim in hazy weather, and the path of a spotlight can be clearly seen in moist air. The quality of synthesized images largely depends on the degree to which the effects of the phenomena mentioned above are taken into account.

For rendering buildings, natural light (i.e., sky light) should be taken into account. And for rendering natural scenes the displaying of the atmosphere (sky color), water color, and objects composed of particles, such as clouds, are important factors. Shading models taking into account scattering/absorption due to particles are categorized as follows;

a) light scattering from particles in the air
the shafts of light caused by spot lights [16][7], light beams passing through gaps in clouds or through leaves [12], the sky color taking account atmospheric scattering [11][10][21], scattered light due to nonuniform density particles such as clouds and smoke [16][24][8][19], the cloud color due to multiple anisotropic volume scattering [13][8][1][18][22], the color of the atmosphere viewed from space [17], and the effect of the radiosity of a participating medium [23]
b) light scattering from particles in water
the color of water affected by particles in the water, such as ponds [9], the color of the sea as viewed from outer space [17], and optical effects such as caustics and shafts of light within the water [25][18]
c) light scattering from particles in objects
Saturn's rings (reflective ice particles) [2], and subsurface scattering such as skin [6] and snow [22]

For calculation of optical effects due to particles, the following components should be considered; i) density distribution and sizes of the particles, ii) phase functions which are dependent on particle sizes, iii) optical paths, and iv) multiple scattering between particles.

For high albedo particles such as in clouds, multiple scattering should be calculated. On this, Kajiya [8] was first to offer a solution.

3 Atmospheric Scattering

There are many optical phenomena due to light scattering/absorption due to particles in the atmosphere as described above; sky color, clouds, smoke, fog, haze, snow, shafts of light (light beams illuminated by light in a dusty room/a foggy night). The characteristics of the scattering depend on the size of the particles. Scattering by small particles such as air molecules is called *Rayleigh scattering*, and scattering by aerosols such as dust is known as *Mie scattering*.

In this section, calculation methods for the spectrum of the sky, clouds, and the atmosphere viewed from outer space are discussed (see references [14], [16] for fog effects, smoke, and shafts of light).

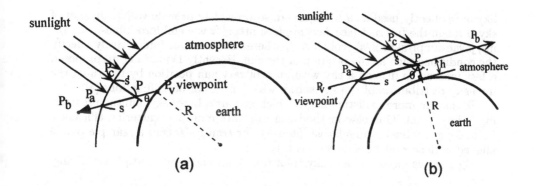

Fig. 1. Optical paths for calculation of sky color and atmosphere.

3.1 Sky Color (Atmosphere viewed from the earth)

Klassen [11] approximated the atmosphere as multiple layers of plane-parallel atmosphere with uniform density; however, this method results in a large error near the horizon. We proposed a spherical-shell atmosphere with continuous variations in density in order to improve accuracy [10].

Let's consider scattering due to air molecules. The light reflected due to Rayleigh scattering, I_p, is given by the following equation;

$$I_p(\lambda) = I_s(\lambda)KF_r(\theta)\rho\frac{1}{\lambda^4}exp(-\tau(PP_c, \lambda)), \tag{1}$$

where I_s is the solar radiation at the top of the atmosphere, λ the wavelength of incident light, K is a constant for the standard atmosphere (molecular density at sea level); $K = 2\pi^2(n^2-1)^2/(3N_s)$, F_r the scattering phase function indicating the directional characteristic of the scattering (given by $3/4(1+cos^2(\theta))$ for Rayleigh scattering), θ the scattering angle (see Fig. 1(a)), n the index of refraction of the air, N_s the molecular number density of the standard atmosphere, and ρ the density ratio. ρ depends on the altitude h ($\rho = 1$ at sea level) and is given by $\rho = exp(\frac{-h}{H_0})$, where H_0 is a scale height. And $\tau(PP_c, \lambda)$ the optical depth from the top of the atmosphere to point P and given by $\tau(PP_c, \lambda) = \int_P^{P_c} c(l)\rho(l)dl$ (l is the integration variable, c is the attenuation coefficient).

As the light scattering from P is also attenuated before reaching viewpoint P_v (see Fig.1(a)), the intensity of the light reaching P_v, I_{pv}, can be obtained by multiplying the attenuation by the intensity at P, that is $I_{pv}(\lambda) = I_p(\lambda)exp(-\tau(PP_v, \lambda))$. So, I_v reaching P_v can be obtained by integrating the scattered light due to air molecules on P_vP_a:

$$I_v(\lambda) = \int_{Pv}^{Pa} I_{pv}(\lambda)ds$$

$$= I_s(\lambda)\frac{KF_r(\theta)}{\lambda^4}\int_{P_v}^{P_a} \rho \, exp(-\tau(PP_c,\lambda) - \tau(PP_a,\lambda))ds. \qquad (2)$$

Sky is frequently used as a background, so we widely provide various images of sky through the Internet (see Java applets: http://www.eml.hiroshima-u.ac.jp/~nis/javaexampl/skycol/skycol2.html), where the user can interactively specify the conditions such as the position of the sun. Recently Dobashi et al. proposed a fast method to display a sky with an arbitrary sun position by recording the intensity distribution of the sky shown with a series of basis functions [4].

To display more realistic images which give us a bluer sky, multiple scattering is important. Our new method can calculate multiple scattering efficiently by making a table of cumulative intensity (or transmittance) at sample points aligned with several light directions [21].

Fig.2 shows an example of sky color taking into account multiple scattering.

3.2 Skylight

Skylight is the light generated by scattering as a result of small particles (air molecules and aerosols) found in the atmosphere.

First, I developed a lighting model for skylight, and solved it as that the geometry of skylight is usually modeled as a huge hemisphere [15]. Kaneda [10] improved on it; in his method the distribution of the sky color is obtained by the method described in section 3.1. This method has can display buildings illuminated by skylight taking into account the sun position. Dobashi [3] applied his model to interior lighting design and calculated accurate intensity due to skylight passing through windowpanes.

3.3 Atmosphere viewed from Outer Space

The color of the earth when viewed from space varies according to the relationship between the view direction and the position of the sun. In the famous words of the astronaut, "the earth was blue". When we observe the earth from relatively close to the atmosphere, the part of the atmosphere surrounding the earth appears as blue, and the part near the boundary of the shadow due to the sun appears red (i.e., sunset). These phenomena are optical effects caused by particles in the atmosphere. They (i.e., the spectrum of the earth, the spectrum of the atmosphere) are concerned with the calculation of optical length and sky light (see Fig. 1(b)).

For these calculations, numerical integrations taking into account atmospheric scattering are required. As shown in Fig.1(b), the light reaching P_v can be obtained as the remainder after scattering and absorption due to air molecules along the intersection line between the ray and the atmosphere, P_bP_a. The intensity of the light scattered at point P in the direction of P_v, I_p, is obtained by Eq.(1). Intensity I_v reaching the viewpoint can be obtained by the modified Eq.(2); the interval$[P_vP_a]$ is exchanged to $[P_aP_b]$ in Eq.(2).

Fig. 3 (color plate) shows an example of the color of the atmosphere viewed from space [17]. In this figure, the color of the sea, direct sunlight, sky light, and atmospheric scattering/absorption are taken into account. This example depicts the beautiful variations in color of the earth and the atmosphere.

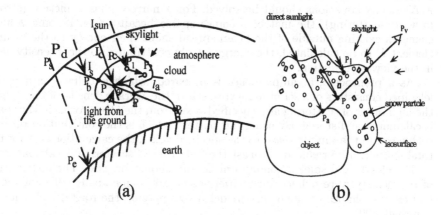

Fig. 4. Intensity calculation for clouds and snow.

3.4 Clouds and Smoke

The color of clouds varies according to the relationship between the viewing direction and the position of the sun. The albedos of clouds are very high, so multiple scattering can not be ignored. Clouds are illuminated by both direct sunlight and sky light affected by atmospheric scattering. Their reflected light from the ground (or the sea) also can not be ignored. Their color is determined by the light path, which is on the light which passes through the atmosphere of scattered light due to cloud particles, again passing through the atmosphere, and reaches the viewpoint (see Fig.4(a)).

The size of particles in clouds/smokes is larger than that of air molecules or of aerosols. The light reflected from clouds depends on the phase function called *Raleigh scattering*. The particles they consist of have strong forward scattering characteristics. This was considered as one of the difficulties due to the intensity calculation in the previous work. We use this characteristic in a positive way.

Let's denote the intensity at point P in direction ω as $I(P, \omega)$, the extinction coefficient per unit length as c, the length of cloud in viewing ray S, the path length from P as s ($s = 0$ at P, P_S ; S is the length of PP_s). Then $I(P, \omega)$ is expressed by

$$I(P, \omega) = I(P_S, \omega)exp(-\tau(PP_S))+$$
$$\int_{s=0}^{S}[c\,\rho(s)exp(-\tau(P(s)P))\frac{1}{4\pi}\int_{4\pi}F(\theta)I(P(s), \omega')d\omega']ds, \qquad (3)$$

where θ is phase angle between ω and ω', and ω' is angular variable for integration over the incoming directions at $P(s)$. As various sizes of particles exist, the phase function $F(\theta)$ is expressed by the linear combination of several phase functions which have strong forward scattering characteristics.

The problem is that I exists in both sides of the equation. To solve this, this space is subdivided into a number of volume elements. If we denote the number of voxels as N, and the number of the discrete directions as M, then MN matrix equations should be solved. For a narrow phase function (strong forward scattering), the matrix becomes sparse because even if some volume elements existing outside of the beam spread shoot their energy to the volume element P to be calculated, the energy hardly contributes to the intensity at P in the viewing direction.

As a preprocess, a sample space is prepared, which is defined as a parallelepiped consisting of a set of some voxels with the average density of the clouds, the high order of scattering at a specified voxel from the other voxels in the space is calculated and stored. By using this pattern which is the contribution ratio at each voxel in the sample space to the specified voxel, the calculation cost for the total space can be reduced. At least the 3rd order of scattering is calculated.

For clouds and smoke, nonuniform density should be used. The distribution of the density were defined by Fourier series [16] or metaballs [19]. For metaball clouds, shapes of clouds are modeled by applying the fractal technique to metaballs [20].

Fig.5 shows an example of clouds taking into account multiple scattering.

4 Subsurface Scattering (snow)

Snow should be treated as particles with a density distribution since it consists of water particles, ice particles, and air molecules. In order to express the material property of snow, the phase functions of the particles must be taken into account, and it is well-known that the fact that snow is white is because of the multiple scattering of light.

We proposed a calculation method for light scattering due to snow particles taking into account both multiple scattering and sky light [22]. The surface of the snow is defined by the isosurfaces of potential fields which are in turn defined by a set of metaballs.

The calculation method is similar to that for clouds [20] described in 3.4, that is, snow is also illuminated by direct sunlight and sky light. The differences are as follows. The representative value of ice grains in snow is 1mm; they are much larger than cloud droplets. That is, the phase function is more narrow. Because deep snow is optically thicker than clouds, the snow is brighter. For rendering snow grains, small primitives are used. We call each of these a sub-ball because the shape of the snow is modeled using metaballs. The number of these sub-balls is very large, so they are defined by their solid texture. Fig.4(b) shows optical paths for the calculation of snow.

Fig.6 shows an example of a snow scene.

5 Light Scattering within Water

In this section, the respective calculation methods for the color water viewed from above and from below the water surface are discussed, as are the calculation methods for caustics and shaft of light due to refracted light from the water surface.

5.1 Color of the water surface

Let's consider the light reaching a viewpoint from the surface of the sea. There are three paths (see Fig. 7(a)): (1) reflected light on the water surface, (2) scattered light due to particles within the water leaving the surface, (3) attenuated light passing through the sea after reaching the bottom.

When the light is incident to the water surface, the light path is divided into reflection and refraction. The relation between these two on the water surface obeys Fresnel's law of reflection. The relation between the incident angle and reflection angle obeys Snell's law. The refracted light is scattered/absorbed by water molecules in the water, and reaches the viewpoint after refracting again at the water surface. The light intensity transmitted in the water, I_v, is given by (see Fig. 7(a))

$$I_v(\theta_{ii}, \theta_{io}, z, \lambda) = \frac{I_i(\lambda)T_i(\theta_{ii}, \theta_{i0})T_o(\theta_{ji}, \theta_{jo})\beta(\delta, \lambda)}{n^2(\cos\theta_{io} + \cos\theta_{ji})c(\lambda)[1 - \omega_0(\lambda)F(\lambda)]}$$
$$\times(1 - exp(-zc(\lambda)[1 - \omega_0(\lambda)F(\lambda)](\sec\theta_{ji} + \sec\theta_{io})), \tag{4}$$

where λ is the wave length, z the depth of the sea, θ_{ii} the angle between the surface normal at point P and the viewing direction (see Fig.7(a) for θ_{io}, θ_{jo}), $I_i(\lambda)$ the irradiance of sunlight just above the water surface, n the refractive index of the water, T_i and T_o the transmittance of the incident light at point S and P, respectively, c the attenuation coefficient of light which expresses the ratio of lost energy of light when the light travels a unit length, β a volume scattering phase function, ω_0 the albedo of water, and F the fraction of the scattering coefficient in a forward direction. Eq. (4) shows that the color of the water depends on the depth, the incident angles and viewing direction. For displaying ponds [9] and the earth [18], the above method is used.

5.2 Optical Effects within Water

In the discussion of optical effects under water, the following three factors must be carefully taken into account; shafts of light and caustics due to refracted light from waves (refracted light from water surface both converges and diverges) and the color of water due to scattering/absorption of particles. It is difficult to calculate them by conventional ray-tracing because light refracts when passing through waves.

Let's consider light arriving from an object or a wave surface; the light is influenced by scattering/absorption of particles along its ray. The scattered intensity from particles at point P on the ray is given by (see Fig. 7(b))

$$I_p(\lambda) = (I_iT(\theta_i, \theta_t)F_p\beta(\phi, \lambda)exp(-c(\lambda)l_s) + I_a)\rho, \tag{5}$$

where λ is the wavelength, I_i the intensity of incident light onto the water surface, T the transmittance of the incident light, θ_i and θ_t the incident angle and the refracted angle, respectively, ρ the density, I_a ambient light, and see section 5.1 for β and c; the color of the water is determined by c and β. F_p is the flux ratio between the intensity at just beneath the water surface and at point P. As the intensities of scattered light from every particle on the ray are attenuated, the intensity at the viewpoint, I_v, is given by

$$I_v(\lambda) = I_q(\lambda)exp(-c(\lambda)L) + \int_0^L I_p exp(-c(\lambda)l)dl, \tag{6}$$

where I_q is the reflected intensity from the object (or refracted light from the wave surface), I_p the scattered light from particles, and l the distance between P

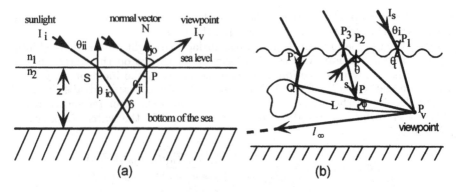

Fig. 7. The optical paths arriving at the viewpoint.

and the viewpoint, L the distance between the object and the viewpoint. When the light does not intersect with any objects, the visible length is set to $L = \infty$.

The optical effects of caustics and shafts of light are calculated as follows. Let's consider subdivided triangle elements from a lattice after meshing a water surface. The refraction vectors are calculated at every lattice point and swept, and the swept volume of a triangle along the refracted vectors is called the *illumination volume*. The intensity of light is inversely proportional to the intersection area between the scan plane and the illumination volume. Then, the intensities of scattered light due to particles can be obtained by applying scan conversion on the illumination volumes. That is, after calculation of the intensity distribution on a scan line, these intensities are stored in an accumulation buffer. The intensity at each pixel is calculated by multiplying the intensity of the triangle and the intersection length between the ray and the triangle.

Fig. 8 shows the shafts of light and caustics on a killer whale in water [18].

6 Conclusion

The previous shading models were mostly concerned with surface shading, and not with volume shading. Volume shading including scattering/absorption play an important role in natural scenes. This paper has discussed algorithms for rendering the optical (scattering) effects such as sky color, shafts of light, caustics, and the color of the water. As shown in the examples, the methods discussed here give us photo-realistic images taking into account scattering characteristic of particles and intensity distribution of light within the water.

Acknowledgment: The author would like to acknowledge Prof. E. Nakamae (formally of Hiroshima Univ.) and Prof. H. Yamashita (Hiroshima Univ.) at whose laboratories some of the images were created. Thanks to Dr. Nelson Max for his review of the draft paper.

References

1. P. Blasi, B.L. Saec, C. Schlick, "A Rendering Algorithm for Discrete Volume Density Objects," *Proc. of EUROGRAPHICS'93*, Vol.12, No.3 (1993) pp.201-210.

2. J.F. Blinn, "Light Reflection Functions for Simulation of Clouds and Dusty Surfaces," *Computer Graphics*, Vol.16, No.3 (1982) pp.21-29.
3. Y. Dobashi, K. Kaneda, T. Nakashima, H. Yamashita, T. Nishita, "Skylight for Interior Lighting Design," *Computer Graphics Forum*, Vol. 13, No. 3, (1994) pp.85-96.
4. Y. Dobashi, K. Kaneda, H. Yamashita, T. Nishita "A Fast Display Method of Sky Color using Basis functions" *Pacific Graphics'95*,(1995) pp.194-208.
5. G.Y. Gardener, "Visual Simulation of Clouds," *Computer Graphics*, Vol. 19, No. 3 (1985) pp.297-303.
6. P. Hanrahan, W. Krueger, "Reflection from Layered Surfaces due to Subsurface Scattering," *Proc. of SIGGRAPH'93*,(1994) pp.165-174.
7. M. Inakage,"Volume Tracing of Atmospheric Environments," *Visual Computer*, (1991) pp.104-113.
8. J.T. Kajiya, B.V. Herzen, "Ray tracing Volume Densities," *Computer Graphics*, Vol.18, No.3 (1984) pp.165-174.
9. K. Kaneda, G. Yuan, E. Nakamae, T. Nishita, "Photorealistic Visual Simulation of Water Surfaces Taking into account Radiative Transfer," *Proc. of CG & CAD'91*,(China) (1991) pp.25-30.
10. K. Kaneda, T. Okamoto, E. Nakamae, T. Nishita, "Photorealistic Image Synthesis for Outdoor scenery Under Various Atmospheric Conditions," *The Visual Computer*, Vol.7 (1991) pp.247-258.
11. R.V. Klassen, "Modeling the Effect of the Atmosphere on Light, " *ACM Transaction on Graphics*, Vol. 6, No. 3 (1987) pp. 215-237.
12. N. Max, "Light Diffusion through Clouds and Haze," *Computer Graphics and Image Processing*, Vol.33, No.3 (1986) pp.280-292.
13. N. Max, "Efficient Light Propagation for Multiple Anisotropic Volume Scattering," *Proc. of the Fifth Eurographics Workshop on Rendering* (1994) pp.87-104.
14. E. Nakamae, K. Harada, T. Ishizaki, T. Nishita, "Montage: The Overlaying of The Computer Generated Image onto a Background Photograph," *Computer Graphics*, Vol. 20, No. 3 (1986) pp. 207-214.
15. T. Nishita, and E. Nakamae, "Continuous tone Representation of Three-Dimensional Objects Illuminated by Sky Light," *Computer Graphics*, Vol. 20, No. 4 (1986) pp. 125-132.
16. T. Nishita, Y. Miyawaki, E. Nakamae, "A Shading Model for Atmospheric Scattering Considering Distribution of Light Sources," *Computer Graphics*, Vol. 21, No. 4 (1987) pp. 303-310.
17. T. Nishita, E. Nakamae, "Display of The Earth Taking into Account Atmospheric Scattering," *Proc. of SIGGRAPH'93* (1993) pp.175-182.
18. T. Nishita, E. Nakamae, "Method of Displaying Optical Effects within Water using Accumulation Buffer," *Proc. of SIGGRAPH'94* (1994) pp.373-379.
19. T. Nishita, E. Nakamae, "A Method for Displaying Metaballs by using Bezier Clipping," *Proc. of Eurographics'94*, Vol.13, No.3 (1994) c271-280.
20. T. Nishita, Y. Dobashi, E. Nakamae, "Display of Clouds Taking into Account Multiple Anisotropic Scattering and Sky Light," *Proc. of SIGGRAPH'96* (1996) pp.379-386.
21. T. Nishita, Y. Dobashi, K. Kaneda, H. Yamashita, "Display Method of the Sky Color Taking into Account Multiple Scattering," *Pacific Graphics'96* (1996) pp.117-132.

22. T. Nishita, H. Iwasaki, Y. Dobashi, E. Nakamae, "A Modeling and Rendering Method for Snow by Using Metaballs," *Computer Graphics Forum*, Vol.16, No.3, (1997) pp.357-364.
23. H.E. Rushmeier, K.E. Torrance, "The Zonal Method for Calculating Light Intensities in The Presence of a Participating Medium," *Computer Graphics*, Vol.21, No.4 (1987) pp.293-302.
24. G. Sakas, M. Gerth, "Sampling and Anti-Aliasing of Discrete 3-D Volume Density Textures," *Proc. of Eurographics'91* (1991) pp.87-102.
25. M. Watt, "Light-Water Interaction using Backward Beam Tracing," *Computer Graphics*, Vol. 24, No. 4 (1990) pp. 377-376.

Editors' Note: see Appendix, p. 325 for colored figures of this paper

A New Change of Variables for Efficient BRDF Representation

Szymon M. Rusinkiewicz
Stanford University

Gates Building, Wing 3B
Stanford, CA 94305
smr@cs.stanford.edu

Abstract

We describe an idea for making decomposition of Bidirectional Reflectance Distribution Functions into basis functions more efficient, by performing a change-of-variables transformation on the BRDFs. In particular, we propose a reparameterization of the BRDF as a function of the halfangle (i.e. the angle halfway between the directions of incidence and reflection) and a difference angle instead of the usual parameterization in terms of angles of incidence and reflection. Because features in common BRDFs, including specular and retroreflective peaks, are aligned with the transformed coordinate axes, the change of basis reduces storage requirements for a large class of BRDFs. We present results derived from analytic BRDFs and measured data.

1 Introduction - BRDF Representation for Computer Graphics

Historically, the reflection models used by computer graphics renderers have been limited. Despite their physical inaccuracy, simple equations such as the Phong lighting model remain popular. True photorealism, however, will require more sophisticated and accurate models of surface properties.

The major difficulty in moving to more sophisticated reflection models has been the difficulty in representing these BRDFs efficiently. The domain of computer graphics requires BRDF representations that are accurate, have high angular resolution, and cover the entire range of possible angles of incidence and reflection. Traditionally, computer graphics systems either relied on analytic models (which were not always available for the exact surface that had to be represented) or had to store enormous quantities of data to represent even relatively simple, smooth BRDFs.

We will examine some of the approaches taken to storing BRDFs in the past, including both analytic models and decompositions into basis functions. We then present an approach for reducing the number of basis functions required to represent a BRDF by reparameterizing the BRDF in terms of the halfangle and a difference angle, rather than the usual angles of incidence and reflection. We show the savings in storage achieved by this transformation for several classes of commonly-encountered BRDFs, including BRDFs with specular, retroreflective, and anisotropic peaks. Note that in this paper we discuss only the directional dependence of BRDFs, not variation with wavelength.

2 Previous Approaches to BRDF Representation

The main efforts in BRDF representation have focused on either analytic formulas that can represent some narrow class of BRDFs, or generic techniques suitable for storing arbitrary four-dimensional functions. We review some of the principal work in both areas.

2.1 Analytic Models - Physically-based and Phenomenological

Most renderers today use BRDFs computed by an analytic formula. Many of these formulas are the result of modeling the properties of a real surface and mathematically computing the amount of light that would be reflected by a surface with those properties. For example, physically-based BRDFs have been derived for primarily specular surfaces (e.g. the Cook-Torrance-Sparrow model [Cook 81]), for rough diffuse surfaces (the Oren-Nayar model [Oren 94]), and for dusty surfaces (the Hapke/Lommel-Seeliger model, developed to model lunar reflectance [Hapke 63]). They range in complexity from simple formulas that ignore many real-world effects to complex models that attempt to account for most actually-observed surface phenomena (e.g. the He-Torrance-Sillion-Greenberg model [He 91]). Because they are derived from physical principles, these models, to a large degree, satisfy the criteria of physical plausibility, such as energy conservation and Helmholtz reciprocity.

An inherent property of physically-based models is that because they start with specific assumptions about microgeometry, they can only predict the reflectance of surfaces that closely match those assumptions. As a result, even the most elaborate theoretical models cannot predict reflectance from surfaces with complex microstructure or mesostructure (e.g. cloth, metallic paint, fur). Phenomenological models have been used successfully to widen the range of representable BRDFs. For example, the Minnaert BRDF [Minnaert 41] was an early empirical formula developed to characterize the reflectance of the moon before more physically-correct models were derived. Lafortune's generalized cosine lobes [Lafortune 97] can represent a wide range of phenomena, including off-specular reflection and retroreflection. As with all phenomenological models, however, there exist behaviors that Lafortune's functions cannot represent (including most commonly-seen kinds of anisotropy). Shade trees [Cook 84] and shading languages such as RenderMan [Hanrahan 90] attempt to generalize phenomenological models even further, by allowing simpler models to be combined in flexible ways.

As we have seen, both physically-based and phenomenological models can only represent certain limited classes of surfaces. Given an arbitrary BRDF, whether it was measured directly or obtained through simulation (as in [Westin 92]), there is no guarantee that any analytic model can represent it. This often does not meet the requirements of computer graphics, since in general one would like to produce realistic renderings of arbitrary surfaces. Thus, despite the simplicity and utility of analytic models, there have been several attempts to look at BRDFs in more general frameworks, in which it is possible to represent exactly any given BRDF.

2.2 Decomposition into Basis Functions

Describing a complex function as a linear combination of some set of basis functions is a widely used technique for representing continuous functions. In the domain of BRDF representation, the most popular classes of basis functions are tensor products of the spherical harmonics, Zernike polynomials, and spherical wavelets.

Spherical harmonics, the spherical analogue of sines and cosines, are a popular choice for representing BRDFs. Because spherical harmonics are compact in frequency space, smooth BRDFs should have fewer fewer nonzero (or at least non-negligible) coefficients than complex ones when expressed in this basis. Westin et al. present an implementation of using spherical harmonics to represent BRDFs, taking advantage of symmetry and reciprocity in the BRDF to reduce storage requirements [Westin 92].

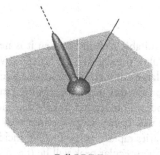 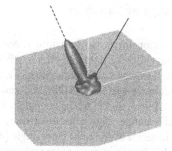

Full BRDF   High-frequency components removed
(Torrance-Sparrow)   (Spherical harmonics through order 8 retained)

Figure 1: Ringing caused by truncation of high-frequency terms. The graphs are goniometric plots of the BRDF as a function of reflectance angle, for a fixed angle of incidence. The directions of incident light and ideal specular reflection are also shown.

An alternative to using a basis of spherical harmonics, which treat functions on a hemisphere as a special case of functions on a sphere, is to use a basis of functions on a disk, then map those functions onto a hemisphere. Koenderink et al. explore this possibility, looking at representing BRDFs in terms of the Zernike polynomials, which form an orthonormal basis of functions on the unit disk [Koenderink 96]. The authors use an equal-area mapping from the disk onto the hemisphere, and enforce reciprocity by taking particular linear combinations of the functions. Thus, the paper develops a representation very similar to that used by Westin et al., but optimized for the hemisphere rather than the sphere.

Both the spherical harmonics and Zernike polynomials require large numbers of basis functions in order to represent quickly-varying BRDFs accurately. The consequence of using too few terms is "ringing" in the BRDF, caused by sharp edges in frequency space (see Figure 1). Moreover, since neither the spherical harmonics nor the Zernike polynomials are compact in space, evaluating a BRDF represented in terms of these functions requires computation time proportional to the total number of nonzero coefficients. Therefore, because large numbers of coefficients are necessary to avoid artifacts, evaluating BRDFs stored in terms of these basis functions is expensive.

Wavelets have been proposed as an alternative set of basis functions for BRDF representation, because they help to reduce evaluation time and storage cost. Because wavelets are localized in space, evaluating the BRDF at a particular pair of angles of incidence and reflection requires computation time proportional to the depth of the coefficient tree (i.e. logarithmic in most cases) rather than time proportional to the total number of nonzero coefficients. In addition, the spatially-localized wavelets can represent the large spikes (e.g. the specular peak) of many common BRDFs more efficiently than spherical harmonics or Zernike polynomials.

Schröder and Sweldens have proposed a basis of wavelets optimized for representing functions on a sphere [Schröder 95], although they did not actually describe an implementation of storing complete BRDFs using these spherical wavelets. Lalonde and Fournier describe a complete implementation, using wavelets on the Nusselt embedding of the hemisphere [Lalonde 97] and a tree-based encoding of the coefficients. Their experience shows that, as expected, using wavelets results in significant BRDF compression.

3 Change of Variables

Decomposition into basis functions is certainly a suitable technique when it is necessary to have the ability to represent arbitrary BRDFs. All of the sets of basis functions we have described, however, share the problem of requiring large numbers of coefficients to describe even moderately specular BRDFs. The main reason for this inefficiency is the fact that a BRDF parameterized in terms of angles of incidence and reflection usually does not have "localized" change. For example, a shiny surface will have a large specular peak whose position in terms of the reflectance angle (θ_o, ϕ_o) varies rapidly as the angle of incidence is changed. Although in most cases *we* know where the specular peak will lie (i.e. mostly in the direction of ideal specular reflection), the usual basis functions do not take advantage of this fact, and therefore cannot represent the peak efficiently. Similarly, the standard methods in general do not require any less storage if the BRDF is isotropic. It is possible to design a basis specialized for storing only isotropic BRDFs, but it is up to the *user* to specify that these, not the generic basis functions, should be used.

We propose an approach for making decomposition of BRDFs in terms of such functions more efficient while retaining the ability to represent (in the limit) arbitrary BRDFs. The idea was inspired by considering the possibility of using basis functions that are more "tuned" to representing common BRDFs. Instead of doing this, however, we propose the opposite approach: transforming BRDFs such that they can be represented more efficiently using the traditional basis functions.

The transformation we propose is a change of variables that causes some features of common BRDFs to lie along the new coordinate axes. Specifically, we propose parameterizing the BRDF in terms of the halfway vector (i.e. the vector halfway between the incoming and reflected rays) and a "difference" vector, which is just the incident ray in a frame of reference in which the halfway vector is at the north pole (see Figure 2). That is, instead of representing the BRDF as $\beta = \beta(\theta_i, \phi_i, \theta_o, \phi_o)$, we regard the BRDF as a function of the halfangle and difference angle: $\beta = \beta(\theta_h, \phi_h, \theta_d, \phi_d)$, where

$$\vec{n} \quad = \quad \text{Surface normal,} \tag{1}$$

$$\vec{t} \quad = \quad \text{Surface tangent (i.e. orientation of an anisotropic surface),} \tag{2}$$

$$\vec{b} \quad = \quad \text{Surface binormal (i.e. } n \times t\text{),} \tag{3}$$

$$\vec{\omega}_i \quad = \quad \text{sph}(\theta_i, \phi_i), \text{ where } \theta_i \text{ is measured relative to } \vec{n} \text{ and } \phi_i \text{ is relative to } t, \tag{4}$$

$$\vec{\omega}_o \quad = \quad \text{sph}(\theta_o, \phi_o), \tag{5}$$

$$\vec{h} \quad = \quad \text{sph}(\theta_h, \phi_h), \tag{6}$$

$$\quad = \quad \frac{\vec{\omega}_i + \vec{\omega}_o}{\|\vec{\omega}_i + \vec{\omega}_o\|}, \tag{7}$$

$$\vec{d} \quad = \quad \text{sph}(\theta_d, \phi_d), \tag{8}$$

$$\quad = \quad \text{rot}_{\vec{b}, -\theta_h} \text{ rot}_{\vec{n}, -\phi_h} \vec{\omega}_i. \tag{9}$$

Note that (θ_h, ϕ_h) are the spherical coordinates of the halfway vector in the $\vec{t}$–$\vec{n}$–$\vec{b}$ frame. The two rotations in equation 9 bring the halfangle $\vec{h}$ to the north pole, and (θ_d, ϕ_d) are the spherical coordinates of the incident ray in this transformed frame.

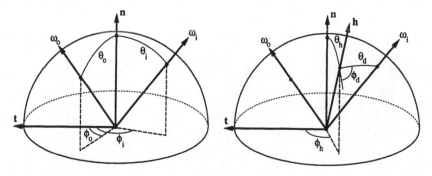

Figure 2: Proposed reparameterization of BRDFs. Instead of treating the BRDF as a function of (θ_i, ϕ_i) and (θ_o, ϕ_o), as shown on the left, we consider it to be a function of the halfangle (θ_h, ϕ_h) and a difference angle (θ_d, ϕ_d), as shown on the right. The vectors marked $\vec{n}$ and $\vec{t}$ are the surface normal and tangent, respectively.

3.1 Properties of BRDFs in the New Coordinates

From the point of view of BRDF representation, the main effect of the proposed change of variables is that it aligns the features of common BRDFs (such as specular and retroreflective peaks) with the new coordinate axes. Thus, representing most BRDFs in terms of basis functions in these new coordinates should require a smaller number of nonzero coefficients than would be required in the untransformed coordinates for equivalent accuracy. The reason for this is that in the new coordinates the BRDFs show strong dependence on each axis individually, but show only weak dependence on combinations of axes. That is, while a BRDF might depend on θ_h or θ_d, most common BRDFs will not have a dependence that is some complex function of both θ_h and θ_d. Therefore, coefficients that correspond to terms that are high-frequency in both θ_h and θ_d will, for most BRDFs, be small. The only large coefficients should be the ones that correspond to variation in only one axis.

Let us now examine how certain BRDFs appear in the transformed coordinates. First, we note that isotropic BRDFs are independent of ϕ_h in this coordinate system. This means that an isotropic BRDF will have basis function coefficients equal to zero for all basis functions that vary with ϕ_h. Therefore, we have automatically reduced the number of nonzero coefficients to a three-dimensional subset of the four-dimensional space. This contrasts with the standard coordinates, where the entire four-dimensional space will be populated with nonzero coefficients even in the case of an isotropic BRDF.

A second property of the new coordinates is that the angles of incidence and reflection become much more symmetric. In particular, the condition of Helmholtz reciprocity becomes a simple symmetry under $\phi_d \rightarrow \phi_d + \pi$. Is is therefore easy to enforce reciprocity in any representation based on this change of variables.

Ideal specular and near-ideal specular peaks are transformed by the change of variables to lie mostly along the θ_h axis. An ideal specular peak is represented as a delta function of θ_h, and is completely independent of the other three variables. Similarly, a simple BRDF such as Blinn's variation [Blinn 77] on Phong's model [Phong 75] is also a function of only θ_h. In general, any BRDF that depends only on $(\vec{n} \cdot \vec{h})$ is independent of three of the four variables in the transformed space. In terms of representation, this means that only a one-dimensional subset of the four-dimensional space of coefficients will be nonzero for such a BRDF.

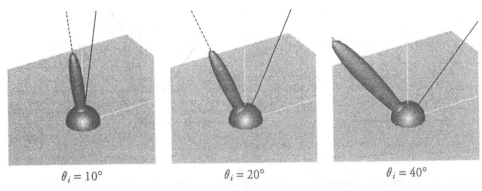

$\theta_i = 10°$ $\theta_i = 20°$ $\theta_i = 40°$

The Cook-Torrance-Sparrow BRDF seen as a function of (θ_o, ϕ_o), for various values of (θ_i, ϕ_i). Note that the position of the peak in space varies considerably.

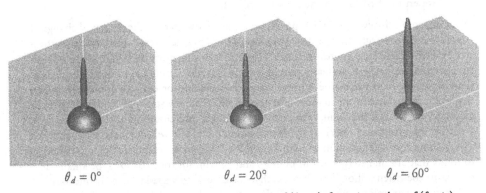

$\theta_d = 0°$ $\theta_d = 20°$ $\theta_d = 60°$

The Cook-Torrance-Sparrow BRDF seen as a function of (θ_h, ϕ_h), for various values of (θ_d, ϕ_d). Note that although the size of the peak changes (as predicted by the Fresnel term), the position and shape of the peak remain constant. The BRDF is therefore approximated very closely by a function of the form $\beta = \beta_1(\theta_h)\beta_2(\theta_d)$, which means that only a small number of basis function coefficients will be nonzero.

Figure 3: Cook-Torrance-Sparrow BRDF in standard and transformed coordinates.

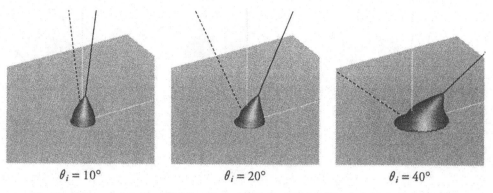

$\theta_i = 10°$ $\theta_i = 20°$ $\theta_i = 40°$

The Hapke/Lommel-Seeliger BRDF seen as a function of (θ_o, ϕ_o), for various values of (θ_i, ϕ_i).

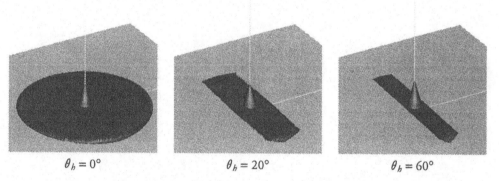

$\theta_h = 0°$ $\theta_h = 20°$ $\theta_h = 60°$

The Hapke/Lommel-Seeliger BRDF seen as a function of (θ_d, ϕ_d), for various values of (θ_h, ϕ_h). Compare this to Figure 3, where we plot slices of constant θ_d. In that figure, it is the specular peak that stays relatively stationary; here it is the retroreflective peak.

Figure 4: Hapke/Lommel-Seeliger BRDF in standard and transformed coordinates.

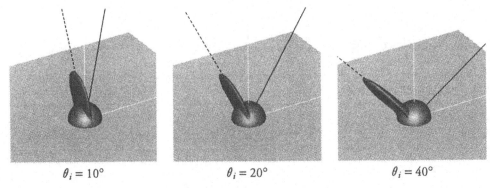

| $\theta_i = 10°$ | $\theta_i = 20°$ | $\theta_i = 40°$ |

Ward's elliptical Gaussian BRDF seen as a function of (θ_o, ϕ_o), for various values of (θ_i, ϕ_i).

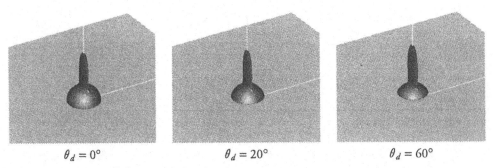

| $\theta_d = 0°$ | $\theta_d = 20°$ | $\theta_d = 60°$ |

Ward's elliptical Gaussian BRDF seen as a function of (θ_h, ϕ_h), for various values of (θ_d, ϕ_d). The BRDF is very closely approximated by a function of the form $\beta = \beta_1(\theta_h)\beta_2(\phi_h)$.

Figure 5: Elliptical Gaussian BRDF in standard and transformed coordinates.

A further generalization of the above is that any analytic model that starts with an isotropic microfacet distribution and ignores masking, shadowing, and interreflectance will be a separable function of only θ_h and θ_d in the transformed coordinates. To prove this, we observe that the halfangle is the direction in which a microfacet's normal must be oriented such that light will reflect from the given direction of incidence in the given angle of reflection. Therefore, the number of microfacets pointed in the "correct" direction must be some function of θ_h. This amount must then be corrected depending on the orientation of the incident ray relative to the microfacet (e.g. to include a Fresnel term). This term will be a function of only θ_d, and therefore the resultant BRDF must be of the form $\beta = \beta_1(\theta_h)\beta_2(\theta_d)$.

More complex BRDFs will include effects such as masking, shadowing, and interreflectance, and therefore will not have completely separable representations. However, in many cases the decomposition in terms of our proposed coordinates will still be fairly simple, and may be nearly separable (see Figure 3, which shows a Torrance-Sparrow BRDF). This means that it is inexpensive to store accurate approximations even for these more realistic BRDFs.

Retroreflective peaks are transformed by our change of variables to be functions of only θ_d: an ideal retroreflective peak would be a delta function around $\theta_d = 0$, and a more diffuse peak is represented by some smoother function of θ_d. Again, in many cases more complicated retroreflective BRDFs come close to being functions of only one variable and have only a weak dependence on the other three. For example, Figure 4 shows the primarily retroreflective Hapke/Lommel-Seeliger BRDF.

Finally, let us consider one particular kind of anisotropy, namely the anisotropy associated with BRDFs that have elliptical specular peaks. These are the kinds of BRDFs predicted by, for example, Ward's elliptical Gaussian BRDF [Ward 92], and are commonly seen on surfaces such as brushed metals. We observe that in transformed coordinates these BRDFs are again largely separable, and their main features can be represented as the product of some function of θ_h (representing the shape of the specular peak) and some function of ϕ_h (representing the anisotropic variation). Figure 5 shows Ward's BRDF in our transformed coordinates.

3.2 Results

We now present some examples of how the proposed change of variables can reduce the cost required to store BRDFs. First, let us look at some data obtained from analytic models (see Figure 6). Three analytic BRDFs were randomly sampled, then least-squares fits to cubic wavelet basis functions were performed in both the standard and transformed coordinates. The fits were done to the logarithm of the BRDF rather than the BRDF itself, which had the effect of minimizing relative, rather than absolute, error. This is a more appropriate error metric for perceptual differences, and so the results should be more applicable to most applications than if the absolute error had been minimized.

Figure 6 shows the required storage for data derived from the analytic BRDFs. The first is the Torrance-Sparrow BRDF[1], a standard glossy-surface model. The second is the Hapke/Lommel-Seeliger BRDF[2], which describes a (mostly retroreflective) dusty surface. The third is Ward's elliptical Gaussian model[3], which describes reflection from an

---

[1] The parameters used were: $m = 0.2$, $\eta = 2.0$ $\kappa_s = 0.5$, and $\kappa_d = 0.5$.

[2] The parameters used were: $g = 0.6$ and forward scattering coefficient = 0.1.

[3] The parameters used were: $\alpha_x = 0.2$, $\alpha_y = 0.5$, $\kappa_s = 0.2$, and $\kappa_d = 0.5$.

20

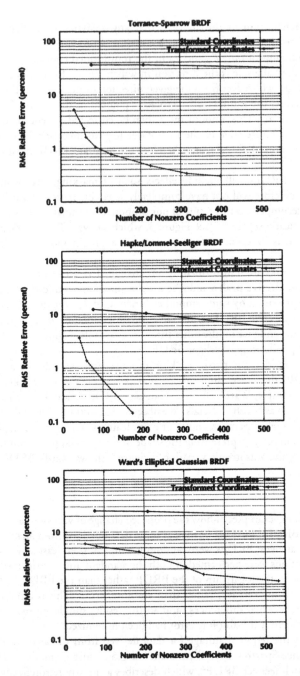

Figure 6: Representation error as a function of the number of nonnegligible wavelet coefficients in standard and transformed coordinates.

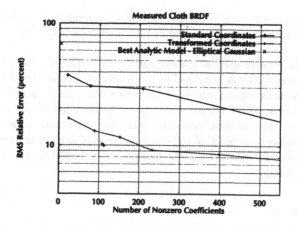

Figure 7: Results of fitting measured cloth BRDF to basis functions. For comparison, the best analytic result is also presented. Note that because the analytic model has a fixed number of parameters (namely, 4), it is plotted as a single point.

anisotropic rough surface. We see that in all cases performing the change of variables reduced storage requirements. Note that all of these BRDFs contained reasonably narrow (specular or retroreflective) peaks. We would expect somewhat smaller storage savings for more diffuse BRDFs.

Figure 7 shows the results obtained for the measured BRDF of a cloth sample. The cloth sample was slightly anisotropic, and exhibited markedly increased reflection towards grazing. For purposes of comparison, we also show a fit to an analytic model that best matches our data.

The cloth BRDF measurements were taken with a four-axis computer-controlled camera gantry that lets a sample be illuminated and photographed from any direction, except for an occlusion that prevents measurements within 3° of retroreflection. Because the data were used without any smoothing, there was about 5% variation among closely-spaced samples because of camera noise and, more importantly, surface irregularities. Therefore, we did not expect to be able to obtain errors below about 5% (using very large numbers of coefficients to try to achieve lower errors would have led to overfitting).

4 Conclusions and Future Work

We have presented an idea for transforming BRDFs such that they can be represented efficiently by standard classes of basis functions. The transformation:

- Reduces the number of basis coefficients necessary to represent a broad range of BRDFs.
- Allows significant savings when storing isotropic BRDFs, as compared to anisotropic ones.
- Exhibits greater savings for increasingly specular BRDFs, since it does a good job of aligning their high-frequency components (namely specular and retroreflective peaks) with the transformed coordinate axes.

Although we have concentrated on applying this transformation to fitting a BRDF to linear combinations of basis functions, it should be possible to apply the same idea to nonlinear fits to arbitrary sets of functions. This should increase compression even further, at the cost of increased running time and possible numerical instability in the nonlinear fitting routines.

In addition, the possible space of transformations on BRDFs should be explored more widely. We have presented only one possible transformation, and surfaces displaying different kinds of phenomena, especially highly anisotropic surfaces, might benefit from different transformations. Most BRDF research in computer graphics has concentrated on diffuse and glossy surfaces, since analytic models for these were the easiest to obtain. As BRDF measurement becomes more practical and common, we plan to measure a greater variety of complicated, "exotic" BRDFs, and explore their properties, including their behavior under this change of variables.

Acknowledgments

I would like to thank my advisor Marc Levoy for his helpful comments on the ideas presented here. I also thank the National Science Foundation for their financial support. Our BRDF-acquisition research is funded by Interval Research and Honda.

References

[Blinn 77]   Blinn, J. "Models of Light Reflection for Computer Synthesized Pictures," *Proc. Siggraph*, 1977.

[Cook 81]   Cook, R. and Torrance, K. "A Reflectance Model for Computer Graphics," *ACM Computer Graphics*, Vol. 15, No. 4, 1981.

[Cook 84]   Cook, R. "Shade Trees," *Proc. Siggraph*, 1984.

[Hanrahan 90]   Hanrahan, P. and Lawson, J. "A Language for Shading and Lighting Calculations," *Proc. Siggraph*, 1990.

[Hapke 63]   Hapke, B. "A Theoretical Photometric Function for the Lunar Surface," *Journal of Geophysical Research*, Vol. 68, No. 15, 1963.

[He 91]   He, X., Torrance, K., Sillion, F., and Greenberg, D. "A Comprehensive Physical Model for Light Reflection," *Proc. Siggraph*, 1991.

[Koenderink 96]   Koenderink, J., van Doorn, A., and Stavridi, M. "Bidirectional Reflection Distribution Function Expressed in Terms of Surface Scattering Modes," *Proc. European Conference on Computer Vision*, 1996.

[Lafortune 97]   Lafortune, E., Foo, S. C., Torrance, K., and Greenberg, D. "Non-Linear Approximation of Reflectance Functions," *Proc. Siggraph*, 1997.

[Lalonde 97]   Lalonde, P. and Fournier, A. "Filtered Local Shading in the Wavelet Domain," *Proc. Eurographics Workshop on Rendering*, 1997.

[Minnaert 41]   Minnaert, M. "The Reciprocity Principle in Lunar Photometry," *Astrophysical Journal*, Vol. 93, 1941.

[Oren 94]   Oren, M. and Nayar, S. "Generalization of Lambert's Reflectance Model," *Proc. Siggraph*, 1994.

[Phong 75]   Phong, B. T. "Illumination for Computer-Generated Pictures," *Communications of the ACM*, Vol. 18, No. 8, June 1975.

[Schröder 95]   Schröder, P. and Sweldens, W. "Spherical Wavelets: Efficiently Representing Functions on the Sphere," *Proc. Siggraph*, 1995.

[Ward 92]   Ward, G. "Measuring and Modeling Anisotropic Reflection," *Proc. Siggraph*, 1992.

[Westin 92]   Westin, S., Arvo, J., and Torrance, K. "Predicting Reflectance Functions from Complex Surfaces", *Proc. Siggraph*, 1992.

Approximating Reflectance Functions using Neural Networks

David Gargan   Francis Neelamkavil

Image Synthesis Group,
Dept. of Computer Science,
Trinity College Dublin
David.Gargan@cs.tcd.ie
Francis.Neelamkavil@cs.tcd.ie

Abstract. We present a new representation for the storage and reconstruction of arbitrary reflectance functions. This non-linear representation, based on a neural network model, accurately captures the spectral and spatial variation of these functions. It is both computationally efficient and concise, yet expressive. We reconstruct the subtle reflection characteristics of an analytic reflection model as well as measured and simulated reflection data

1 Introduction

The physically accurate rendering of any scene requires that we appropriately model the interaction of light with the surfaces involved. In real scenes this interaction can often be quite complex and varied. The appearance industry [18] illustrates this with the many terms, such as gloss, lustre, haze and sparkle, that it uses to classify different types of interaction that occur at a microgeometric level[1]. In computer graphics we attempt to capture these effects using a specific shading model, or rather when using the rendering equation [9], this behaviour is embodied in the bi-directional reflectance distribution function (BRDF).

Often a particular shading model provides the only visual cues that an observer uses to gain some intuitive understanding of the texture and construction of a particular surface. Traditionally we have attempted to capture important appearance classifications, such as diffuse, glossy or ideal specular, and combine these to produce a shading model. This limits the range of surfaces that we can accurately render. A general method of representing the bi-directional reflectance distribution function is required, coupled with an adequate global illumination algorithm to reproduce the appearance of a diverse range of real world surfaces.

Unfortunately there is a delicate balance between expressiveness and efficiency. In general it may be relatively simple to capture one or two effects, such as diffuse and directional diffuse components, in a simple mathematical model. If we pay proper attention to both reciprocity and energy conservation [15] then this physically plausible model will adequately reproduce these effects. As the expressiveness increases, to include backscattering for instance, so does the complexity of our

---

[1] Geometry at a scale which is too small to be visibly discernable[29]

underlying model. This increase in complexity often leads to increased computational and storage costs.

If BRDF data is either sampled or precomputed, and an appropriate storage and reconstruction method chosen, then this leads to a generally expressive formulation; however efficiency and accuracy both become even more complicated issues.

In this paper we present a general method for storing and reconstructing BRDF data. This new method exploits the non-linear function approximation capabilities of neural networks. The representation encodes the entire BRDF concisely and allows computationally efficient reconstruction.

2 Previous Work

The reflectance equation [26], equation 1, illustrates the vital role the BRDF, f_r, plays in the rendering of any computer-generated image. Different rendering algorithms use various numerical approaches to solve this integral. Algorithms such as simple local shading or raytracing will restrict the domain of integration, whereas classical radiosity makes the assumption that the BRDF is constant with respect to direction. Regardless of the illumination algorithm used it should be clear that it is the BRDF which will inevitably have the greatest influence on the final appearance of a rendered surface. For this reason much research has been focused on generating appropriate representations of the BRDF.

$$L_r(\mathbf{x}, \omega) = \int_{\Omega_{in}} f_r(\omega, \omega') L_i(\mathbf{x}, \omega')(-\omega'.n) d\sigma(\omega') \qquad (1)$$

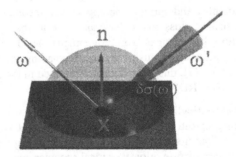

Fig. 1 Variables used in the reflectance equation

We can classify shading model research using the following categories

- analytic solutions derived from a theoretical basis.
- analytic solutions derived from empirical observation.
- analytic solutions with an intuition based parameterisation.
- BRDF data[2] storage and reconstruction methods.

---

[2] The data may be sampled or more often precomputed.

As mentioned previously the delicate balance between efficiency[3], accuracy, expressiveness and complexity are weighted differently in each of the shading models presented to date. The first classification restricts the expressiveness of the BRDF to a certain class of microgeometries to facilitate a framework for theoretical analysis. The majority of models in this classification attempt to produce physically accurate representations often increasing the complexity of the model. Examples of these models are [1][3][21][8][19][7].

The second classification balances accuracy, efficiency and ease of implementation [28][24]. Often their parameterisation contains some measurable parameters, such as surface roughness or indices of refraction, which are required to evaluate the BRDF. Accuracy is accounted for by fitting the model to observed data.

The third class of model attempts to provide intuitive parameter based models that encompass many different microgeometries. This is usually done by first identifying and modelling the most important features of common BRDF's and then weighting them appropriately and intuitively. In the earlier models [20] physical plausibility is ignored and validation of the model is done on a purely aesthetic basis. More recent efforts have produced energy conserving reciprocal versions of this type of model [15] [12] [26].

Unfortunately finding a single equation which matches all materials is probably impossible[5]. For this reason expressiveness has been sacrificed and the trend in current research is to concentrate on simplicity, efficiency and physical correctness and the focus is now on the representation of known BRDFs. Since the BRDF is either physically measured or simulated the accuracy of the BRDF representation depends primarily on the accuracy of the measuring equipment/ simulation. The efficiency of the representation depends on the storage and reconstruction scheme employed. Previous researchers have used combinations of linear basis functions such as spherical harmonics [8] [2] [29], adaptive geodesic spheres [6], and wavelets [25] [14], as well as non-linear basis functions like generalised cosine lobes [13].

3 Neural Networks as Universal Function Approximators

Artificial Neural Networks (ANNs) are well known for their classification and approximation abilities. Our aim in this section is not to provide a comprehensive taxonomy of neural network research to date. Instead we will focus on the most popular universal approximation method i.e. backpropagation. We begin with an introduction to linear neural networks and then extend this discussion to non-linear models. The aim here is to give the reader some intuitive understanding of how these models work as approximators.

3.1 Linear Approximators : The ADALINE / Perceptron

Figure 2 illustrates the simplest form of neural network, a single perceptron or ADALINE. The inspiration for this model comes from a biological model of the

---

[3] Both computational and storage efficiency

brain. The unit, an activation function and basis function combined, is a discrete analogue of a single brain neuron. The weights, w_j represent the synapses connecting neurons, a heavier weight implies a stronger synaptic connection. An entire interconnected set of these units forms a neural network. The inputs i_j may come from other units and similarly the output may lead to other units (figure 3). Neural networks are normally classified according to the interconnection method used.

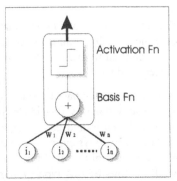
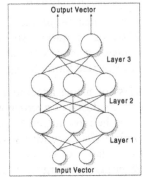

Fig. 2 The Perceptron/ADALINE   **Fig. 3** A multilayer feedforward network

The basis function describes how weights and inputs combine before application to the activation function. Equation 2 represents the most common form of basis function, the linear basis function.

$$u(i_1..i_n, w_1..w_n) = \sum_{j=1}^{n} w_j i_j \qquad (2)$$

Another commonly used basis function is the radial basis function (Equation 3) which uses the Euclidean distance between the weight vector and input vector.

$$u(i_1..i_n, w_1..w_n) = \sum_{j=1}^{n} (i_j - w_j)^2 \qquad (3)$$

The activation function shown in Figure 2 is a simple step function, which is typically used for classification problems. Some alternatives, commonly used for approximation are shown in Figure 4.

It should be obvious from the above formulation that a single input perceptron with a linear basis function and linear activation function can represent any line that passes through the origin. Often an additional bias, essentially a dummy input fixed at 1.0 is included to remove this origin fixed restriction (Figure 5).

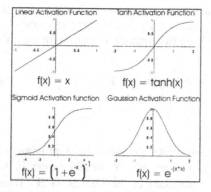

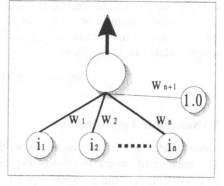

Fig. 4 Common Activation Functions   **Fig. 5** Bias weight

3.2 Approximation using Single Layer Feed-Forward Networks

Given a set of k input vectors $\mathbf{i_k}$ and a set of corresponding training values $\mathbf{t_k}$ the aim of the learning phase is to minimise the error between the desired output, given by our training values, and the network output $\mathbf{o_k}$. An error metric is defined over the input and weight space as

$$E_k(\mathbf{i_k}, \mathbf{w}) = \left| \mathbf{t_k} - \mathbf{o_k}(\mathbf{i_k}, \mathbf{w}) \right|^2 \tag{4}$$

Each step, or epoch, in the learning cycle involves presenting the network with our inputs, calculating the error and adjusting the weights appropriately[4]. In order to update the weights we need to find the gradient of the error function with respect to a particular weight and then adjust the weight in the direction opposite to the gradient. More formally we seek the partial derivative of the error function with respect to a particular weight. Expanding our vector notation we adjust the weight j for a specific training pattern using

$$w_j \leftarrow w_j - \eta \frac{\partial E}{\partial w_j} \tag{5}$$

where η is a chosen constant known as the learning rate. The partial derivative can be determined by applying the chain rule as follows:

$$\frac{\partial E}{\partial w_j} = \frac{\partial E}{\partial o} \frac{\partial o}{\partial u} \frac{\partial u}{\partial w_j} \tag{6}$$

This implies for training pattern k we have

$$\frac{\partial E}{\partial w_j} = -(t_k - o_k) f'(u_k) i_k \tag{7}$$

---

[4] At the beginning of training weights are usually assigned a random value in the same range of values that the activation function produces.

where f' represents the derivative of the activation function. It is clear that the activation function must be differentiable in order to apply this learning algorithm and the previous step function is not applicable. Using the sigmoid activation function the weight update rule is as follows

$$w_j \leftarrow w_j + \eta(t_k - o_k)o_k(1 - o_k)i_k \qquad (8)$$

3.3 Non-linear Approximation

Figure 3 depicts a 3 layer neural network appropriate for non-linear approximation. The number of layers is usually defined as the number of weight layers in the network. This specific network approximates functions of the form $\mathbf{R}^2 \rightarrow \mathbf{R}^2$. Using multiple layers of units with a linear activation function has no benefit [11]. By definition the composition of linear functions is separable so we can always form an equivalent single layer network. If we use non-linear activation functions such as the sigmoid, tanh or Gaussian function the network will perform non-linear approximation.

3.3.1 Backpropagation

Backpropagation [23] allows the error in the output layer to be transferred to the hidden layers and in turn the weights may be adjusted appropriately. During each epoch the input signal is fed forward, the output is generated and the error is calculated. This error is then propagated back through the entire network adjusting the weights along the way.

Consider the penultimate weight layer $\mathbf{w}_2$. In order to minimise the error we must calculate the partial derivative of the error with respect to the weights in this layer. The change in the error is simply the sum of the changes to each of the output units. In our example we have

$$\frac{\partial E}{\partial w_{2j}} = \sum_{i=1}^{2} \frac{\partial E}{\partial o_{3i}} \frac{\partial o_{3i}}{\partial u_{3i}} \frac{\partial u_{3i}}{\partial o_{2j}} \frac{\partial o_{2j}}{\partial u_{2j}} \frac{\partial u_{2j}}{\partial w_{2j}} \qquad (9)$$

This may in turn be applied recursively to each of the hidden layers.

4 Learning BRDF data

In this section we discuss how to exploit the non-linear function approximation capabilities of neural networks to represent BRDF data. We reproduce the BRDFs given by data from three independent sources. An analytic model, measured data and simulated data.

4.1 Data Preparation

The most important phase of approximation using neural networks is the choice of representation for input and output data. The type of activation function will always dictate the range of output values reproducible by the network. All of the most common non-linear activation functions presented earlier have an output range of either [0,1] or [-1,1]. After an appropriate choice of activation function has been made the output data is scaled to the appropriate output range[5].

A far more important issue is the appropriate choice of representation for the input data. Figure 6 shows the variation of the normalised Phong BRDF for several incoming angles. As can be seen the shape of the BRDF is very similar for each of the given angles. Figure 7 shows the same function, but this time plotted in polar co-ordinate space. As can be seen the shape of the function varies greatly at high incident angles. We therefore need to find some parameterisation of the hemisphere (see Figure 8) which minimises the variation due to the parameter space chosen. In general we find the XY parameter space, or projected hemispherical space to be the most efficient. Intuitively this is because directions that are close to each other in world space remain close to each other in projected space.

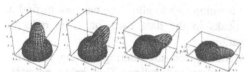

Fig. 6 The normalised phong BRDF plotted for
several incoming angles

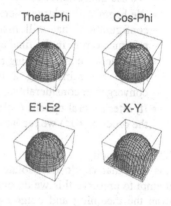

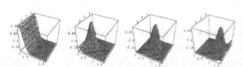

Fig. 7 The normalised phong BRDF plotted in polar
co-ordinate space.

Fig. 8 Parameterisations of the
hemisphere from [17]

4.2 Network Size and Interactive learning

The networks used are standard backpropagation networks with two or three weight layers[6], linear basis functions and sigmoid activation functions. Typically the storage of a 5 dimensional BRDF (fully anisotropic and spectrally varying) requires on the order of 50 - 200 weights, depending on its characteristics. Isotropic BRDFs will be learned with fewer weights, as the input space is only three-dimensional. The complexity of the BRDF will also greatly influence the number of units required. Extremely smooth BRDFs require far fewer units than highly complex ones. A perfectly diffuse BRDF could be represented by a single bias weight.

---

[5] This is not strictly necessary as a neural network which contains an output layer with linear activation functions is capable of learning any n to m dimensional real valued mapping, $R^n \rightarrow R^m$. Convergence is often far better for non-linear outputs.

[6] This is based on the evidence presented by Kung [11]. He claims that most real world data is best approximated using two or three layers in the network.

In general we begin training with a gross underestimation of the number of units required and gradually repeat the training with more and more units until a satisfactory RMS error has been reached.

The motivation for this is well founded. A neural network with too many hidden units will tend to overfit the data presented. Overfitting is a dangerous problem; although the sample values may appear to be adequately represented the network may vary rapidly between samples, thus violating both reciprocity and energy conservation. By using a minimum number of hidden units together with a sufficiently large sample set (on the order of tens of thousands of samples) we minimise the risk of overfitting.

During training we have found the following ideas useful:

- We begin with a large sample set and relatively small network, typically a single hidden layer network with 5 hidden units.
- We use the continuous wrong update method as a training strategy[7].
- We begin with a large tolerance and learning rate to allow the network to find a configuration that will match the peaks in the data early on. This reduces min/max errors in the final approximation.
- The tolerance and learning rate are gradually decreased during training. In doing so we are mimicking a form of simulated annealing. This helps network convergence considerably.
- If, after several epochs ($<10^2$), the total RMS error is not sufficiently small (on the order of 10^{-2}) we repeat the training with a larger network.

4.2.1 Reciprocity

It is vital that we do not violate reciprocity when approximating our BRDFs. In an attempt to preserve this we do one of two things. Firstly we take the original data and swap the incoming and outgoing directions, effectively doubling the dataset's size. The network is then trained to an appropriate error using this new dataset. Alternatively we can use the reciprocal data as test data. During training the reciprocal data is fed into the network and training continues with the original data until the RMS error in the reciprocal data is acceptable.

4.2.2 Analytic Model Fit

To validate our representation we develop a simple BRDF by combining several normalised Phong lobes and a constant diffuse term. The model is specifically designed to exhibit strong forward and exaggerated backward scattering. The data presented to the network for training is a randomly generated set of 10,000 samples from this BRDF. The samples are rescaled to match the output range [0,1] and a theta-phi parameterisation. Figure 11 depicts a scene rendered with both the theoretical model and the neural network representation. The neural network used has a total of 51 weights (3 inputs, 10 hidden units and a single output unit[8]), and learning was stopped after the RMS error reached 0.017. The walls of the room immediately behind the viewer are deliberately red and green to illustrate the exaggerated backscattering.

---

[7] The 'continuous wrong' training strategy involves adjusting the weights only for samples that lie outside a certain tolerance.

[8] We assume the BRDF has no spectral variation.

4.3 Measured Data Fit

We choose 3 sets of measured reflectance data [4] and use our neural network techniques to learn these. The original database contains only 205 samples for each surface type, therefore it is vital that we do not overfit the data. Each sample in its original form gives radiance values for the red, green and blue camera phosphors. Using the data provided we reconstruct the entire spectrum and resample using the Meyer [16] colour space before learning. The 3 sets chosen, leather, velvet and wood represent a range of surface properties. Both the leather and velvet exhibit low anisotropy, whereas the wood has high anisotropy. Figure 9 shows the RMS error for the networks for the first 200 epochs of training. One particular interesting feature is how the wood BRDF oscillates about several local minima before finally settling down. These minima are due to the anisotropy in the dataset. The following table shows the network parameters used to fit the data.

Table 1. Network Configuration and RMS Error for leather, velvet and wood datasets.

Dataset	Network Configuration	RMS Error (after 2000 epochs)
Leather	4 input, 10 hidden & 4 output units	0.01081
Velvet	4 input, 8 hidden & 4 output units	0.00812
Wood	4 input, 12 hidden & 4 output units	0.0203

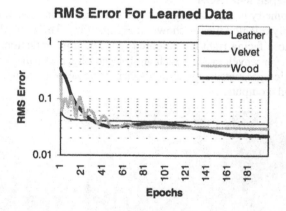

Fig. 9 RMS error for the first 200 epochs of training

Figures 12(a) and 12(b) contain the velvet and leather BRDF's respectively. Although the data will inevitably contain noise due to error in the measurement process the networks will produce smoothly varying BRDFs.

4.4 Simulated Data Fit

We have built a virtual gonioreflectometer, which enables us to generate samples of a BRDF from a given microgeometry. Since we are only required to generate a set of samples we do not need to store the entire BRDF in memory at once. In fact we choose a random incoming direction and perform a Monte Carlo simulation caching the interaction of light with the microgeometry for that specific direction. Once this simulation is complete we generate a set of samples to be learned later by the network. This entire process is repeated for several hundred incoming directions. Although this simulation is slow (on the order of 5-12 hours on a Pentium P200) it needs to be performed only once for a given BRDF. We choose three microgeometries and have generated BRDFs for these.

4.4.1 Anisotropic Brushed Metal

The microgeometry, similar to that used by Poulin [21], consists of a set of parallel ideally specular cylinders, shown in Figure 10(a). Figure 13(a) shows two spheres rendered using the learned BRDF. Note the characteristic anisotropic highlights from both the light source and the diffuse interreflection. The BRDF, which is assumed to have no spectral variation, was learned using a two layer neural network with 8 hidden units. The learned BRDF consists of 41 weights and a single scaling coefficient, which we store as 42 double precision numbers.

4.4.2 Specular Fibre Model

This microgeometry, shown in figure 10(b), consists of ideal specular cylinders arranged randomly in a plane. The BRDF produces results similar to that of silk or satin (see Figure 13(b)). The neural network consisted of 4 input units, 10 hidden units and a single output unit.

4.4.3 Spectrally Dependent BRDF

The final microgeometry is deliberately arranged to have a strong spectral variation. It consists of randomly oriented tiles as shown in Figure 10(c). Those tiles facing left are coloured green, those facing right are coloured red. The gonioreflectometer produces values for the 4 Meyer wavelengths, which are learned directly by the network. Figure 13(c) shows the microgeometry applied to a sphere. The network consists of 4 inputs, 10 hidden units and 4 outputs.

Fig.10 (a) Brushed metal microgeometry, (b) Specular fibre microgeometry (c) Spectrally varying microgemoetry

5 Rendering

We currently use a naïve Monte Carlo path tracer to visualise our BRDFs. The incoming and outgoing directions are converted to the appropriate parameterisation, used when training the network, and then applied in a single feed forward pass. The resulting radiance is rescaled appropriately using the same scaling constant which was used to normalise the data before training.

6 Conclusions & Future Work

Our initial investigation of neural networks for BRDF representation has produced promising results. The representation shows itself to be a concise and efficient one capable of faithfully encoding general BRDFs. Clearly only the network weights together with a scaling factor are needed to reconstruct the BRDF. All of the BRDFs presented in this paper are stored in less than 1k of memory. During rendering retrieving a BRDF value from the network simply involves doing a single feed forward pass. For each layer in the network (we use only 2) with i inputs and j units the retrieval cost is simply i*j multiplications, (i+1)*j additions and j evaluations of the activation function. Many issues remain open for future research.

The choice of a linear basis function and sigmoid activation function was based purely on evidence suggested by Kung [11]. A radial basis function coupled with a Gaussian activation function deserves further attention.

By generating samples of the same microgeometry at different scales it should be possible to use a single network which will capture the spatial variation of a BRDF across a surface. We should be able model effects such as crushed velvet. Across the surface the microgeometry changes gradually from perpendicular to parallel fibres. By generating samples, again from a virtual gonioreflectometer, we should be able to build a higher dimensional BRDF, which includes an additional parameter representing the angle of fibres. Some work on spatially varying reflection has previously managed to capture appearance characteristics like speckle [10]

One final area, which we are currently investigating, is the generation of reflected directions from a neural network BRDF. The aim is to use the BRDF as a probability density function and use this when importance sampling in Monte Carlo rendering. Initial analytic attempts have proved fruitless. The network is separable among the input parameter space, however the sigmoid activation function we use is non-integrable, and therefore we cannot build the required cumulative distributions. Our alternative is to first generate the BRDF samples and then numerically build the inverted cumulative probability distributions. We can then learn these distributions using the network.

References

1. Blinn, James. "Models of Light Reflection for Computer Synthesised Pictures," *Computer Graphics*, 11(4), July 1977
2. Cabral, Brian, Max Nelson and Springmeyer Rebecca. "Bi-directional Reflection Functions from Surface Bump Maps," *Computer Graphics*, 21(4), July 1987
3. Cook, Robert L. and Torrence, Kenneth E., "A Reflectance Model for Computer Graphics," *Computer Graphics*, 15(3), August 1981

Editors' Note: see Appendix, p. 325 for colored figures of this paper

34

4. Dana ,K.J., Nayar S.K., vam Ginneken Bram and Kooenderink Jan J., "Reflectance and Texture of Real-World Surfaces", Technical Report, *http://www.cs.columbia.edu/CAVE/curet/html*

5. Glassner., Andrew, S., " Principles of Digital Images Synthesis," The Morgan Kaufmann Series in Computer Graphics and Geometric Modelling. 1995

6. Gondek, Jay S., Meyer G.W., and Newmann J.G., "Wavelength Dependent Reflectance Functions" *Computer Graphics*, 28(4), July 1994

7. He, Xiao D., Torrence Kenneth E., Sillion, François X. and Greenberg Donald P., "A Comprehensive Physical Model for Light Reflection," *Computer Graphics*, 25(4), July 1991

8. Kajiya, James T., "Anisotropic Reflectance Models," *Computer Graphics, 19(4), August 1985*

9. Kajiya, James T., "The Rendering Equation," *Computer Graphics, 20(4), August 1986*

10. Krueger, Wolfgang. "Intensity Fluctuations and Natural Texturing," *Computer Graphics*, 22(4), August 1988

11. Kung., S.Y. Digital Neural Networks. Prentice Hall, 1992

12. LaFortune, Eric P. and Willems, Yves D., "Using the modified Phong Reflectance Model for Physically Based Rendering," Report CW 197, Department of Computing Science, K.U. Leuven, November 1994.

13. LaFortune, Eric P., Foo, Sing-Choong, Torrence, Kenneth E. and Greenberg, Donald P., "Non-Linear Approximation of Reflectance Functions," *Computer Graphics, 31(4), August 1997.*

14. Lalonde, Paul, "Representation and Uses of Light Distribution Functions," PhD. Thesis, University of British Columbia, December 1997.

15. Lewis, Robert R., "Making Shaders More Physically Plausible," In *Proceedings of Fourth Eurographics Workshop on Rendering*, Paris 1993

16. Meyer, G.W. "Wavelength Selection for Synthetic Image generation," *Computer Vision, Graphics and Image Processing*. No 48, 1998

17. Neumann, L. "Photosimulation : interreflection with arbitrary reflectance models and illumination," *Computer Graphics Forum* 8(1) March 1989

18. National Institute of Standards and Technology, United States Department of Commerce. Report of Workshop on Advanced Methods and Models for Appearance of Coatings and Coated Objects, March 1997

19. Oren, Michael and Nayar, Shreek., "Generalisation of Lambert's Reflectance Model," *Computer Graphics*, 22(4), August 1994, 28(4), July 1993

20. Phong, Bui-Tuong. "Illumination for Computer Generated Images." *Communications of the ACM*, 18(6) 1975.

21. Poulin, Pierre and Fournier, Alain. "A Model for Anisotropic Reflection" *Computer Graphics*, 24(4), August 1990

22. Rosenblatt, F., "The perceptron : A probabilistic model for information storage and organisation in the brain," *Psychology Review* Vol 65, 1958

23. Rummelhart, D. E.,et. al. "Parallel Distributed Processing (PDP) : Exploration in the Microstructure of Cognition," volume 1. The MIT Press, Cambridge MA, 1986

24. Sanford, B. and Robertson D., "Infrared Reflectance Properties of aircraft Paints," *Proceedings of the 1985 Meeting of the IRIS Speciality Group on Targets, Backgrounds and Discrimination*, February 1985

25. Schröder, Peter and Sweldens, Wim. "Spherical Wavelets: Efficiently Representing Functions on the Sphere," *Computer Graphics*, 29(4), August 1995

26. Shirley, Peter, Wang, Changyaw and Zimmerman Kurt. "Monte Carlo Techniques for Direct Lighting Calculations,"

27. Strauss, P. S., "A realistic lighting model for computer animators," *IEEE Computer Graphics & Applications*, 10(11) November 1990.

28. Ward, Gregory J., "Measuring and Modelling Anisotropic Reflection," *Computer Graphics*, 26(2), July 1992

29. Westin, Stephen H., Arvo, James R. and Torrance Kenneth E., "Predicting Reflectance Functions from Complex Surfaces," *Computer Graphics*, 26(2), July 1992

A new Form Factor Analogy and its Application to Stochastic Global Illumination Algorithms

Robert F. Tobler[1], László Neumann[2], Mateu Sbert[3], and Werner Purgathofer[1]

[1] Vienna University of Technology, Austria
[2] Maros u. 36, H-1122 Budapest, Hungary
[3] Universitat de Girona, Spain

Abstract. A new form factor analogy, that has been derived from results of integral geometry, is introduced. The new analogy is shown to be useful for stochastic evaluation of the local form of the rendering equation used in various Monte Carlo methods for calculating global illumination. It makes it possible to improve importance sampling in these methods, thereby speeding up convergence. A new class of bidirectional reflection distribution functions that directly benefits from the analogy and permits exact evaluation and calculation of correctly distributed vectors for Monte Carlo integration is presented.

1 Introduction

Nusselt's analogy is a well known geometric relation that expresses the form factor as a ratio of two areas. A patch j is projected onto the hemisphere above the point $\mathbf{P}_i$. Now this spherical patch is projected orthogonally down onto the disk underneath the hemisphere. The form factor from the differential area around the center of the hemisphere to patch j, F_{ij} is the ratio of the area of the twice projected patch and the area of the disk (see figure 1), i.e. $F_{ij} = \frac{A'_j}{A_D}$. This analogy can be used for cosine-

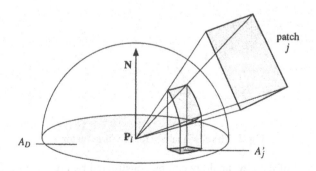

Fig. 1. Nusselt's analogy for the form factor.

weighted hemispherical sampling and thus for the calculation of form-factors. In this paper we will present a new analogy together with its applications to stochastic global illumination algorithms.

2 Integral geometry and analogies for the form factor

A very interesting result about the differential form factor between two spherical patches i and j on the surface of the same sphere, that is related to Integral Geometry can be found or derived from results of various authors ([6], [10], [7]). Namely, that this form factor is $F_{ij} = \frac{A_j}{A_S}$, where A_S is the area of the sphere. This can be seen very easily considering figure 2. Since the two angles θ are equal and $r = 2R \cdot \cos\theta$, the expression for the form factor becomes

$$F_{ij} = \frac{1}{A_i} \int_{A_i} \int_{A_j} \frac{\cos\theta_i \cos\theta_j}{\pi r^2} V_{ij} dA_i dA_j = \frac{1}{4\pi R^2 A_i} \int_{A_i} \int_{A_j} dA_i dA_j = \frac{A_j}{A_S} \qquad (1)$$

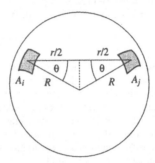

Fig. 2. Geometry for the form factor between patches in a sphere.

This new way of thinking about the form factor can be represented in a nice analogy. Consider projecting a patch j onto a sphere that just touches the surface at point $\mathbf{P}_i$. Now the form factor F_{ij} from the differential area with the touching point as center to the patch j is the fraction of the area of the total sphere that is covered by the projection, i.e. $F_{ij} = \frac{A'_j}{A_S}$ (see figure 3).

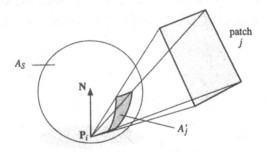

Fig. 3. The new analogy for the form factor.

Another proof of the new analogy can be constructed by considering a differential area on the hemisphere (see figure 4):

$$dA_H = \sin\theta \, d\theta \, d\phi \qquad (2)$$

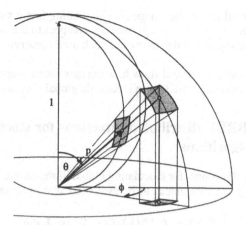

Fig. 4. A proof of the equivalency of the two analogies by differential areas.

Projecting this differential area onto the disk of Nusselt's analogy adds a cosine factor, and the result is the well known cosine-weighted differential area used in the rendering equation:

$$dA_D = \cos\theta\,\sin\theta\,d\theta\,d\phi \tag{3}$$

If we project the differential area from the hemisphere onto the sphere of the new analogy, two factors have to be taken into consideration: the perspective shortening $p^2 = \cos^2\theta$, and the stretching of the differential area on the sphere, due to the tilt of the projection direction with respect to the normal: $t = \frac{1}{\cos\theta}$. Thus the projected differential area on the sphere is given by

$$dA_S = p^2 \cdot t \cdot \sin\theta\,d\theta\,d\phi \tag{4}$$

which is exactly the same as the cosine-weighted differential area dA_D from equation (3). Thus the disk in Nusselt's analogy is an area-preserving mapping of the sphere in the new analogy.

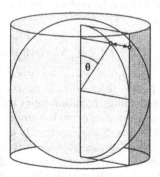

Fig. 5. An area-preserving mapping of a sphere onto a cylinder.

In the following chapters we will make use of another well-known area-preserving mapping of a sphere [5](see figure 5): if you project each point of the sphere horizontally outward from the axis onto a cylinder, the area does not change. Again this can be proven using differential areas: the perspective shortening factor $\sin\theta$ is cancelled out by the stretching due to the tilt of the normal with respect to the projection direction. Maling [5] gives a survey of the derivations of such area-preserving projections of the sphere.

In the following sections we will show how the new analogy and the cylinder mapping can be used for evaluating BRDFs in stochastic global illumination methods.

3 Generating BRDF-distributed directions for stochastic global illumination algorithms

In Monte Carlo type algorithms for calculating global illumination, the following local form of the rendering equation needs to be evaluated at various sampling points in a scene:

$$L^{out}(\mathbf{V}) = \int_\Omega L^{in}(\mathbf{L})\, f_r(\mathbf{L}, \mathbf{V})\, \mathbf{N} \cdot \mathbf{L}\, d\omega \qquad (5)$$

For sake of simplicity, the point at which the evaluation takes place, has been omitted. $L^{out}(\mathbf{V})$ denotes the radiance leaving the point in direction $\mathbf{V}$, $L^{in}(\mathbf{L})$ denotes the radiance coming from direction $\mathbf{L}$, $f_r(\mathbf{L}, \mathbf{V})$ denotes the BRDF, and $\mathbf{N}$ is the normal vector.

Speeding up the stochastic evaluation of this integral can be achieved by sampling the direction $\mathbf{L}$ according to the known parts of the integrand and summing up the results of evaluating the remaining parts of the product for each sampled direction.

Standard methods assume an unknown distribution of incoming radiance L^{in}, thus the optimal sampling scheme for these methods distributes $\mathbf{L}$ according to $f_r(\mathbf{L}, \mathbf{V})\,\mathbf{N} \cdot \mathbf{L}$. This distribution can, however, not be evaluated exactly for arbitrary BRDFs.

3.1 Cosine-distributed directions

In radiosity algorithms all surfaces are assumed to be perfect lambertian reflectors, thus $f_r(\mathbf{L}, \mathbf{V}) = \frac{1}{\pi}$ and the directional sampling schemes need only take the cosine term $\mathbf{N} \cdot \mathbf{L}$ into account. The usual way of generating randomly distributed directions that follow the lambertian law is based on Nusselt's analogy for the form factor. This is due to the fact that $\mathbf{N} \cdot \mathbf{L}\, d\omega$ in equation (5) can be replaced by the cosine weighted differential area from equation (3) yielding:

$$L^{out}(\mathbf{V}) = \int_\Omega L^{in}(\mathbf{L})\, f_r(\mathbf{L}, \mathbf{V})\, \sin\theta\, \cos\theta\, d\theta\, d\phi \qquad (6)$$

First a uniformly distributed point within a disc is generated. This can be done using a rejection method based on generating a uniform point within a square and discarding points that lie outside the inscribed disc or with a direct method which is based on generating points $(\sqrt{r}, \varphi)$, where r and φ are uniformly distributed. In the second step this point is projected onto a hemisphere over the disc, and so the resulting vector from the origin of the hemisphere to the projected point is then distributed according to the cosine distribution. Shirley et al. [9] compiled more detailed descriptions of these methods.

Of course, due to equation (4), the new analogy can also be easily used to generate random directions that follow the lambertian law. First a uniformly distributed point on the surface of a sphere is generated. Again this can be done using a rejection method: generate a uniformly distributed point within a cube, reject it, if it is outside the inscribed sphere, and then normalize its corresponding vector to get a uniformly distributed point on a unit sphere. Alternatively a direct method can be used that is based on the area-preserving mapping of the sphere to a cylinder (see figure 5): generate the points on the cylinder and project them back to the sphere. In the second step the normalized, uniformly distributed direction vector is added to the normalized normal vector of the surface to obtain a random direction according to the lambertian law.

The methods based on our new analogy are faster since there is no need of a local coordinate system. Thus the directions can be directly generated in the global coordinate system, foregoing the need of a transformation.

3.2   Directions distributed according to general BRDFs

For general BRDFs the sampling of directions should proceed according to $f_r(\mathbf{L}, \mathbf{V})\, \mathbf{N} \cdot \mathbf{L}$. In order to make this possible BRDFs could be defined as functions over $S^2 \times S^2$, i.e. $f_r(\mathbf{L}, \mathbf{V}) = f_{r,S}(\mathbf{L}_S, \mathbf{V}_S)$, where $\mathbf{L}_S$ and $\mathbf{V}_S$ are elements of S^2 (the surface of the sphere) and defined by the projection of the original vectors $\mathbf{L}$ and $\mathbf{V}$ onto the sphere in the new analogy (see figure 6). Note that the sphere can have any scale; we use $r = 1$ for the following derivations. Calculating the normalized vectors $\mathbf{L}_S$ and $\mathbf{V}_S$ is very simple,

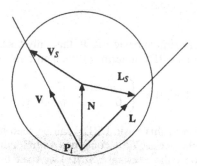

Fig. 6. Projecting the original vectors of the BRDF, L and V onto the sphere of the new analogy.

assuming that the normal vector $\mathbf{N}$ and the vectors $\mathbf{V}$ and $\mathbf{L}$ are normalized as well:

$$\mathbf{L}_S = 2(\mathbf{N} \cdot \mathbf{L})\mathbf{L} - \mathbf{N}$$
$$\mathbf{V}_S = 2(\mathbf{N} \cdot \mathbf{V})\mathbf{V} - \mathbf{N} \tag{7}$$

A suitable scheme for storing such BRDFs could be a four dimensional extension of the spherical wavelet approach introduced by Schröder and Sweldens [8]. Alternatively, both points could be projected from the surface of the sphere to the area-preserving cylinder, and the r and ϕ coordinates on this cylinder could be used to define the BRDF. In both cases importance sampling according to the function on the sphere or cylinder will automatically incorporate the cosine term due to the construction of the sphere according to the new analogy.

4 A new class of BRDFs suited for stochastic global illumination algorithms

Here we introduce a class of BRDFs for which the distribution of directions L according to $f_r(\mathbf{L}, \mathbf{V})$ $\mathbf{N} \cdot \mathbf{L}$ in equation (5) can be performed exactly. For this purpose we will define the BRDF in terms of the vectors $\mathbf{V}_S$ and $\mathbf{R}_S$, both from S^2. $\mathbf{V}_S$ is calculated according to equation (7), but $\mathbf{R}_S$ is now the reflection vector of $\mathbf{L}_S$.

Again the normalized $\mathbf{R}_S$ can be calculated very easily:

$$\mathbf{R}_S = 4(\mathbf{N} \cdot \mathbf{L})^2 \mathbf{N} - 2(\mathbf{N} \cdot \mathbf{L})\mathbf{L} - \mathbf{N} \tag{8}$$

We now define the new BRDF based on the following metric in S^2:

$$m = m_S(\mathbf{R}_S, \mathbf{V}_S) = 1 - \mathbf{R}_S \cdot \mathbf{V}_S \tag{9}$$

If we define our BRDF exclusively using m, we automatically gain reciprocity by the simple fact that we could just as well reflect the vector $\mathbf{V}_S$ instead of the vector $\mathbf{L}_S$ without a change in the value of m. Note that we can use $\mathbf{L}_S$ instead of $\mathbf{R}_S$ to obtain retro-reflective BRDFs.

4.1 Basic BRDF

We start by defining a very simple basic BRDF that only takes on the values of a constant or zero. We define this BRDF in terms of the metric introduced in equation (9):

$$f_z(\mathbf{L}, \mathbf{V}) = f_{z,S}(m) = \begin{cases} const(z) & \text{if } m < z \\ 0 & \text{otherwise} \end{cases} \quad z, m \in [0, 2] \tag{10}$$

A graphical representation of this basic BRDF can be seen in figure 7. The value of the BRDF is constant if the vector $\mathbf{V}_S$ is contained within a disk around $\mathbf{R}_S$, otherwise it is zero. In order to choose the constant $const(z)$ we scale the basic BRDF so that its

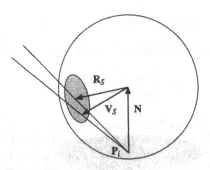

Fig. 7. A geometric definition of the basic BRDF.

albedo (i.e. the part of the incoming energy it reflects back into the hemisphere) is equal to 1. The parametrization of the basic BRDF with parameter z has been chosen so that the area of the spherical disk where the basic BRDF is non-zero is linearly dependent on z. This can be easily seen by using an area-preserving map of the sphere to a cylinder with its axis aligned to $\mathbf{R}_S$. Figure 8 gives a graphical representation of this relation.

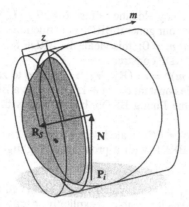

Fig. 8. The area of the disk used by the basic BRDF.

Since the albedo of the basic BRDF is linearly dependent on the area covered by the disk, the constant necessary for setting the albedo to 1 can be easily derived as:

$$const(z) = \frac{2}{\pi \cdot z} \qquad (11)$$

4.2 Combining basic BRDFs

A simple generalization of these basic BRDFs would be a linear combination of a finite number of such BRDFs. This, however, is only the discrete special case of combining infinitely many basic BRDFs using arbitrary weighting functions. For a simple construction of such a combination we rewrite the basic BRDF as follows:

$$f(m) = \frac{2}{\pi} \cdot pdf(m) \qquad (12)$$

where $pdf(m)$ is a probability density function over the interval $[0, 2]$:

$$pdf(m) = \begin{cases} \frac{1}{z} & \text{if } m < z \\ 0 & \text{otherwise} \end{cases} \quad z, m \in [0, 2] \qquad (13)$$

We can now conveniently replace $pdf(m)$ with an arbitrary probability density function defined in $[0, 2]$ without compromising our BRDF construction, so that the albedo of the BRDF remains exactly 1 for all directions V!

Calculating a randomly distributed vector according to this new BRDF construction is very simple. Taking the given vector and calculating the reflected vector within the sphere $\mathbf{R}_S$ we can construct the area-preserving cylinder aligned with $\mathbf{R}_S$ according to figure 8. On this cylinder we can generate a point in polar coordinates using:

$$m = F^{-1}(u)$$
$$\phi = 2\pi v \qquad (14)$$

where u and v are uniform random variables in $[0, 1]$ and $F(m) = \int pdf(m)dm$. Projecting this point back onto the sphere yields the correctly distributed vector.

A few examples for useful probability density functions are the following:

the constant function $\frac{1}{2}$: this yields the diffuse BRDF $f_r(\mathbf{L}, \mathbf{V}) = \frac{1}{\pi}$ which turns out to be a special case of our general BRDF construction

the Dirac delta function at $m = 0$: this yields the BRDF for a perfect mirror, another special case of our BRDF construction

a power function $c \cdot (1 - m)^n = c \cdot (\mathbf{R}_S \cdot \mathbf{V}_S)^n$**:** clamped to zero for $\mathbf{R}_S \cdot \mathbf{V}_S < 0$ and scaled with a suitable constant $c = n + 1$, this yields a BRDF that can be thought of as an analog to the Phong BRDF, but defined with the vectors $\mathbf{R}_S$ and $\mathbf{V}_S$ instead of $\mathbf{R}$ and $\mathbf{V}$.

an exponential function $c \cdot e^{-k \cdot m}$**:** scaled with a suitable constant $c = k/(1 - e^{-2k})$, this yields another BRDF with a pronounced highlight, but it has a non-zero diffuse component

The freedom to choose any $pdf(m)$ which can be evaluated and for which equations (14) can be calculated, makes it possible to explicitly specify the intensity drop-off for the highlight.

4.3 Properties of the new BRDFs

The new class of BRDFs is highly suited for use in Monte Carlo global illumination algorithms due to the simple evaluation procedures. There are a number of advantages when these new BRDFs are used:

- The generation of correctly distributed reflection direction is performed using a mapping of two 1-dimensional uniform random values (see equation (14)). This makes it possible to use quasi-Monte Carlo methods [3], thereby significantly speeding up the convergence of these algorithms.
- Due to the new construction it is possible to define probability density functions containing both a constant term and a non-diffuse component. Therefore correctly distributed reflection directions for such BRDFs can be generated in one step without the need for handling the diffuse part of the function separately.
- Due to the way the new BRDFs are defined, they have an albedo of 1 for every incoming direction. This facilitates exact control of the reflected energy using a reflectance coefficient.

We applied the new BRDFs using the indicated power function as pdf to generate a few example surfaces. Figure 9 (see also appendix) shows three objects with the exponents set to (from left to right) $n = 2, 8$, and 64. An unsolved detail of the BRDF is its behaviour at grazing angles. As the incident angle varies from normal to grazing incidence, the shape of the highlight varies from perfectly circular, to elliptic (see the right highlight on the sphere in figure 9) and finally in the limit case the reflection directions are distributed in a near 1-dimensional distribution in all grazing directions. However, the grazing angle represents a problem in a number of BRDF constructions.

5 Directional importance in global illumination algorithms

A number of Monte Carlo algorithms for global illumination use directional importance information to speed up the integration ([4], [2]). This corresponds to having information about the distribution of $L^{in}(\mathbf{L})$ in equation 5.

We can use our new analogy to sample the directions according to $L^{in}(\mathbf{L})\mathbf{N} \cdot \mathbf{L}$. This can be achieved by storing the importance information (i.e. our current estimate of

Fig. 9. Examples using the power function as pdf (see also appendix).

the incoming radiance) in a data structure based on the new analogy. We propose two possible data structures for this kind of importance scheme.

If we use the area-preserving mapping to a cylinder, we can use a two dimensional array, or better any hierarchical two dimensional data structure for storing the current estimate of the incoming radiance for any direction (see figure 10). The second scheme

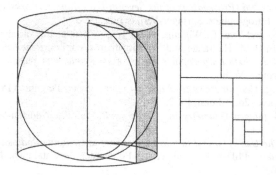

Fig. 10. A k-d-tree based storage scheme for directional importance.

uses the segmentation of the surface of the sphere into a number of spherical triangles. If estimates of incoming radiance are stored for each of these spherical triangles, Arvo's method of generating uniformly distributed points within each of these spherical triangles [1] can be used to generate corresponding cosine-distributed directions for importance sampling.

6 Conclusion and future work

We have shown a new analogy for the form factor, and demonstrated its applicability in the field of stochastic global illumination algorithms. The new formulation leads to a number of improvements that speed up stochastic global illumination algorithms. A new class of BRDFs based on this analogy has been presented, which is perfectly suited for quasi-Monte Carlo algorithms. Currently we are investigating the new class of BRDFs for use in fitting for arbitrary BRDFs and the use of the new analogy for more complex BRDF definitions, in order to make these more suitable for stochastic algorithms as well. We are also working on modifications to the model in order to overcome the problem at grazing angles.

Acknowledgements

This work was partly performed under the "Human Capital and Mobility" project *Online graphics systems for the creation of photorealistic images for complex scenes*, contract number *ERBCHGECT92009*, providing "Acces to Supercomputing Facilities for European Researchers".

References

1. James Arvo. Stratified sampling of spherical triangles. In Robert Cook, editor, *SIGGRAPH 95 Conference Proceedings*, Annual Conference Series, pages 437–438. ACM SIGGRAPH, Addison Wesley, August 1995. held in Los Angeles, California, 06-11 August 1995.
2. Philip Dutre and Yves D. Willems. Importance-Driven Monte Carlo Light Tracing. In G. Sakas, P. Shirley, and S. Müller, editors, *Photorealistic Rendering Techniques (Proceedings of the Fifth Eurographics Workshop on Rendering)*, pages 185–194, Wien–New York, June 1994. Springer-Verlag.
3. Alexander Keller. Quasi-monte carlo radiosity. In X. Pueyo and P. Schröder, editors, *Rendering Techniques '96 (Proceedings of the Seventh Eurographics Workshop on Rendering)*, pages 101–110. Eurographics, Springer-Verlag, June 1996.
4. Eric P. Lafortune and Yves D. Willems. A 5D Tree to Reduce the Variance of Monte Carlo Ray Tracing. In P. M. Hanrahan and W. Purgathofer, editors, *Rendering Techniques '95 (Proceedings of the Sixth Eurographics Workshop on Rendering)*, pages 11–20, Wien–New York, 1995. Springer-Verlag.
5. Derek H. Maling. *Coordinate systems and map projections*. Pergamon Press, Oxford–New York–Seoul–Tokyo, 2nd edition, 1992.
6. Luis A. Santaló. *Integral Geometry and Geometric Probability*. Addison-Wesley, New York, 1976.
7. Mateu Sbert. *The Use of Global Random Directions to Compute Radiosity – Global Monte Carlo Techniques*. PhD thesis, Universitat Politècnica de Catalunya, Barcelona, Spain, November 1996.
8. Peter Schröder and Wim Sweldens. Spherical Wavelets: Texture Processing. In P. M. Hanrahan and W. Purgathofer, editors, *Rendering Techniques '95 (Proceedings of the Sixth Eurographics Workshop on Rendering)*, pages 252–263, Wien–New York, 1995. Springer-Verlag.
9. Peter Shirley, Chang Yaw Wang, and Kurt Zimmermann. Monte carlo techniques for direct lighting calculations. *ACM Transactions on Graphics*, 15(1):1–36, January 1996.
10. Robert Siegel and John R. Howell. *Thermal Radiation Heat Transfer*. Hemisphere Publishing Corporation, Washington, D.C., 3rd edition, 1992.

Editors' Note: see Appendix, p. 326 for colored figure of this paper

An ambient light illumination model.

S. Zhukov*, A.Iones*, G.Kronin**

Creat Studio,
*Applied Mathematics Dept., St.-Petersburg State Technical University (Russia),
**Calculus Dept., St.-Petersburg State University (Russia)
{zh,iones,kronin}@creatstudio.spb.ru
www.CreatStudio.com

Abstract. In this paper we introduce an empirical ambient light illumination model. The purpose of the development of this model is to account for the ambient light in a more accurate way than it is done in Phong illumination model, but without recoursing to such expensive methods as radiosity. In our model we simulate the indirect diffuse illumination coming from the surfaces of the scene by direct illumination coming from the distributed pseudo-light source. The estimation of indirect illumination is based on the concept of obscurance coefficients that resemble the integrated weighted form-factors computed for some vicinity of a given point. The same idea is used to account illumination of a given point (patch) from light sources. This illumination is computed as a sum of direct illumination calculated using the standard local reflection model and empirically estimated indirect illumination based on the same obscurance concept.

Keywords: illumination model, ambient light, radiosity, form-factor, obscurance.

1. Introduction

The principal goal of realistic image synthesis is to develop the methods that would allow to visualize three-dimensional scenes realistically. To achieve this goal a number of illumination models have been elaborated. These models range from local illumination models that are easy to implement and fast to compute to much more complex global methods that allow the generation of photorealistic-quality images [Watt92, Foley92]. The latter methods range from classic schemes [Goral84, Whitt80] to more modern approaches [Hanr91, Cohen93, Sill94, Sbert93, Sbert96]. The other way to create realistic images is based on the use of more ad hoc techniques (e.g., procedural shaders in MentalRay® or RenderMan® [Upst92]) that strive to recreate the most appealing lighting effects and may sacrifice the physical realism for this goal.

In this paper we introduce an empirical ambient light illumination model. Our model simulates the indirect diffuse illumination coming from the surfaces of the scene by direct illumination coming from the distributed pseudo-light source. The use of this model enables to account for ambient light without recoursing to the expensive radiosity method. Due to the nature of indirect illumination simulation, it is possible to reproduce appealing and realistic images without explicit light source setting at all. This feature enables us to quickly preview the scenes that look appealing and realistic.

The idea of the method lies in computing the *obscurance* of a given point. Obscurance is a geometric property that reflects how much a given surface point is open. For a given scene, obscurances for each patch can be computed only once and stored into file. This will enable to quickly recompute the lighting in the environment with moving light sources.

The computation of obscurances involves form-factor determination in the vicinity of a given point. Due to the locality of the method, it is much faster than radiosity, while the generated images are realistic and look similar to the ones generated by radiosity after a number of iterations.

The same idea is used to account illumination of a given point (patch) from light sources. This illumination is computed as a sum of direct illumination calculated using the standard local reflection model and empirically estimated indirect illumination based on the same obscurance concept.

2. Observations leading to the obscurance illumination model.

Most local illumination models use ambient light to account for secondary diffuse reflections in the environment. These models usually assume that ambient light is constant over the whole scene. Clearly, this is just a rough approximation. E.g., in a room lit with the diffuse light the illumination over the surface of the walls is not constant – it is normally darker near the room corners. A similar effect is a shadow under the car standing on the snow in a cloudy day. These lighting (darkening) effects in the obscured areas can be modeled with the use and at the expense of radiosity method, since this global illumination model takes into account secondary diffuse reflections.

Our goal is to develop an illumination model that would enable us to reproduce the darkening effects in the obscured areas by empirically accounting for indirect illumination without recoursing to the expensive radiosity solution. We simulate indirect illumination by a direct illumination of a specific distributed ambient light source. The darkening in the obscured areas is achieved by measuring the geometric obscurance in these areas.

The rest of this paper is organized as follows. We start with the 'practical' explanation of our illumination model, showing how it works and how to compute the illumination using it. Then we give some physical foundations of the model. Finally, we show some results.

3. Obscurance illumination model.

As it was stated above, our illumination model is based on a more accurate empiric accounting for the indirect illumination than it is done for ambient light in Phong illumination model and does not involve expensive computations like the ones in radiosity. Similarly to radiosity, our model is view independent and is based on subdividing the environment into discrete patches.

Our primary observation is that in the environments illuminated by diffuse light it is usually darker in obscured areas. Our illumination model is based on the notion of *obscurance*. Roughly speaking, *obscurance* measures the part of the hemisphere obscured by the neighbor patches. E.g., near a corner of a room the obscurance of patches is higher than on the plane open parts. From the physics of light transport point of view, obscurance simulates the lack of secondary light ray reflections coming to the specific parts of the environment that makes them darker. This is unlike radiosity where secondary reflections add the intensity.

3.1. Obscurance illumination model without light sources.

Let's assume firstly that there are no specific light sources in the environment, in other words, the environment is lit by a perfectly diffuse light coming from everywhere. We will better elaborate this notion in the following sections.

Definition. Let P be a point on the surface in the scene, x belongs to the unit hemisphere hS^2 centered in P, aligned with the surface normal in P and lying in the outer part of the surface. The function $L(P,x)$ is defined as follows:

$$L(P,x) = \begin{cases} dist(P,C), \text{where } C \text{ is the first intersection point of ray } Px \text{ with the scene} \\ +\infty, \text{otherwise (ray } Px \text{ does not intersect the scene)} \end{cases} \tag{1}$$

Note, that for properly constructed scenes any ray outcoming from the point P on a surface always hits a frontface before a backface.

We assume that the more patch is open, the more its intensity is. In other words, the farther the intersection point is in a given direction, the more energy is coming from this direction to the patch. Therefore the intensity of a given patch can be approximated as follows:

$$I(P) = \frac{2}{\pi} \times I_A \times \iint_{x \in hS^2} \rho(L(P,x)) \cos \alpha \, dx , \tag{2}$$

where:

$\rho(L(P,x))$ –   an empirical mapping function that maps the distance $L(P,x)$ to the first obscuring patch in a given direction to the energy coming from this direction to patch P

α –   the angle between the direction x and patch normal

I_A –   ambient light power - a global constant for the whole environment

Let's discuss the meaning of the mapping function $\rho()$. Firstly, this function is a monotone increasing function of L. Indeed, we assume the farther the patch is obscured in a given direction, the more light is coming from it. Secondly, this function is up bounded. This reflects the observation that normally the lighting of a given point is primarily affected by its neighborhood. This is especially true for the environments without bright light sources that may affect the illumination at large distances. Therefore, we conclude that the function $\rho()$ has the shape shown schematically on Figure 1. The function $\rho(L) = 1$ for $L > L_{max}$. The shape and the meaning of the mapping function $\rho()$ is discussed more formally in the following sections.

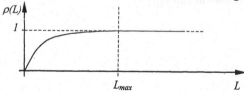

Figure 1. An example plot of a typical function $\rho(L)$

Since $\rho(L) = 1$ for $L > L_{max}$, we can take into account only the intersections within some neighborhood of a given point – if the ray Px does not intersect any patch at the

48

distance less than L_{max}, it may be assumed that it does not intersect any patch at all and the energy coming from this direction is constant.

Definition. Let P be a point on the surface in the scene. *Obscurance[1]* of a point P is defined as follows:

$$w(P) = \frac{2}{\pi} \times \iint_{x \in hS^2} \rho(L(P,x)) \cos \alpha \, dx. \tag{3}$$

So defined, the obscurance is the weighted average length of a chord originating from the point P (the length of chords is measured between P and the first intersection with a patch in the scene). Clearly, for any surface point P: $0 \leq w(P) \leq 1$. E.g., for a point standing in the middle of the base of a large hemisphere $w(P) = 1$. The function $w(P)$ reflects the local geometric properties of point P. Obscurance value 1 means that the patch is fully open, 0 - fully closed (this may happen only in the degenerate cases). Typical obscurance values for different patches are illustrated on Figure 2. As follows from its definition, patch obscurance resembles patch's average form-factor (see Section 5).

From the equations (2) and (3) we conclude that the intensity in point P can be computed as follows:

$$I(P) = I_A * w(P). \tag{4}$$

The equation (4) actually determines the illumination model working in the absence of light sources. This model can be called *locally-global* since it enables to directly compute the intensity for a given patch, but the computation involves the analysis of the local environment geometry.

On Figure 2 we show a snapshot of a real scene the illumination for which is computed using equation (4) (there are no specific light sources). Notice that a number of pseudo-shadow effects are clearly visible. We claim that the scene looks similar to the view in a cloudy day.

The use of obscurances facilitates helps to outline the surface profile without light source setting at all. Indeed, in practice the task of light source position selection is fairly delicate and laborious, since it requires a number of iterations to produce appealing results (that is the goal in many applications, e.g. in video games). So the purpose of actual illumination calculation is to show shadows and color light spots.

Obscurance is an intrinsically geometric property - it does not depend on light sources, so for a given scene it can be computed only once and stored for future use to recompute the lighting from the moving light sources, or while stuffing the light sources during the environment editing.

---

[1] The name 'obscurance' is a bit confusing, since it actually denotes the 'openness' of a patch, but we prefer to stick to the earlier introduced name [Zhuko98].

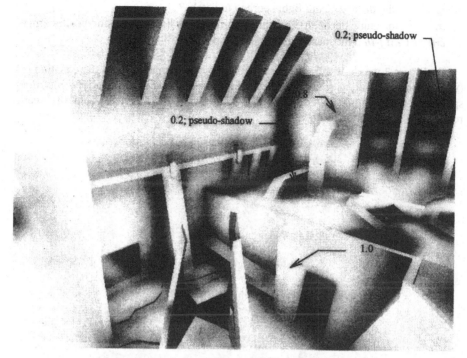

Figure 2. Obscurances of different patches.

An empiric approach called 'accessibility shading' resembling the use of obscurances was exploited in [Mill94], where similar geometric properties of the surfaces were used to recreate specific visual effects such as aging of materials.

3.2. Obscurance illumination model accounting for light sources.

The goal of our illumination model is to account indirect illumination in an easy manner. To compute the illumination of a given patch caused by a given light source, we separate the illumination to the direct and indirect terms. We compute the direct illumination using the standard local diffuse illumination model. The indirect illumination from the light sources is empirically accounted (weighted) with the help of obscurance coefficient as well. Specifically, we use the following equation (the extension of the equation (4)):

$$I(P) = (I_A + I_S) * w(P) + I_S,$$ (5)

where:

I_A - ambient light intensity

I_S - the sum of direct intensities coming from visible light sources,

$$I_S = \sum_{j=1}^{N} \delta(j) \frac{I_j}{r_j^2} \cos\alpha_j,$$ (6)

where:

N - the number of light sources in the environment

I_j - the intensity of j-th light source

r_j - the distance from light source j to patch P
α_j - the angle between direction toward the j-th light source and patch normal
$\delta(j)$ - 1 - if j-th light source is visible from patch P, 0 - otherwise

On Figure 3 we show a fragment of the same scene rendered using the equation (5) (with a number of light sources added). Notice that a number of shadows and light spots appeared, but the overall impression does not differ much from the scene rendered with obscurances only.

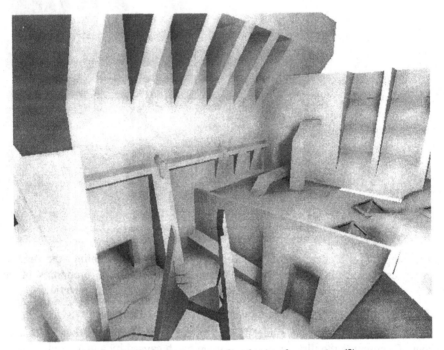

Figure 3. The same scene rendered using the equation (5).

4. Physical justification of the model.

We start the justification of our model by first considering the environment illuminated by purely diffuse light without specific light sources. In our model, to simulate the diffuse light we assume that the environment is filled with the emitting non-absorbing gas with constant emittance λ per unit volume. Indirect illumination from the surfaces is modeled using direct illumination coming from this gas. Clearly, the more the visible gas volume is from a given patch A, the more this patch is illuminated.

It is well known that the perceived intensity (brightness Y) is usually estimated as a gamma-corrected value of the actual intensity [Poynt93]:

$$Y = (I)^{1/\gamma}, \text{ where } \gamma \approx 2.2 \dots 2.5 \tag{7}$$

Let's compute now the actual illumination of the patch P formally. The differential intensity $I_{P, dV}$ coming from a volume element dV to patch P is equal to (Figure 4):

$$I_{P, dV} = \lambda \frac{1}{4\pi r^2} \cos\alpha \, dV \,, \tag{8}$$

where:

α - is the angle between patch normal and the direction toward the volume element

r - the distance between the patch and volume element.

In equations (2) and (3), x was a surface element of a unit hemisphere hS^2; $dx = \sin\varphi \, d\varphi \, d\psi$. Computing the intensity of patch P due to direct illumination of the gas and taking into account that $dV = r^2 \sin\varphi \, d\varphi \, d\psi \, dr$ in polar coordinates, obtain:

$$I = \lambda \iiint\limits_{overvisiblevolume} \frac{1}{4\pi r^2} \cos\alpha \, dV = \lambda \int_0^{2\pi} d\psi \int_0^{\frac{\pi}{2}} d\varphi \int_0^{L(\varphi,\psi)} dr \frac{1}{4\pi r^2} r^2 \cos\left(\frac{\pi}{2} - \varphi\right) \sin\varphi =$$

$$= \frac{\lambda}{4\pi} \int_0^{2\pi} d\psi \int_0^{\frac{\pi}{2}} d\varphi L(\varphi,\psi) \sin^2\varphi = \frac{\lambda}{4\pi} \iint\limits_{x \in hS^2} L(x) \cos\alpha \, dx \tag{9}$$

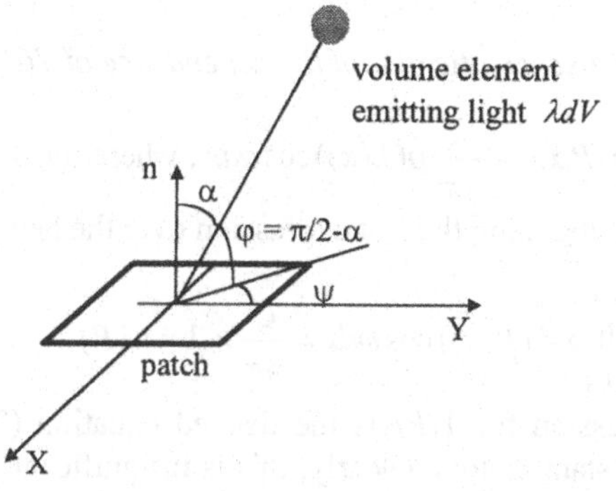

Figure 4. Illumination of a patch by a gas volume element

Note, that equation (9) is similar to equation (2). The difference is that in the equation (9) the distance $L(x)$ is not weighted by a function $\rho(L)$. In (9) the expression $I(P,x) = \frac{\lambda}{4\pi} L(x) \cos\alpha \, dx$ shows the differential intensity coming from the given direction x. Due to the equation (7), unlike $I(P,x)$ the perceived intensity $Y(P,x)$ is non-linearly dependent on L. The more the distance L, the less is the relative influence of the newly accounted volume elements to the perceived intensity. On Figure 5 schematic plots of $I(P,x)$ and $Y(P,x)$ as a function of distance L are shown.

52

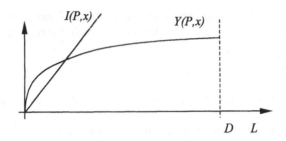

Figure 5. A schematic plot of the dependence of I(P,x) and Y(P,x) of L. D is the diameter of the environment.

In practice, we approximate the dependence of $Y(P,x)$ of the distance L by the function $\rho(L)$ plotted schematically on Figure 6.

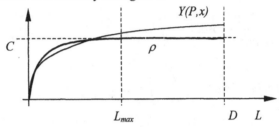

Figure 6. A schematic plot of the dependence of Y(P,x) of L.

Therefore, obtain: $Y(P,x) = \dfrac{\lambda}{4\pi}\rho(L(x))\cos\alpha\,dx$, where the shape of function $\rho(L)$ is shown on Figure 6. Integrating the last expression over the hemisphere, obtain:

$$Y(P) = \frac{1}{4\pi} \times \lambda \times \iint_{x \in hS^2} \rho(L(P,x))\cos\varphi\,dx = \frac{C}{4\pi} \times \lambda \times w(P). \tag{10}$$

The obtained expression for $Y(P)$ is the desired equation (2) for $I(P)$, the only difference is in the constant factors. Clearly, this is insignificant since to compute the lighting for a given scene it is sufficient to use just some constant factor (dependent either on I_A (2) or on λ in (10)). The final remark is that usually the intensities computed using radiosity method are gamma-corrected before the actual rendering, while in our model this gamma-correction is intrinsic.

To justify the equation (5), we suppose that with the presence of light sources affects the luminance of gas in the neighborhood of patch P. Approximately (if light sources are relatively far from patch P), the intensity of the gas in the neighborhood of patch P can be computed using the equation (6). Therefore, we accurately account for the direct illumination and approximate the indirect light using the lit gas. Indirect illumination from the invisible light sources can be accounted by changing the function $\delta(j)$ in equation (6): $\delta(j) = 1$, if the source is visible, $\delta(j) = \beta$, $0 \le \beta \le 1$, if the source is not visible.

Definitely, the physical foundation of the obscurance illumination model is empiric, and the final justification of the usefulness of the chosen approach is that it produces nice illumination results by simulating the subtle indirect illumination effects without recoursing to the expensive methods.

5. Computing the lighting using the obscurance illumination model

As it was stated above, the obscurance coefficient is similar to integrated weighted form factors. Indeed, for a given patch Q illuminating the patch P the differential form-factor dF_{PQ} is defined as:

$$dF_{PQ} = \frac{1}{\pi r^2} \cos\varphi \cos\phi dS_Q,$$ (11)

where:

φ, ϕ - the angles between the line PQ and respective patch normals
dS_Q - the differential area of patch Q
r - the distance between patches

Clearly, $dx = \frac{1}{r^2} \cos\phi dS_Q$, and hence:

$$w(P) = \int_{over\, all\, visible\, patches\, Q} \rho(dist(P,Q)) dF_{PQ}$$ (12)

This last equation indicates the computation method for the obscurance of a given patch. To compute the obscurance, one may compute the form factors for the given patch with all the visible patches in the neighborhood of patch P using, e.g., the hemicube method, and then integrate the form factors weighted with the function $\rho(L)$.

As we noted above, the function $\rho(L)$ differs from 1 only in the range $[0..L_{max}]$. Therefore we may compute the form factors for patches only in the neighborhood of a given patch P. Practically, in our sample environments it is sufficient to take L_{max} equal to 15-25 patch sizes, while D, the diameter of the environment, is approximately 500-700 patch sizes. A slight modification of the hemicube method is needed to take into account the 'absent patches', i.e. the patches that are not found in the L_{max}-neighborhood of patch P.

The locality of the method makes it much faster than the conventional radiosity. Besides, since in obscurance illumination model it is valid that no patch is found in a given direction, the method works perfectly in partially-open environments.

6. Results and conclusions

The illumination model described in the paper has been practically implemented in a proprietary lighting tool. Our goal was to compute the illumination for the highly complex environments like the ones shown on Figure 3 and in Appendix. In such environments there can be up to 10000 polygons, so computing lighting using conventional radiosity would be too slow since it takes many iterations to set up the lights in the environment properly. The lighting data was stored in light map textures and used for real-time visualization to modulate the ordinal textures of the scene (practical issues of lightmapping are described in [Zhuko97, Zhuko98]). Even though we tuned the tool specifically for this kind of scenes, we also computed the illumination of more usual 'table and chair' scenes like the one on the figures below. Notice, there is no much difference in the general outlook in Figure 7 and Figure 8.

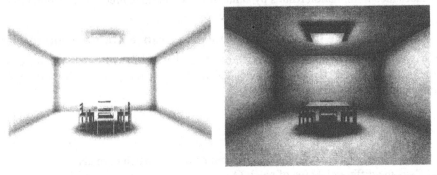

Figure 7. A 'table and chair' scene computed with pure obscurances only (left) and using the full model (equation (5)) with a point light source (right).

Figure 7 (left) was computed using the obscurances only using the equation (4)); Figure 7 (right) was computed using the full model (equation (5)) with one light source. In our computations we subdivided the scene into 14000 patches. On Pentium-166 it took less than 1 minute to compute the illumination using the illumination model proposed. The scene on Figure 8 was computed using Monte Carlo radiosity [Sbert96]. The scene was subdivided into 2000 patches and it took 10 minutes to compute it on a similar machine [Sbert98].

Figure 8. The same scene with distributed light source rendered using Monte-Carlo radiosity. (Scene is courtesy of Peter Shirley. Image is courtesy of Mateu Sbert, University of Girona, Spain)

In the Appendix we show some more screenshots of the scenes illuminated using the model described.

Finally, we briefly summarize the basic advantages of the proposed model:

Obscurance illumination model enables to simulate (non-constant) ambient light distribution in the environment. Besides, it makes possible to reproduce appealing and realistic images and preview scenes without explicit light source setting at all.

- The use of the suggested model enables as well to account the indirect illumination from light sources without the use of expensive radiosity method or ray tracing.
- Once the obscurance coefficients for the given scene are computed, it is easy to recompute the lighting while scene editing or for moving light sources.
- The complexity of the suggested algorithm is much less than the radiosity algorithm since only the geometry in the vicinity of a patch of interest is accounted during the computation.
- The illumination model works well in partially open environments.

Overall, obscurance illumination model can serve as a good alternative to radiosity for the applications where it is possible to sacrifice the physical accuracy of the computations, but gain in speed and ease of previewing and rendering the scenes without explicit light source setting.

References

[Cohen93] Cohen M., Wallace J.: Radiosity and Realistic Image Synthesis, Academic Press Professional, CA, USA, 1993

[Foley92] Foley J., van Dam A., Feiner S., Hughes J.: Computer Graphics. Principles and Practice. Addison-Wesley, 1992.

[Hanr91] Hanrahan P., Salzman D., Aupperle L.: A Rapid Hierarchical Radiosity Algorithm. SIGGRAPH'91 Conference Proceedings Vol.25, July 1991, pp.197-206

[Goral84] Goral C., et al.: Modeling the interaction of light between diffuse sources, ACM SIGGRAPH'84 Conf. Proc.

[Poynt93] Poynton C.: ""Gamma" and its disguises: The nonlinear mappings of intensity in Perception, CRTs, Film and Video", SMPTE J., 12, 1993, pp. 1099-1108

[Mill94] Miller G.: "Efficient Algorithms for Local and Global Accessibility Shading", SIGGRAPH'94 Conference Proceedings, 1994, pp.319-326

[Sbert93] Sbert M.: An integral geometry based method for fast form-factor computation, Computer Graphics Forum, Vol. 12, N 3, 1993

[Sbert96] Sbert M. et al.: "Global multipath Monte Carlo algorithms for radiosity", The Visual Computer, 12:47-61, 1996.

[Sbert98] Private communication

[Sill94] Sillion F., Puech C.: "Radiosity and Global Illumination." Morgan Kaufmann, SF, US, 1994

[Upst92] Upstill, S: "The RenderMan Companion. A Programmer's Guide to Realistic" Computer Graphics, Addison-Wesley, 1992

[Watt92] Watt A., Watt M.: "Advanced Animation and Rendering Techniques. Theory and Practice." Addison-Wesley, 1992.

[Whitt80] Whitted T.: "An Improved Illumination Model for Shaded Display", Comm. Of ACM Vol. 6, No. 23, 1980

[Zhuko97] Zhukov S., Iones A., Kronin G.: "On a practical use of light maps in real-time application", in Proc. of SCCG'97 Conference (Bratislava, Slovakia)

[Zhuko98] Zhukov S., Iones A., Kronin G.: "Using Light Maps to create realistic lighting in real-time applications", in Proc. of WSCG'98 Conference (Plzen, CZ), pp. 464-471

Editors' Note: see Appendix, p. 326 for colored figures of this paper

Computing the Approximate Visibility Map, with Applications to Form Factors and Discontinuity Meshing

A. James Stewart and Tasso Karkanis

Dynamic Graphics Project
Department of Computer Science
University of Toronto

Abstract. This paper describes a robust, hardware–accelerated algorithm to compute an approximate visibility map, which describes the visible scene from a particular viewpoint. The user can control the degree of approximation, choosing more accuracy at the cost of increased execution time. The algorithm exploits item buffer hardware to coarsely determine visibility, which is later refined. The paper also describes a conceptually simple algorithm to compute a subset of the discontinuity mesh using the visibility map.

Key words:
approximate visibility, visibility map, discontinuity meshing,
hardware assisted, occlusion culling, form factor, item buffer

1   Introduction

The visibility map is a data structure that describes the projection of the visible scene onto the image plane. It is a planar graph in which the vertices, edges, and faces are annotated with the corresponding vertices, edges, and faces of the scene. The visibility map provides more than just the set of visible surfaces; it provides their arrangement on the image plane. This paper describes a fast and robust algorithm to compute an approximation of the visibility map.

Previous algorithms have computed the exact visibility map. They are somewhat complicated to implement, due to the large number of "special case" input configurations that must be handled. They are also prone to robustness failure upon encountering geometric degeneraces and very small features. As most algorithm implementors will report, it is very difficult to make a geometric algorithm that is immune to this sort of failure.

This paper presents an algorithm to compute an approximate visibility map, which is approximate in the sense that some small features may be missing or distorted. The user can control the degree of approximation, choosing more accuracy at the cost of increased execution time.

The new algorithm can exploit graphics hardware to accelerate the computation, using an item buffer [1] to determine visibility at a coarse scale. The algorithm is more robust than previous object space algorithms because it operates on the discrete pixels of the item buffer, thus avoiding most of the small features and numerical uncertainty that tend to plague object space algorithms.

Many visibility algorithms have been developed. That of Watkins [2] was the first widely used visible surface algorithm in computer graphics, but it has been all but abandoned in favor of hardware Z–buffering which, although less elegant, is typically much faster. Various other approaches have been taken, including the "cookie cutter" approach of Atherton, Weiler, and Greenberg [3], the use of BSPs by Fuchs, Kedem, and Naylor [4], a randomized algorithm by Mulmuley [5], a plane sweep algorithm by McKenna [6], and the classic Z–buffer technique of Catmull [7].

The new approach differs in that it builds an *approximate* visibility map which contains a subset of the visible scene features. The visibility map is not discretized (as are the results of Watkins and Catmull); it is, rather, represented with floating–point coordinates on the image plane.

2 Visiblity Map Algorithm

The new algorithm to compute an approximate visibility map is conceptually simple:

1. The scene is rendered into an item buffer in which each pixel contains the unique identifier of the face visible in that pixel (Figure 1a).
2. The item buffer is traversed to build a coarse, rectilinear planar graph whose vertical and horizontal edges separate pixels containing different faces (Figure 1b).
3. The rectilinear graph is "relaxed" to produce straighter edges and a more correct positioning of its features. The topology of the graph can change to become closer to that of the exact visibility map.

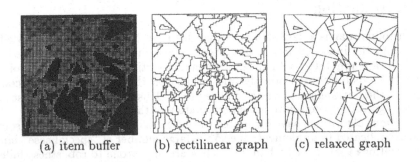

(a) item buffer   (b) rectilinear graph   (c) relaxed graph

Fig. 1. The algorithm first renders the scene into an item buffer, then produces a coarse, rectilinear graph, and finally "relaxes" the rectilinear graph to produce a more accurate visibility map.

2.1 Step 1: Rendering into the item buffer

In the first step, each face of the scene is given a unique color and is drawn with the standard Z–buffer technique [7]. Upon completion, each pixel of the item buffer stores the color of the scene face visible through that pixel. An item buffer pixel is said to **contain** a scene face if it stores the color of that face. Each pixel also stores the depth of the corresponding face.

To distinguish pixels on a face boundary from those on the interior, the line segments bounding each face are also drawn. Each line segment of a particular face is assigned a unique bit and the logical "or" of these bits is accumulated in the alpha channel of each pixel on the face boundary. (If a face has more than eight edges, multiple passes are required.) This information is later used to help determine the correspondence between scene edges and rectilinear graph edges.

2.2 Step 2: Building a Rectilinear Graph

In the second step, a rectilinear planar graph is built. The item buffer is traversed and horizontal or vertical edges are placed between pixels that contain different faces.

The horizontal and vertical edges must be connected with vertices, which are of two types: A NORMAL vertex has exactly two adjacent edges, while a T vertex has three and is used where one edge becomes blocked by another on the image plane. The **blocked** edge corresponds to a scene edge that disappears from view. The other two **blocking** edges at the T vertex correspond to the single scene edge under which the blocked edge disappears. If fours edges meet at a vertex, they are decomposed into NORMAL and T vertices as described below.

Care must be taken to meaningfully connect the horizontal and vertical edges with the appropriate types of vertices. Figure 2a shows four pixels separated by three edges. Each pixel contains a face and is labelled with a depth, where '1' denotes the closest face and '3' denotes the most distant. The edge separating '2' and '3' bounds face 2 and becomes blocked when it passes below face 1, so the three edges must be joined by a T vertex.

Other cases are shown in Figure 2: If exactly two edges meet at a point, they are joined by a NORMAL vertex. If four edges meet at a point, they are either split into two pairs of joined edges, or two of the four edges become blocked at two T vertices. The algorithm simply iterates over all positions in the item buffer and, at each position, connects the edges according to the rule shown in the figure.

Once the graph is connected, more vertices are added which correspond to the visible scene vertices: A list is made of scene faces that appear in at least one pixel. For each such face, each of its scene vertices is projected onto the image plane. If the projection falls within a pixel containing that face then a graph vertex, which corresponds to the scene vertex, is added to the rectilinear graph.

As shown in Figure 3, vertices are projected outward from the centroid of the projected scene face until they meet a graph edge (this assumes that scene faces are convex). Each new vertex causes a graph edge to be split into two edges. More than one vertex of the face can project to the same edge, which would be split into multiple pieces. Each new vertex is labelled with the corresponding scene vertex, while the two graph edges adjacent to the vertex are labelled with the scene edges adjacent to the scene vertex. Let the notation "SCENE vertex" (as opposed to "scene vertex") denote a *graph* vertex created and labelled in this manner. There are now three types of graph vertices: NORMAL, T, and SCENE.

After all vertices have been projected, the known edge labels are propagated throughout the graph: The label on one edge can be propagated to an adjacent, unlabelled edge provided (a) they are not separated by a SCENE vertex and (b) they are not separated by a T vertex at which the edge being propagated is blocked.

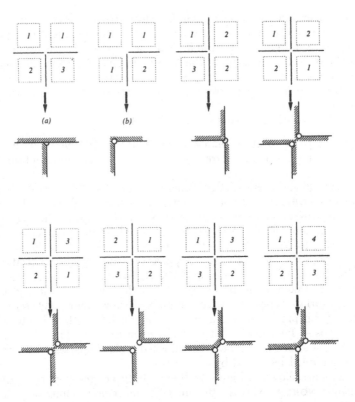

Fig. 2. The different ways in which edges are joined in the rectilinear graph. NORMAL vertices appear as open circles and T vertices appear as open semicircles, with the blocked edge on the semicircle. Symmetric cases are not shown. A depth number is repeated if and only if a particular face occupies more than one pixel. The hashing to one side of each edge indicates the closer scene face, which is bounded by that edge.

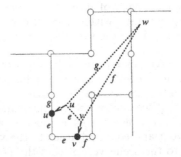

Fig. 3. The projection of a scene face, uvw, into the item buffer. Each scene vertex that appears in a pixel containing uvw is projected onto a graph edge on the pixel boundary. A new graph vertex is created and labelled with the corresponding scene vertex. The adjacent graph edges are labelled with the corresponding scene edges. SCENE vertices appear as solid circles (recall that NORMAL vertices appear as open circles).

Occasionally, a graph edge cannot be labelled in this manner. This occurs, for example, on a chain of edges whose ends are blocked by two T vertices; the SCENE vertices that would normally bound the chain are hidden. To label such a graph edge, consider that it separates two pixels, one of which contains a closer scene face. Two labelling strategies are used: If the pixel containing the closer face is identified with a unique scene edge in the item buffer, that scene edge labels the graph edge. Otherwise, each scene edge of the closer scene face is projected onto the image plane. A particular projected edge labels the graph edge if (a) it passes within one pixel of the graph edge, (b) no other projected edge passes within one pixel, and (c) it is oriented within 90 degrees of the graph edge. New labels are propagated as described above.

Upon completion of this step, the algorithm has produced a rectilinear graph that is connected and planar, and that has its faces, edges, and vertices labelled with the corresponding faces, edges, and vertices of the scene. Note that NORMAL vertices are not labelled, and T vertices are labelled with their blocked and blocking edges.

2.3   Step 3: Relaxing the Graph

The final steps consist of straightening the graph edges and moving the graph vertices to their correct locations on the image plane. A vertex of the graph is said to be **settled** if it is at its correct location (i.e. the location of its projection) on the image plane. A T vertex is correctly located at the intersection of the projections of its blocked and blocking scene edges.

Define a **multi–edge** to be a maximal–length chain of graph edges, each labelled by the same scene edge. A multi–edge represents one segment of the projection of a single scene edge. Its endpoints can be SCENE vertices or, if the scene edge disappears from view at some point, T vertices. See Figure 4.

The relaxation algorithm repeatedly cycles through the graph vertices, settling them whenever possible. Two conditions must be satisfied before a vertex may be settled:

- The interior vertices of a multi–edge cannot be settled until the endpoints have themselves been settled (otherwise, the straight line segment joining the unsettled endpoints would be undefined). This imposes a partial order on the multi–edges, in which one multi–edge is **deeper** than another if one of its endpoints is a T vertex on the other. Multi–edges must be processed in order of increasing depth.
- For a given multi–edge with settled endpoints, the straight line segment joining the endpoints in the image plane may be parameterized between 0 and 1. Each interior T vertex of the multi–edge corresponds to some parameter, which is that of the image plane intersection of the straight line segment and the projection of the T vertex's blocked scene edge. Interior T vertices cannot be settled until their parameters are **consistently ordered**: Each parameter must be in the range $[0, 1]$ and parameters must increase monotonically as the multiedge is traversed from the tail endpoint (with parameter 0) to the head endpoint (with parameter 1).

Relaxation. To relax the graph, the algorithm iterates through the following steps until nothing remains to be settled. Each step is explained below.

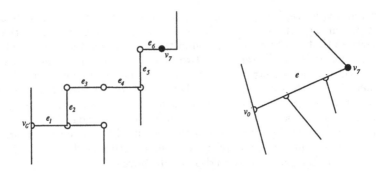

Fig. 4. *Left:* A multi–edge consisting of graph edges $e_1, e_2, \ldots e_6$, bounded by graph vertices v_0 (T) and v_7 (SCENE). Each e_i is labelled with the same scene edge. *Right:* The relaxed multiedge, e. Only after v_0 and v_7 are settled may the multi–edge be straightened and its five interior vertices be settled.

1. For each multiedge with settled endpoints but unsettled interior vertices, attempt to order the interior vertices consistently by their parameters. Consistently ordered vertices are said to be **viable**.
2. For each unsettled graph vertex, v, attempt to settle it. If v is a T vertex, it must be viable before being settled.

It is possible that the algorithm terminates with remaining **unsettlable** vertices, which cannot move to their correct positions without violating the planarity of the graph. At that point, these vertices are frozen in their (incorrect) positions. The quality of an approximate visibility map can be measured by its percentage of unsettlable vertices.

Settling vertices. To settle a vertex, the vertex must be moved to its correct location in the image plane. When a vertex moves, its attached edges are pulled along with it.

In attempting to settle a vertex, the algorithm first determines whether the moving vertex intersects any graph edge on the way to its correct location. The algorithm also determines whether any of the attached edges intersects a graph vertex as it is being pulled along. These tests need only to be performed in at most three graph faces immediately adjacent to the moving vertex.

If no intersection occurs, the vertex is moved to its correct location and is considered to be settled. If an intersection occurs, the movement would cause the graph to become non–planar, so the attempt fails and the vertex remains at its current position. The algorithm thus guarantees that the graph remains planar.

Consistent ordering on multi–edges. Interior vertices of a multi–edge are consistently ordered if their parameters increase monotonically from tail to head, and if all parameters lie in the range $[0, 1]$.

In attempting to settle a multi–edge, the algorithm searches for inconsistencies in the order of interior vertices. Where inconsistences are found, the topological structure of the graph is modified to eliminate them.

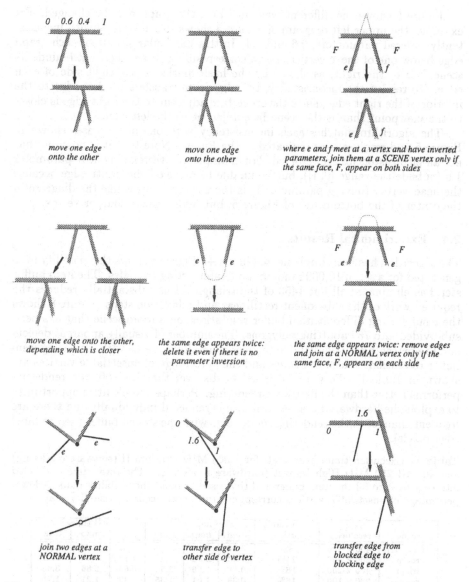

Fig. 5. Different actions are taken to resolve inconsistencies on a multi–edge. The diagrams show the adjacent scene faces with hash marks and sometimes show the parameter values at the T vertices. In all cases, there is either a parameter inversion (first and second rows) or a parameter is outside the range [0, 1] (third row)

Figure 5 shows the different ways in which the graph may be changed. For example, the upper left diagram of Figure 5 shows two T vertices with inconsistently ordered parameters, 0.6 and 0.4. In this particular situation, each graph edge below one of the T vertices has a corresponding scene edge that bounds its scene face to the right, as shown by the hash marks on the right side of each edge. To resolve the inconsistency, the algorithm transfers the left edge to the interior of the right edge, since the scene face adjacent to the right edge is closer to the viewpoint than is the scene face adjacent to the left edge.

The algorithm matches each inconsistency with one of the cases shown in Figure 5 and resolves it as indicated in the Figure. Note that it is possible that not all inconsistencies are resolved. For example, a T vertex may have parameter 1.6 (outside the range $[0, 1]$), but be unable to move off the multi–edge because the head vertex (having parameter 1) is too convex. This is like the diagram in the center of the bottom row of Figure 5, but with a much sharper vertex.

2.4 Experimental Results

The algorithm has been implemented in C++. Figure 6 shows the visibility map generated for a set of 10,000 random, non–intersecting triangles. The item buffer step has eliminated all but 1453 of the triangles; this substantially reduces the required work of the subsequent rectilinear and relaxation steps. Figure 8 shows the results using different item buffer resolutions on a room scene that has been subdivided into 784 abutting polygons. The number of visually apparent defects and the percentage of unsettlable vertices generally decrease with increasing item buffer resolution. Execution times and the percentage of unsettlable vertices are shown in Table 1. We were surprised to discover that the software rendering performs better than the hardware rendering. Perhaps there's little opportunity to exploit the hardware because each face is rendered individually and there are frequent changes in the rendering mode (e.g. with the stencil buffer used to label edge pixels).

Table 1. Execution times in seconds for a 300 MHz Pentium II (software rendering) and an SGI 250 MHz High Impact (hardware rendering). The total time is divided into that used to fill the item buffer and that used to build the visibility map. A lower percentage of unsettlable vertices corresponds to a more accurate visibility map.

scene	resolution	unsettlable vertices	Pentium times			SGI times		
			item	map	total	item	map	total
10,000 tri	400 × 400	8%	3.57	6.92	10.49	6.06	11.91	17.97
room	800 × 400	14%	1.53	3.94	5.47	1.12	7.35	8.47
room	400 × 200	18%	0.60	1.96	2.56	0.63	2.85	3.48
room	200 × 100	16%	0.34	1.01	1.35	0.49	1.21	1.70
room	100 × 75	25%	0.26	0.53	0.79	0.46	0.66	1.12

3 Application to Form Factors

The point–to–patch form factor describes the fraction of energy leaving a patch that arrives at a point, and is computed by integrating over the visible area of the patch. Determining visibility is the most expensive operation in computing a form factor. The hemicube method [8] is typically used to determine form factors, but can suffer aliasing problems because visibility is sampled on a

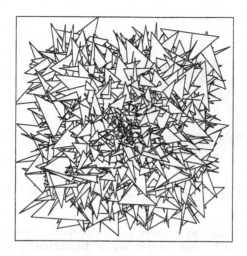

Fig. 6. The approximate visibility map of a scene of 10,000 randomly positioned triangles, using a 400 × 400 item buffer. Only 1453 triangles appear in the item buffer, which substantially reduces the subsequent processing time.

relatively coarse, regular grid. Other algorithms use ray tracing to sample the visibility [9, 10].

The approximate visibility map would seem ideally suited for form factor computation. After rendering into the item buffer, the relaxation step should produce a more accurate visibility map and better form factors.

However, approximate visibility performs poorly in computing form factors. It takes substantially longer and, in some cases, actually has a larger RMS normalized error than the hemicube method, as shown in Table 2. The large RMS error occurs in the approximate visibility map when small faces are "pulled larger" in order to maintain topological constraints.

Table 2. Comparison of the hemicube and the approximate visibility map in computing form factors, with hardware rendering.

resolution	method	mean error	RMS error	time (sec)
100 × 100	hemicube	0.011703	0.148934	0.08
100 × 100	visibility	0.017119	0.494695	1.71
400 × 400	hemicube	-0.000557	0.026423	0.50
400 × 400	visibility	-0.006068	0.037054	4.50

4 Application to Discontinuity Meshing

Discontinuity meshing is used to increase the accuracy and speed of radiosity algorithms. The work of Campbell and Fussell [11] and Heckbert [12] in this area has spurred a number of meshing algorithms, including those of Lischinski, Tampieri, and Greenberg [13, 14], Drettakis and Fiume [15], and others. A more general structure, the Visibility Skeleton [16], may also be used to compute the discontinuity mesh.

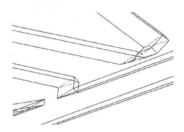

Fig. 7. *Left:* Discontinuity mesh. *Right:* Closer view of some segments.

Discontinuity meshing has received little attention recently, in part because of the unavailability and sheer complexity of discontinuity meshing programs. These programs must treat many special, complex geometric configurations and are prone to robustness failure in the presence of degenerate configurations or numerical uncertainty.

This section describes a very simple discontinuity meshing algorithm that computes *only a subset* of the segments of the discontinuity mesh: in particular, it determines those segments defined by a source vertex and scene edge (called EV_s) and some of the segments defined by a source edge and two scene edges (called EEE_s). It does not determine any segments that do not involve a source vertex or edge.

Two observations from the literature are exploited:

- The EV_s segments can be determined by computing the visibility map at each source vertex and projecting the edges of the map into the scene [12].
- The endpoints of the discontinuity segments can be tagged by the scene features that define them. Endpoints with matching tags can be joined with a discontinuity segment [17, 16].

The algorithm does exactly that. It computes the approximate visibility map from each source vertex and projects the edges of the map onto the faces of the scene to form EV_s segments. Each segment endpoint is tagged with the source vertex and two scene edges that define it. Where there are two vertices tagged by adjacent source vertices and identical scene edges, an EEE_s segment is identified. Each such segment is projected onto the appropriate faces with a two-dimensional visibility calculation [12, 15].

Figure 7 shows the computed mesh for a simple scene of 104 polygons and a triangular light source. Using software rendering on a 300 MHz Pentium, the mesh was computed in 9.5 seconds.

This simple meshing algorithm has some attractive properties: It avoids the slow, error–prone geometric computations of other meshing algorithms; it is very easy to implement, given the approximate visibility code; and the item buffer step very efficiently eliminates most occluded faces, which don't contribute to the mesh. Some potential problems exist with the algorithm: If the approximate visibility map does not contain an edge visible from the source, then the corresponding EV_s segment will not appear in the mesh. Using a higher–resolution

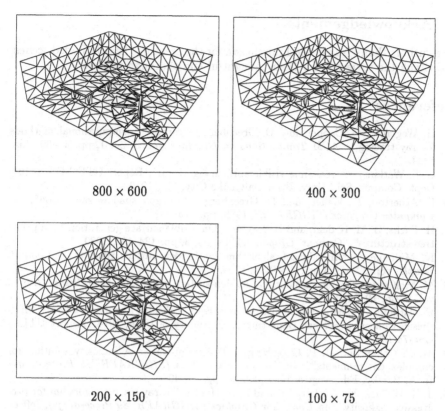

800 × 600   400 × 300

200 × 150   100 × 75

Fig. 8. Various item buffer resolutions are used for a 784–polygon scene.

item buffer helps to avoid this. Spurious segments can occur in the mesh if the approximate visibility map contains incorrect T vertices. This is avoided by ignoring unsettlable T vertices.

5 Summary

This paper has introduced an algorithm to compute the approximate visibility map. The user can control the accuracy of the approximate visibility map by choosing a higher resolution item buffer, at the cost of increased execution time.

There are two principal advantages to using the item buffer: It culls many of the small or hidden objects from consideration, and it provides the algorithm with a well defined graph as an initial approximation to the visibility map. Planarity of the final visibility map is guaranteed because no planarity–violating alterations are permitted during the relaxation phase.

The approximate visibility map is not useful for form factors, but it is useful for a limited form of discontinuity meshing. The authors hope that other applications, such as occlusion culling, may also find a use for approximate visiblity.

6   Acknowledgements

The authors thank Nasser Keshmirshekan for the creating the subdivided scenes, and thank the reviewers for their thoughtful and thorough reviews.

References

1. H. Weghorst, G. Cooper, and D. Greenberg, "Improved computational methods for ray tracing", *ACM Transactions on Graphics*, vol. 3, no. 1, pp. 52–69, Jan. 1984.

2. G. S. Watkins, "A real-time visible surface algorithm", Report UTEC-CS-70-101, Dept. Comput. Sci., Univ. Utah, Salt Lake City, UT, 1970.

3. P. Atherton, K. Weiler, and D. Greenberg, "Polygon shadow generation", in *Computer Graphics (SIGGRAPH)*, 1978, pp. 275–281.

4. H. Fuchs, Z. M. Kedem, and B. Naylor, "On visible surface generation by a priori tree structures", *Comput. Graph.*, vol. 14, no. 3, pp. 124–133, 1980.

5. K. Mulmuley, "An efficient algorithm for hidden surface removal", *Comput. Graph.*, vol. 23, no. 3, pp. 379–388, 1989.

6. M. McKenna, "Worst-case optimal hidden-surface removal", *ACM Trans. Graph.*, vol. 6, pp. 19–28, 1987.

7. E. Catmull, *A subdivision algorithm for computer display of curved surfaces*, Ph.d. thesis, Computer Science Department, University of Utah, 1974, Report UTEC-CSc-74-133.

8. M. F. Cohen and D. P. Greenberg, "The Hemi-Cube: A radiosity solution for complex environments", in *Computer Graphics (SIGGRAPH '85 Proceedings)*, B. A. Barsky, Ed., August 1985, vol. 19, pp. 31–40.

9. J. R. Wallace, K. A. Elmquist, and E. A. Haines, "A ray tracing algorithm for progressive radiosity", in *Computer Graphics (SIGGRAPH '89 Proceedings)*, Jeffrey Lane, Ed., July 1989, vol. 23, pp. 315–324.

10. P. Schroder, "Numerical integration for radiosity in the presence of singularities", in *Proceedings of the Fifth Eurographics Workshop on Rendering*, 1993, pp. 177–184.

11. A. T. Campbell and D. S. Fussell, "Adaptive mesh generation for global diffuse illumination", in *Computer Graphics (SIGGRAPH)*, 1990, pp. 155–164.

12. P. Heckbert, *Simulating Global Illumination Using Adaptive Meshing*, Ph.D. thesis, CS Division (EECS), Univ. of California, Berkeley, June 1991.

13. D. Lischinski, F. Tampieri, and D. P. Greenberg, "Discontinuity meshing for accurate radiosity", *IEEE Computer Graphics Graphics & Applications*, pp. 25–39, November 1992.

14. D. Lischinski, F. Tampieri, and D. P. Greenberg, "Combining hierarchical radiosity and discontinuity meshing", in *Computer Graphics (SIGGRAPH '93 Proceedings)*, 1993, pp. 199–208.

15. G. Drettakis and E. Fiume, "A fast shadow algorithm for area light sources using backprojection", in *Computer Graphics (SIGGRAPH)*, 1994, pp. 199–208.

16. F. Durand, G. Drettakis, and C. Puech, "The Visibility Skeleton: A powerful and efficient multi-purpose global visibility tool", in *Computer Graphics (SIGGRAPH)*, 1997, pp. 89–100.

17. G. Drettakis, *Structured sampling and reconstruction of illumination for image synthesis*, Ph.D. thesis, Computer Science Department, University of Toronto, January 1994.

Ray Tracing of Subdivision Surfaces

Leif P. Kobbelt   K. Daubert   H-P. Seidel

Computer Science Department, University of Erlangen-Nürnberg
Am Weichselgarten 9, 91058 Erlangen, Germany

Abstract. We present the necessary theory for the integration of subdivision surfaces into general purpose rendering systems. The most important functionality that has to be provided via an abstract geometry interface are the computation of surface points and normals as well as the ray intersection test. We demonstrate how to derive the corresponding formulas and how to construct tight bounding volumes for subdivision surfaces. We introduce envelope meshes which have the same topology as the control meshes but tightly circumscribe the limit surface. An efficient and simple algorithm is presented to trace a ray recursively through the forest of triangles emerging from adaptive refinement of an envelope mesh.

1 Introduction

The general concept of subdivision techniques for the construction and representation of free-form surfaces is gaining more and more attention in computer graphics and related fields [19, 22]. The efficiency of subdivision algorithms and the flexibility with respect to the topology and connectivity of the control meshes makes this approach suitable for many applications such as surface reconstruction [2, 4] and interactive modeling [23]. The close connection to multi-resolution analysis of parametric surfaces provides access to the combination of classical modeling paradigms with hierarchical representations of geometric shape.

So far subdivision surfaces have been used mainly in the context of geometric modeling, i.e., issues like the asymptotic behavior of the scheme [6, 24] and the (discrete) fairness of the resulting meshes were investigated [11, 12]. Meanwhile the related mathematical theory has reached a state of maturity which allows the programmer to choose among the many schemes proposed in the literature [2, 5, 10, 13]. In this paper we do not investigate such properties but we address an important issue for the integration of subdivision schemes into a wider range of potential applications: While the subdivision methods have become a standard in *surface design*, the connection to *rendering applications* is still based on raw triangle data exchange.

There has been a considerable amount of work on the integration of higher order basic shapes like spline surfaces into the generic setup of sophisticated rendering algorithms [1, 7, 9, 15, 17, 20, 21]. Most of the approaches derive a more or less tight, preferably convex, bounding volume for each patch. The size of this bounding volume provides an upper bound on the spatial extent of the object such that ray intersection tests can be implemented much more efficiently by discarding rays according to simple tests against the bounding volume. If the bounding volume aligns to the local geometry of a patch then its shape can be used as an oracle to rate the local flatness, i.e., the approximation error if the true geometry would be replaced by a planar face.

In this paper we will present the basic prerequisites which are necessary to transfer the generic bounding volume technique to subdivision surfaces. The mathematical difficulties emerge from the fact that in general there is no explicit description for the limit surface and hence possible bounds have to be derived from the coefficients of the underlying refinement equation (i.e., from the coefficients of the subdivision masks).

We start by finding points and normal vectors on the limit surface corresponding to the initial control vertices. The triangles of the initial control mesh imply a decomposition of the limit surface into triangular patches with these limit points at their corners. For each patch we compute a bounding prism by sweeping the chord triangle spanned by the three corners in the (triangle-) normal direction.

Based on the individual bounding prisms, we build envelope meshes for the composite surface by moving the limit points in (point-) normal direction until the bounding prisms are completely contained. This provides a continuous polyhedral hull. When a given ray intersects one of the envelope triangles, we perform local subdivision to obtain a better piecewise linear approximation of the limit surface. Since the neighboring triangles in a mesh data structure can be found in $O(1)$, we can formulate an efficient recursive scheme that traces the ray through the hierarchy of envelope triangles. The recursion stops when a local flatness criterion is met, indicating that the intersection with the chord triangle does not deviate from the true solution by more than a prescribed ε. The algorithm has been integrated as a new geometric primitive into a general purpose ray tracing tool to compute the pictures shown in the result section.

Throughout the paper we will explain the theoretic concepts and general methods in the context of *univariate* subdivision. Transferring the results to the bivariate setting is rather obvious but depending on the vertices' valences several special cases have to be considered. To make the reproduction of the results as easy as possible, we apply the corresponding formulas explicitly to Loop's subdivision scheme [13].

2 Limit points and normals for subdivision surfaces

Let a control polygon $P_0 = [\mathbf{p}_i^0]$ with $\mathbf{p}_i^0 \in \mathbf{R}^3$ be given which represents the curve

$$P(t) = \sum_i \mathbf{p}_i^0 \, \phi(t-i) \qquad \text{with} \qquad \phi(t) = \sum_j \alpha_j \phi(2t-j). \qquad (1)$$

It is well known that the subdivision rule

$$\mathbf{p}_i^1 := \sum_j \alpha_{i-2j} \, \mathbf{p}_j^0 \qquad (2)$$

generates a new control polygon $P_1 = [\mathbf{p}_i^1]$ such that

$$P(t) = \sum_i \mathbf{p}_i^1 \, \phi(2t-i).$$

provides a *refined* representation of the same curve. Notice that (2) actually combines *two* rules triggered by the parity of i. By iterating the refinement rule we obtain a sequence of polygons $P_m = [\mathbf{p}_i^m]$, $m = 0, 1, \ldots$ which — depending on the coefficients α_j — converges to a smooth limit curve P_∞.

We are interested in computing points on the limit curve P_∞ directly from the control points $\mathbf{p}_i^m$ on some level m without going through the iterative refinement. The following technique has become standard in the analysis of subdivision schemes: Construct a local subdivision matrix and transform it into its basis of (generalized) eigenvectors.

Let the coefficients $[\alpha_j]_{j=0}^{2n}$ define a univariate subdivision scheme. If we rewrite the polygon P_m as a *vector* then the subdivision step $P_m \to P_{m+1}$ corresponds to the multiplication of P_m by a suitable matrix

$$\widetilde{S} = \begin{pmatrix} \ddots & & \ddots & & \ddots & & \\ \alpha_1 & \cdots & \alpha_{2j+1} & \cdots & \alpha_{2n-1} & 0 & 0 \\ \alpha_0 & \cdots & \alpha_{2j} & \cdots & \cdots & \alpha_{2n} & 0 \\ 0 & \alpha_1 & \cdots & \alpha_{2j+1} & \cdots & \alpha_{2n-1} & 0 \\ 0 & \alpha_0 & \cdots & \alpha_{2j} & \cdots & \cdots & \alpha_{2n} \\ & & \ddots & & \ddots & & \ddots \end{pmatrix}.$$

Due to the fixed bandwidth of $\widetilde{S}$ we observe that the non-zero coefficients of $2n-1$ successive rows form a square matrix S. From an algorithmic point of view this means that the $(n-1)$-neighborhood of $\mathbf{p}_{2i}^{m+1}$ in P_{m+1} is completely determined by the $(n-1)$-neighborhood of $\mathbf{p}_i^m$ in P_m.

Since the control vertices $\mathbf{p}_i^m$ represent the function $P(t)$ with respect to the functional basis $\phi(2^m t - i)$ it is natural to associate the vertex $\mathbf{p}_i^m$ with the parameter value $t_i^m = i 2^{-m}$. Hence, through all subdivision levels, the vertices $\mathbf{p}_{i2^r}^{m+r}$ correspond to the same parameter value and this sequence of vertices converges for $r \to \infty$ to the point $P(t_i^m)$ on the limit curve.

The sub-polygons of $2n-1$ successive vertices $\mathbf{p}_i^m$ and $\mathbf{p}_i^{m+1}$ correspond to the nested parameter intervals $t_i^m + [1-n, n-1]\, 2^{-m}$ and $t_i^m + [1-n, n-1]\, 2^{-(m+1)}$ respectively and since the same subdivision mask S is applied in every step, it is obvious that the limit point $P(t_i^m)$ is determined by the $(n-1)$-neighborhood of the vertex $\mathbf{p}_i^m$ through

$$P(t_i^m)\,[1,\ldots,1]^T \;=\; \lim_{r\to\infty} S^r\, [\mathbf{p}_{i-n+1}^m, \ldots, \mathbf{p}_{i+n-1}^m]^T.$$

The direct computation of $P(t_i^m)$ requires the decomposition of $S = V^{-1} D V$ into a diagonal matrix D and a transform V into the basis of eigenvectors[1]. The convergence of the iterative scheme implies that the dominant eigenvalue of S be $\lambda_1 = 1$ and the affine invariance of the subdivision operator indicates that the corresponding eigenvector is $[1,\ldots,1]$. Therefore the coefficients $l_{1-n}, \ldots, l_{n-1}$ such that

$$P(t_i^m) = \sum_j l_j\, \mathbf{p}_{i+j}^m$$

can be read off that row of V which is associated with this eigenvector. This is obvious since components of the input vector which belong to eigenspaces of smaller eigenvalues fade out during the iteration of S.

If the subdivision scheme generates C^1 curves then there exists a *difference scheme* $\widetilde{S}'$ which maps the divided differences $\triangle \mathbf{p}_i^m = 2^m(\mathbf{p}_{i+1}^m - \mathbf{p}_i^m)$ of P_m to the divided differences of P_{m+1} [6]. For repeated subdivision the differences converge to the derivative of the limit function $P'(t)$. The limit point analysis applied to the difference scheme hence provides limit tangent vectors.

Due to the simple relation between $\widetilde{S}$ and $\widetilde{S}'$ it turns out that the eigenvector $[1,\ldots,1]$ of S' with eigenvalue $\lambda_1 = 1$ corresponds to the eigenvector $[1-n,\ldots,n-1]$ of S with eigenvalue $\lambda_2 = \frac{1}{2}$ (*constant differences*). Hence, just as the eigenvector for the dominant eigenvalue $\lambda_1 = 1$ allows us to compute the limit point, the eigenvector for the subdominant eigenvalue $\lambda_2 = \frac{1}{2}$ determines the limit tangent at $P(t_i^m)$ since it describes the line which is asymptotically approached by the sequence:

$$P'(t_i^m)\,[1-n,\ldots,n-1]^T \;=\; \lim_{r\to\infty} 2^r S^r\, \big[\mathbf{p}_{i-n+1}^m - P(t_i^m), \ldots, \mathbf{p}_{i+n-1}^m - P(t_i^m)\big].$$

If the subdivision scheme converges to a C^1 limit then the modulus of all other eigenvalues $\lambda_3 \geq \ldots \geq \lambda_{2n-1}$ is less than $\frac{1}{2}$. The rate by which the deviation of the sub-polygon $[\mathbf{p}_{i-n+1}^m, \ldots, \mathbf{p}_{i+n-1}^m]$ from a straight line fades out is $|\lambda_3|$. As expected (by Taylor's theorem) the local flattening rate $1/\lambda_3$ is higher than the contraction rate $1/\lambda_2 = 2$ if the limit curve is smooth.

In the bivariate setting, we have two partial derivatives of first order and accordingly the local subdivision matrix S of a refinement scheme which generates C^1 limit surfaces

---

[1] In general D is the Jordan normal form of S but for convergent subdivision schemes with smooth limit surfaces, the leading eigenvalues have algebraic multiplicity one [16, 24].

has a double subdominant eigenvalue $\lambda_2 = \lambda_3$. The components of the input mesh which lie in the corresponding eigenspaces span the tangent plane at the limit point.

Example: Consider the three directional grid spanned by $(1,0)$, $(0,1)$, and $(1,1)$. The quartic box spline M_{222} defined on this grid satisfies the refinement equation [3]

$$M_{222}(u,v) = \sum_{i,j=0}^{4} \alpha_{i,j} \, M_{222}(2u-i, 2v-j) \quad \text{with} \quad [\alpha_{i,j}] = \frac{1}{16} \begin{pmatrix} 0 & 0 & 1 & 2 & 1 \\ 0 & 2 & 6 & 6 & 2 \\ 1 & 6 & 10 & 6 & 1 \\ 2 & 6 & 6 & 2 & 0 \\ 1 & 2 & 1 & 0 & 0 \end{pmatrix}$$

Hence, the corresponding subdivision rules are

$$\mathbf{p}_{i,j}^{m+1} := \begin{cases} \frac{1}{16}(10\mathbf{p}_{l,k}^m + \mathbf{p}_{l,k-1}^m + \mathbf{p}_{l,k+1}^m + \mathbf{p}_{l-1,k}^m + \mathbf{p}_{l+1,k}^m + \mathbf{p}_{l-1,k-1}^m + \mathbf{p}_{l+1,k+1}^m) & i=2l, \quad j=2k \\ \frac{1}{16}(2\,\mathbf{p}_{l,k-1}^m + 6\,\mathbf{p}_{l,k}^m + 6\,\mathbf{p}_{l+1,k}^m + 2\,\mathbf{p}_{l+1,k+1}^m) & i=2l+1, j=2k \\ \frac{1}{16}(2\,\mathbf{p}_{l-1,k}^m + 6\,\mathbf{p}_{l,k}^m + 6\,\mathbf{p}_{l,k+1}^m + 2\,\mathbf{p}_{l+1,k+1}^m) & i=2l, \quad j=2k+1 \\ \frac{1}{16}(2\,\mathbf{p}_{l+1,k}^m + 6\,\mathbf{p}_{l,k}^m + 6\,\mathbf{p}_{l+1,k+1}^m + 2\,\mathbf{p}_{l,k+1}^m) & i=2l+1, j=2k+1 \end{cases} \quad (3)$$

See Figure 1 for several generations of a recursively refined triangle mesh approximating the box-spline basis function and Fig. 2 for a geometric interpretation of the rules.

Fig. 1. Uniformly subdivided regular triangle meshes converging to the quartic box-spline M_{222}.

Several authors have generalized these rules to meshes with arbitrary connectivity [13, 23]. To do this we have to leave the formal setup of refinement rules of the type (3) since regularly indexing the vertices is no longer possible. Due to the small support of the refinement masks in Loop's scheme (cf. Fig. 2), there is no need to modify the "edge"-rules: An inner edge is always adjacent to two triangles. Hence the generalization can be restricted to the definition of alternative "vertex"-rules

$$\mathbf{p}^{m+1} := \frac{\alpha(k)}{\alpha(k)+k}\mathbf{p}^m + \frac{1}{\alpha(k)+k}\sum_i \mathbf{p}_i^m$$

for vertices $\mathbf{p}^m$ with valence $k \neq 6$ and $\mathbf{p}_i^m$ being the direct neighbors of $\mathbf{p}^m$ in P_m. A good choice leading to overall C^1 limit surfaces is [24]

$$\alpha(k) = k\frac{1-\beta(k)}{\beta(k)}, \qquad \beta(k) = \frac{5}{8} - \frac{(3+2\cos(2\pi/k))^2}{64}.$$

Notice that this rule coincides with the original "vertex"-rule (3) if the valence of $\mathbf{p}^m$ is 6.

Remark: At the boundary of *open* triangle meshes, we cannot apply the above masks since some of the neighboring vertices are missing. We avoid this problem by treating the boundary of a mesh as a closed polygon and applying univariate subdivision. By doing this we additionally guarantee that no internal control vertex influences the shape of the boundary curve. This is important if we want to generate creases or join two separate subdivision surfaces along a common curve in a C^0 fashion [10].

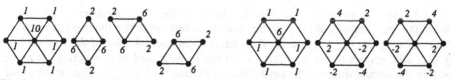

Fig. 2. Geometric notation for the refinement rules (3) and for the limit rules (limit point and the two partial derivatives).

The support of the refinement rules implies that the 1-ring neighborhood of a vertex $\mathbf{p}^{m+1}$ only depends on the 1-ring neighborhood of $\mathbf{p}^m$ (cf. Fig. 2). Hence, the local subdivision matrix S for a regular vertex with valence 6 is

$$
S = \frac{1}{16}\begin{pmatrix}
10 & 1 & 1 & 1 & 1 & 1 & 1 \\
6 & 6 & 2 & 0 & 0 & 0 & 2 \\
6 & 2 & 6 & 2 & 0 & 0 & 0 \\
6 & 0 & 2 & 6 & 2 & 0 & 0 \\
6 & 0 & 0 & 2 & 6 & 2 & 0 \\
6 & 0 & 0 & 0 & 2 & 6 & 2 \\
6 & 2 & 0 & 0 & 0 & 2 & 6
\end{pmatrix} = V^{-1} D V
$$

$$
= \begin{pmatrix}
1 & 0 & 0 & 0 & 0 & 1 & 0 \\
1 & 1 & -1 & 0 & 1 & 0 & -1 \\
1 & 0 & -1 & -1 & -1 & -3 & 1 \\
1 & -1 & 0 & 1 & 0 & 0 & -1 \\
1 & -1 & 1 & 0 & 1 & 0 & 1 \\
1 & 0 & 1 & -1 & -1 & -3 & -1 \\
1 & 1 & 0 & 1 & 0 & 0 & 1
\end{pmatrix}
\frac{1}{16}\begin{pmatrix}
16 & 0 & 0 & 0 & 0 & 0 & 0 \\
0 & 8 & 0 & 0 & 0 & 0 & 0 \\
0 & 0 & 8 & 0 & 0 & 0 & 0 \\
0 & 0 & 0 & 4 & 0 & 0 & 0 \\
0 & 0 & 0 & 0 & 4 & 0 & 0 \\
0 & 0 & 0 & 0 & 0 & 4 & 0 \\
0 & 0 & 0 & 0 & 0 & 0 & 2
\end{pmatrix}
\frac{1}{12}\begin{pmatrix}
6 & 1 & 1 & 1 & 1 & 1 & 1 \\
0 & 2 & -2 & -4 & -2 & 2 & 4 \\
0 & -2 & -4 & -2 & 2 & 4 & 2 \\
-6 & -1 & -1 & 5 & -1 & -1 & 5 \\
-6 & 5 & -1 & -1 & 5 & -1 & -1 \\
6 & -1 & -1 & -1 & -1 & -1 & -1 \\
0 & -2 & 2 & -2 & 2 & -2 & 2
\end{pmatrix}
$$

The coefficients of the linear combination for the limit point and the tangents are given by the first three rows of V (cf. Fig 2 for the explicit masks). The tangent masks yield the two partial derivatives at the limit point, and the normal vector can be obtained by their cross product. In [8] we give a complete table of the limit mask coefficients for valences $k = 3, \ldots, 12$.

3 Oriented bounding volumes

To derive bounding volumes for subdivision curves and surfaces we have to know the *ranges* of the basis functions $\phi(t - i)$ over the considered interval. This is not straight forward since we just know the coefficients of the refinement equation and do not have an explicit parameterization in terms of polynomials or such. We will first explain how to compute oriented bounding boxes before we present a simple iterative procedure which yields tight estimates for the actual bounds.

Let a curve $P(t)$ be given by a linear combination of scalar valued basis functions $\phi(t - i)$ as in (1). Its range with respect to some direction $\mathbf{n}$ for $t \in [a, b]$ can be estimated by

$$
\sum_i (\mathbf{n}^T \mathbf{p}_i)_+ \min_{t \in [a,b]} \phi(t - i) + \sum_i (\mathbf{n}^T \mathbf{p}_i)_- \max_{t \in [a,b]} \phi(t - i) \leq P(t),
$$

$$
P(t) \leq \sum_i (\mathbf{n}^T \mathbf{p}_i)_+ \max_{t \in [a,b]} \phi(t - i) + \sum_i (\mathbf{n}^T \mathbf{p}_i)_- \min_{t \in [a,b]} \phi(t - i)
$$

where

$$
(\mathbf{n}^T \mathbf{p}_i)_+ := \max\{\mathbf{n}^T \mathbf{p}_i, 0\}, \qquad (\mathbf{n}^T \mathbf{p}_i)_- := \min\{\mathbf{n}^T \mathbf{p}_i, 0\}.
$$

Example: For the cubic B-spline it is known that

$$
N(\{0, 1, 2, 3, 4\}) = \{0, \tfrac{1}{6}, \tfrac{4}{6}, \tfrac{1}{6}, 0\}
$$

and N is monotonic on each integer interval such that the extremal values occur at the uniform knots. Consider the spline curve $P(t) = \sum_i \mathbf{p}_i N(t - i)$ over the interval $t \in$

74

$[j, j+1]$. Here the curve is completely determined by the control vertices $\mathbf{p}_{j-1}, \ldots \mathbf{p}_{j+2}$ due to the finite support of N. With the limit point rule we obtain the two points $\mathbf{q}_j := P(j)$ and $\mathbf{q}_{j+1} := P(j+1)$ on the limit curve. The chordal error of the straight line $\overline{\mathbf{q}_j \, \mathbf{q}_{j+1}}$ with respect to the arc $P([j, j+1])$ can be estimated by computing the range of P in the normal direction $\mathbf{n}_j$ perpendicular to the chord $\overline{\mathbf{q}_j \, \mathbf{q}_{j+1}}$. Due to the affine invariance we can shift the control vertices by $-\mathbf{q}_j$ and obtain the control vertices' normal distances $r_i := \mathbf{n}_j^T (\mathbf{p}_i - \mathbf{q}_j)$ for $i = j-1, \ldots, j+2$. The normal range

$$[l_j, u_j] = \frac{1}{6}\left[(r_j^+ + r_{j+1}^+ + r_{j-1}^- + 4\,r_j^- + 4\,r_{j+1}^- + r_{j+2}^-),\ (r_j^- + r_{j+1}^- + r_{j-1}^+ + 4\,r_j^+ + 4\,r_{j+1}^+ + r_{j+2}^+)\right]$$

defines a rectangular box which is aligned to the chord $\overline{\mathbf{q}_j \, \mathbf{q}_{j+1}}$ and completely contains the arc $P([j, j+1])$.

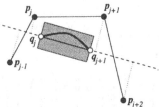

Fig. 3. The gray bounding box is spanned by shifts of the chord $\overline{\mathbf{q}_j \, \mathbf{q}_{j+1}}$ (hollow dots) in normal direction $\mathbf{n}_j$. The actual normal range is computed by a weighted sum of the normal distances r_i of the control vertices $\mathbf{p}_i$ from the supporting line of the chord (dashed line).

In the bivariate setting the procedure to estimate the chordal approximation error is exactly the same: given a submesh which defines one segment of the limit surface (corresponding to one triangle in the control mesh) we use the limit masks to obtain points on the surface spanning a chordal triangle T. As in the univariate case we derive the normal distances r_i of all involved control vertices $\mathbf{p}_i$ from the supporting plane of T and compute a weighted sum according to the ranges of the associated basis functions. The result is an orthogonal triangular prism with the top and bottom face being shifted versions of the chordal triangle T. The possibility that the patch might intersect the quadrilateral sides of the prism will be addressed in Section 4.

For irregular meshes we have to consider the different special cases that occur at extraordinary vertices since the ranges of the basis functions do depend on the local connectivity of the mesh. For the sake of simplicity we assume that the mesh has been uniformly subdivided once before the bounding boxes are to be computed. This reduces the number of special cases since each extraordinary vertex with valence $\neq 6$ has only regular direct neighbors (cf. Fig. 4). Since there is no explicit parameterization for the limit surface of Loop's subdivision scheme, we have to find a reliable numerical algorithm to estimate the true ranges. It turns out that this already is a first application of the bounding volume technique in itself.

The control mesh defining a triangular *Loop-patch* is shown in Fig 4. To approximate the basis function corresponding to one of the vertices, we assign $z = 1$ to it and zero z-values to all other vertices (*Dirac-mesh*). The iterative refinement of such a mesh will approach the specific basis function as depicted in Fig. 1 but it won't provide a reliable bound on the actual range. Notice that we are only interested in the basis function's range over the center triangle.

Let us start with a very coarse over-estimation of the true ranges, e.g. $[-1, 1]$. We apply the subdivision rules to the Dirac-mesh in Fig. 4 and after several subdivision steps we compute the chord aligned bounding boxes for every triangle. Since the mesh has become locally flat, the chord aligned boxes are also flat and provide a much tighter

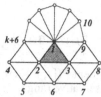 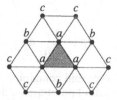

Fig. 4. The mesh on the left contains all vertices that have influence on the surface patch corresponding to the center triangle. Each vertex is associated with a specific basis function. On the right the vertices are symmetrically labeled according to the ranges of their basis functions over the center triangle ($a = [\frac{1}{12}, \frac{1}{2}]$, $b = [0, \frac{13}{96}]$, $c = [0, \frac{1}{12}]$).

bound for the range of the current basis function. Once we have done this for all involved basis functions, we can iterate the whole procedure to obtain even better bounds in every step. Fig. 4 shows the exact ranges for the 12 basis functions whose support covers the center triangle in the regular setting. For a complete table with min/max range values of the $k + 6$ basis functions over a triangle being adjacent to a valence $k = 3, \ldots, 12$ vertex cf. [8].

4 Bounding envelopes

Each triangle of a subdivision surface's control mesh corresponds to a patch segment of the limit surface. When uniformly subdividing the mesh we generate a sequence of meshes whose triangles can be grouped as a forest of quad-trees with each triangle of the original mesh being a root node. In Section 3 we showed how to compute a bounding prism for each triangular sub-patch (on any subdivision level). However, using the individual bounding prisms directly for ray intersection tests is quite complicated since each is aligned to the particular normal direction of the underlying chord triangle.

Further, when we derived the bounding prisms, we did not address the problem that bounding the range in the direction perpendicular to the chordal triangle is not sufficient since the triangular patch might intersect one of the prism's quadrilateral faces. Hence we would have to enlarge the prism to guarantee inclusion. Moreover, testing whether a ray intersects the prism is not trivial since several special configurations have to be checked.

We therefore introduce a pre-processing step where we combine all chord aligned bounding prism to build a global *bounding envelope*. This is a continuous triangle mesh having the same topology and complexity as the control mesh and which tightly circumscribes the limit surface. Obviously for open meshes we have to compute *two* envelopes: one covering the front side and one for the back side.

Each local refinement operation of the control mesh induces a corresponding refinement of the envelope meshes. As the refinement proceeds the envelopes quickly approach the limit surface. In the next section we will explain a simple ray intersection procedure which traces recursively through the forest of mesh triangles thereby testing as few prisms as possible. By using the envelope structure to navigate, we exploit the topological coherence in the mesh, i.e., the fact that a triangle's neighbor can be found with $O(1)$ complexity. The *continuity* of the hull on each subdivision level guarantees that no intersection is missed.

Again, we explain the general envelope construction in the univariate setting for the sake of simplicity. So far we have derived an individual bounding box for each *segment* by estimating the limit curve's range $[l_j, u_j]$ in the direction $\mathbf{n}_j$ perpendicular to the chord $\overline{\mathbf{q}_j \, \mathbf{q}_{j+1}}$. In order to obtain continuous bounding polygons we use the limit tangent rule to compute a (normalized) normal vector $\tilde{\mathbf{n}}_j$ for each limit point $\mathbf{q}_j$. The

four points

$$\mathbf{q}_j + \frac{l_j}{\mathbf{n}_j^T \tilde{\mathbf{n}}_j} \tilde{\mathbf{n}}_j, \quad \mathbf{q}_j + \frac{u_j}{\mathbf{n}_j^T \tilde{\mathbf{n}}_j} \tilde{\mathbf{n}}_j, \quad \mathbf{q}_{j+1} + \frac{l_j}{\mathbf{n}_j^T \tilde{\mathbf{n}}_{j+1}} \tilde{\mathbf{n}}_{j+1}, \quad \mathbf{q}_{j+1} + \frac{u_j}{\mathbf{n}_j^T \tilde{\mathbf{n}}_{j+1}} \tilde{\mathbf{n}}_{j+1}$$

define a minimal trapezoid containing the curve segment $P([j, j + 1])$. Since two segments meet at each limit point $\mathbf{q}_j$, we define the conservative range

$$[\tilde{l}_j, \tilde{u}_j] = \left[\min\{\frac{l_{j-1}}{\mathbf{n}_{j-1}^T \tilde{\mathbf{n}}_j}, \frac{l_j}{\mathbf{n}_j^T \tilde{\mathbf{n}}_j}\}, \max\{\frac{u_{j-1}}{\mathbf{n}_{j-1}^T \tilde{\mathbf{n}}_j}, \frac{u_j}{\mathbf{n}_j^T \tilde{\mathbf{n}}_j}\} \right]$$

for each vertex of the control polygon. Hence, for a given control polygon $[\mathbf{p}_i^m]$ we obtain a *limit polygon* $[\mathbf{q}_i^m]$, an *upper envelope* $[\mathbf{q}_i + \tilde{u}_i \tilde{\mathbf{n}}_i]$ and a *lower envelope* $[\mathbf{q}_i + \tilde{l}_i \tilde{\mathbf{n}}_i]$.

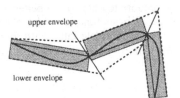

Fig. 5. By computing the ranges $[\tilde{l}_j, \tilde{u}_j]$ for each vertex we find an envelope for the chord based bounding boxes. The envelope's boundaries have the same "topology" as the control polygon (left). In the bivariate case we also compute normal ranges with respect to vertices instead of triangles and replace the orthogonal prism (center) by more general bounding volumes (right). Although each individual bounding object becomes more complicated, the continuity of the bounding envelope enables a simple recursive tracing procedure.

A closed envelope in the bivariate setting is obtained by replacing the orthogonal prisms of Section 3 by more general prisms that are generated by moving each vertex $\mathbf{q}_j$ of the chord triangle T in the corresponding vertex normal direction $\tilde{\mathbf{n}}_j$ (cf. Fig 5). As in the univariate case we compute a normal range for each vertex by taking the minimum and maximum of the *projected* normal ranges of all adjacent triangles (and their orthogonal prisms). After this procedure we end up with three topologically equivalent meshes (derived from the original control mesh), one being a chordal approximant to the limit surface and two meshes (outer and inner envelope) which can be used as bounding volume.

5 Efficient ray intersection

Free form surfaces are usually defined as a collection of individual patches. The standard way to implement a ray intersection test with such objects is to compute a simple bounding volume for each patch. These bounding volumes can be used as a middle layer of a bounding volume hierarchy which is propagated towards the root by enclosing neighboring bounding volumes into a circumscribed larger volume and towards the leaves by subdividing the patches and computing bounding volumes for the sub-patches.

Tracing a ray through such a bounding volume hierarchy means traversing a tree data structure with the descent being controlled by ray intersection tests for the current node's volume. If a leaf is reached, a numerical procedure like Newton iteration is applied to the explicit polynomial parameterization of the patch.

Since subdivision surfaces do not have an explicit parameterization, we propose to base the recursive ray tracing procedure mainly on the information provided by the envelope meshes. Many techniques have been suggested to accelerate the ray intersection tests with a collection of triangles [1]. The most effective one is to build a hierarchical space partition (e.g., BSP-trees) for the whole mesh and incrementally trace the ray through this decomposition. We do use this technique in our algorithm but only to find those triangles where the ray enters the outer envelope corresponding to the *initial* control mesh. Once the entry triangle is found, we trace the ray through the hierarchy of adaptively refined envelope triangles. The navigation is controlled by simple ray–triangle tests. The efficiency of the tracing procedure results from the fact that chord aligned bounding boxes converge much faster than axis-aligned ones (cf. Sect. 6). Another advantage of this strategy is that no redundancy is introduced by overlapping bounding prisms or ambiguous assignment of bounding prisms to a space partition.

To simplify the explanation we assume that all necessary local refinement operations have already been performed when the tracing path intersects a certain triangle. In our actual implementation we used a *lazy-evaluation* mechanism which computes the control vertices of the refined meshes on demand while keeping the overall data structure consistent (i.e. adjacent leaf-triangles may only differ by one generation) and caching all computed information for future requests.

The tracing algorithm has to handle all special cases of the ray intersecting a triangle of the envelope mesh (on a certain refinement level) but intersecting the surface itself at a neighboring patch (if it intersects at all). The most important feature that we exploit for the tracing algorithm is that the envelopes corresponding to the same subdivision level form a continuous surface and the ray cannot pass through without hitting at least one triangle. In one tracing step we either descend in the forest of refined envelope triangles or we proceed the search in a neighboring tree if no intersection can be found below the current one.

We base the local decision on *intersection masks* specifying the seven possible spatial configurations between a ray and a triangle (cf. Fig. 6). For a triangle $T = \triangle(A, B, C)$ and a ray with origin O and direction $\mathbf{r}$ we distinguish the configurations according the signs of the coefficients α, β, and γ in the unique linear combination

$$\mathbf{r} = \alpha(A - O) + \beta(B - O) + \gamma(C - O).$$

Only the $+++$-case reports an intersection of the ray with T. All other cases indicate failure but they provide useful information about where to search for an intersection, i.e. which neighboring triangle to check next.

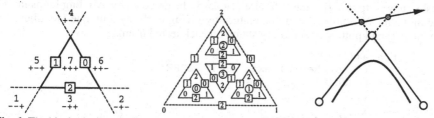

Fig. 6. The bits in the intersection mask define a partition of the triangle's supporting plane. They indicate in which direction the tracing algorithm has to proceed (left). The indexing of triangle corners and edges is inherited from the parent triangle to preserve the orientation (center). A ray passing the envelope triangle mesh without intersection does intersect the associated supporting planes in reverse order. Translated into the intersection masks of the corresponding triangles this configuration would cause a loop in the tracing algorithm (right).

In each step we check if one of the current node's children is intersected, too. If so, we proceed the search with the descendents of this child. If not, we use the intersection masks obtained while testing the children to decide which neighbor of T to test next. The following pseudo code implements the algorithm

```
trace(T)
    for child_0 to child_3
        mask[i] = test_intersect(child_i)
        if mask[i] = '+++'
            trace(child_i)
    lookup neighbor(mask[3],mask[sibling(mask[3])])
    trace(T->neighbor)
```

Cf. Fig. 6 for our indexing convention of a triangle and its four children. If none of the child triangles is intersected by the ray the tracing is controlled by two tables neighbor and sibling. The $child_3$ is the center child and if the corresponding intersection mask has two negative signs then the next neighbor of T is uniquely determined. If only one of the signs is negative, then we use the corresponding sibling's intersection mask to determine where to proceed. If the sibling's mask has two negative signs then we proceed (by convention) in the counter clockwise direction to make the decision deterministic. sibling is indexed by mask[3] and neighbor is row indexed by mask[sibling(mask[3])] and column indexed by mask[3].

```
int sibling[7] = {0,0,0,2,0,1,0};
int neighbor[49] = {
    -1, -1, -1, -1, -1, -1, -1,    /* 2 1 0         */
    -1,  0,  0,  0,  0,  0,  0,    /* - - +  :  1 */
    -1,  1,  1,  1,  1,  1,  1,    /* - + -  :  2 */
    -1,  1,  0, -1,  1,  1,  0,    /* - + +  :  3 */
    -1,  2,  2,  2,  2,  2,  2,    /* + - -  :  4 */
    -1,  2,  0,  2,  0, -1,  0,    /* + - +  :  5 */
    -1,  2,  2,  2,  1,  1, -1 };  /* + + -  :  6 */
```

Don't-care-cases in sibling are set to zero and -1s in neighbor indicate configurations where no intersection occurs. Notice that the case mask[i] = 7 cannot occur since in this case the algorithm descends recursively and the procedure to determine the next neighbor is not called.

Failure of the recursive algorithm (i.e. no intersection occurs) can be detected as follows: The generic situation of a ray passing the object without intersection is depicted in Fig. 6. Such configurations are characterized by the fact that the intersection mask of a triangle T_1 indicates the search to be continued in triangle T_2 while T_2's intersection masks suggest to test T_1. Hence failure can easily be detected by avoiding loops in the tracing path. We implemented this feature by giving each ray a unique identification (an integer) and putting a temporary stamp on each tested triangle:

```
trace(T)
    if touched(T) exit on FAILURE
    touch(T)
    ...
```

6 Approximation tolerance

To eventually compute an intersection point of the ray with the surface, we have to define a stopping criterion to determine whether a bounding prism is sufficiently small in the sense that replacing the true surface geometry by a linear approximant does not lead to an error larger than some prescribed ε. We approximate the location of the true intersection point by computing the intersection of the ray with a chord triangle. For the orthogonal bounding prisms obtained by shifting the chordal triangle in normal direction the error can be estimated by

$$E \leq \min\left\{l, \frac{h}{|\mathbf{n}^T\mathbf{r}|}\right\} \tag{4}$$

with l being the longest edge of the chord triangle, h being the height of the prism, $\mathbf{n}$ the normal vector and $\mathbf{r}$ the direction of the ray (cf. Fig 7).

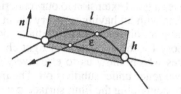 

Fig. 7. The maximum distance between the true intersection point and its approximation depends on the height h of the bounding prism *and* the relative direction of the ray. For very steep angles the maximum edge length of the supporting chord triangle becomes the dominant bound (left). Since the stopping criterion for the subdivision depends on the direction of the ray, the refinement tends to be much finer close to the contour (right).

This reveals the significant advantage of geometry aligned boxes over axis-aligned ones: Subdivision of axis-aligned bounding boxes approximately halves the size of the box in every step, hence going down one level in the bounding volume hierarchy provides one more bit in the precision of the result. In our case however, the *height* of the boxes shrinks much faster than the length of the triangle edges. This behavior can easily be explained by looking at the eigenstructure of the subdivision matrix S. The first eigenvalue $\lambda_1 = 1$ and the associated eigenvector is responsible for the local convergence to a point on the limit surface. The second and third subdominant eigenvalues $\lambda_2 = \lambda_3$ and the associated (two dimensional) eigenspace define the tangent plane. The fourth eigenvalue λ_4 controls the flattening rate by which the deviation of the local sub-mesh from a plane configuration reduces.

In the above example of Loop's scheme in the vicinity of a regular vertex, we have $\lambda_4 = \frac{1}{4}$ and hence one subdivision step approximately bisects the edges of the chord triangles but their height is reduced by the factor 4. As a consequence the tracing algorithm turns out to be significantly faster, especially if high precision is required.

The reciprocal factor $|\mathbf{n}^T \mathbf{r}|$ implies a certain degree of view dependency in the refinement during ray tracing. If the ray intersects almost perpendicularly, the refinement can stop as soon as the height of the bounding prism is smaller than the prescribed tolerance ε. The bounding prisms for triangles close to the visual contour of the object have to be refined much further. Fig. 7 shows the adaptive refinement resulting from rendering a simple convex object by a ray tracing algorithm with fixed tolerance for the intersection tests. To avoid numerical instabilities, we consider intersection tests with sufficiently small bounding prisms (according to (4)) where the ray does not hit the interior of the associated chord triangle as failure.

7 Results and conclusions

Loop subdivision surfaces were integrated as a new geometric primitive into a generic architecture for rendering algorithms [18]. The color plates show several pictures rendered with a simple ray tracer. We chose one very simple example (the initial control mesh being an octahedron) to demonstrate that intersection tests at the contours can be evaluated in a stable manner (upper left). The reflectivity of the material was set to a high value, making the surrounding room's striped wallpaper fully visible on the resulting surface. The smooth reflection lines demonstrate the surface's C^2-continuity.

The initial triangle mesh defining the cat model in the lower right image consists of 728 faces. As in the image of the reflecting octahedron the material properties were set to full reflection to obtain clearly visible reflection lines. The image on the right shows a raytraced subdivision surface with a control mesh consisting of 1374 triangles. A texture was added to define the head's color. Again the smooth shape of the highlights indicate curvature continuity of the surface.

In practice we found subdivision surfaces a good extension to our rendering system, as the control meshes are easy to manipulate without having to worry about continuity issues. Even meshes only roughly outlining an object's shape produce good looking surfaces. We presented the necessary theory for the derivation of tight chord aligned bounding volumes for subdivision surfaces which allow us to effectively handle ray intersection inquiries due to their fast convergence under subdivision. The introduction of continuous envelope meshes completely containing the limit surface gives rise to an efficient algorithm that enumerates all potentially intersected envelope triangles.

References

1. W Barth, W. Stürzlinger, *Efficient ray tracing for Bezier and B-spline surfaces*, Comput Graph 17:423-430
2. E. Catmull, J. Clark, *Recursively generated B-spline surfaces on arbitrary topological meshes*, CAD 10 (1978), 350–355
3. C. de Boor, K. Höllig, S. Riemenschneider, *Box Splines*, Springer Verlag, 1993
4. D. Doo, M. Sabin, *Behaviour of Recursive Division Surfaces Near Extraordinary Points*, CAD 10 (1978), 356–360
5. N. Dyn, J. Gregory, D. Levin, *A Butterfly Subdivision Scheme for Surface Interpolation with Tension Control*, ACM ToG 9 (1990), 160-169
6. N. Dyn, *Subdivision Schemes in Computer Aided Geometric Design*, Adv. Num. Anal. II, W.A. Light ed., Oxford Univ. Press, 1991, 36–104.
7. A. Fournier, J. Buchanan *Chebyshev polynomials for boxing and intersections of parametric curves and surfaces*, Comp Graph Forum (Eurographics 94 issue), 127–142
8. http://www9.informatik.uni-erlangen.de/~kobbelt/RT_of_SS.html
9. J. Kajiya, *Ray tracing parametric patches* SIGGRAPH 82 proceedings, 245–254
10. L. Kobbelt, *Interpolatory Subdivision on Open Quadrilateral Nets with Arbitrary Topology*, Comp Graph Forum (Eurographics '96 issue), C407–C420
11. L. Kobbelt, *Discrete Fairing*, Proceedings of the 7th IMA Conference on the Mathematics of Surfaces, Information Geometers, 1997, 101 – 131
12. L. Kobbelt, S. Campagna, J. Vorsatz, H-P. Seidel, *Interactive Multi-Resolution Modeling on Arbitrary Meshes*, to appear in SIGGRAPH 98 proceedings
13. C. Loop, *Smooth Subdivision for Surfaces Based on Triangles*, University of Utah, 1987
14. M. Lounsbery, T. DeRose, J. Warren *Multiresolution Analysis for Surfaces of Arbitrary Topological Type*, ACM ToG 16 (1997), 34–73
15. T. Nishita, T. Sederberg, M. Kakimoto *Ray tracing trimmed rational surface patches* SIGGRAPH 90 proceedings, 337–345
16. U. Reif, *Neue Aspekte in der Theorie der Freiformaflächen beliebiger Topologie*, Universität Stuttgart, 1993
17. M. Sweeney, R. Bartels, *Ray tracing free-form B-spline surfaces*, IEEE Comput Graph Appl 6 (1986), 41–49
18. Slusallek P, Seidel H-P, *Vision - an architecture for global illumination calculations*, IEEE Trans Vis Comput Graph 1 (1995), 77–96
19. E. Stollnitz, T. DeRose, D. Salesin *Wavelets for Computer Graphics*, Morgan Kaufmann Publishers, 1996
20. D. Toth, *On ray tracing parametric surfaces*, SIGGRAPH '85 proceedings, 171–179
21. C. Yang, *On speeding up ray tracing of B-spline surfaces*, CAD 19 (1987), 122–130
22. D. Zorin, P. Schröder, W. Sweldens, *Interpolating subdivision for meshes with arbitrary topology*, SIGGRAPH 96 proceedings, 189–192
23. D. Zorin, P. Schröder, W. Sweldens, *Interactive Multiresolution Mesh Editing*, SIGGRAPH 97 proceedings, 259–269
24. D.Zorin C^k *Continuity of Subdivision Surfaces*, California Institute of Technology, 1996

Editors' Note: see Appendix, p. 327 for colored figures of this paper

Acquiring Input for Rendering at Appropriate Levels of Detail: Digitizing a Pietà

Holly Rushmeier, Fausto Bernardini, Joshua Mittleman and Gabriel Taubin

IBM TJ Watson Research Center, Yorktown Heights, NY 10598 USA

Abstract: We describe the design of a system to augment a light striping camera for three dimensional scanning with a photometric system to capture bump maps and approximate reflectances. In contrast with scanning an object with very high spatial resolution, this allows the relatively efficient and inexpensive acquistion of input for high quality rendering. This system is being used in a project to digitize a Michelangelo Pietà in Florence, Italy.

1 Introduction

All of the input required to accurately render an object could be derived from a highly detailed description of the object's geometry and material properties. A key to efficient rendering is not to work directly with such a detailed description, but with appropriate representations for various length scales – geometric models, surface maps and reflectances [6]. One approach is to derive the appropriate representations for the various scales from a highly detailed description. In this paper we suggest a step towards acquiring input at appropriate levels of detail directly, to increase the efficiency of acquistion and the subsequent processing required.

The motivation for this work is an ongoing project to create a digital model of the Michelangelo Pietà located in the Museum of the Opera del Duomo, in Florence, Italy (Fig.1a) This work was initiated by art historian Jack Wasserman, who is preparing a book [1] on this sculpture. A geometric model, on a relatively coarse (approximately 5 mm resolution) scale is needed to examine questions about the composition of the figures, to determine the shape left when Michelangelo removed several pieces that have subsequently been reattached, and to speculate on the shape of a piece that is still missing. Some of these issues will be examined with calculations of distance and volume. However, much of the study will be performed by rendering the sculpture in configurations and views specified by the art historian. These rendering requirements dictate the accuracy of the base model.

Finer resolution (on the order of 1 mm or less) descriptions are needed to render the appearance of the work for a CD-ROM to accompany the book, and to examine issues such as what type of tools were used based on fine tool marks. While photographs, panoramic views or light fields would be efficient means for rendering the sculpture as it now appears in the museum, this type of acquisition and rendering would not be adequate for this application. The goal is not to simulate a visit to the museum, but to render the sculpture in detail in alternative locations, such as a reconstruction of the tomb where the work was

---

[1] To be published by Princeton University Press

originally destined to stand. The sculpture is also to be rendered with altered geometry to see it as Michelangelo saw it with parts removed. To perform these renderings we wish to acquire the fine level geometry and reflectance. Surface maps, such as bump or height maps, are needed to render small details such as tool marks. Reflectance variations are required to show where there are cracks that have been repaired, and to show the extent of a yellow patina that was applied to a portion of the sculpture.

There are a number of challenges in this project. One is the sheer size of the sculpture, which stands 2.25 m high. Based on our initial scans, attempting to scan this in detail at even a spatial resolution of 0.5 mm would require on the order of 100 million points. Furthermore, the sculpture has a complicated topology with cavities formed by the overlapping limbs of the figures. No large piece of equipment that simply rotates about the sculpture would be adequate to capture the interior of these cavities. Finally, obviously samples can not be removed to measure reflectance in a separate laboratory. All measurements must be taken in the museum with minimal physical contact. Based on a review of previous projects to scan three dimensional works of art (e.g. [3],[9]) we decided to design a system around an existing multiview camera for three dimensional scanning.

(a)   (b)

Figure 1a: A photograph of the Pietà in the Museum of the Opera del Duomo, 1b: An example of the geometry captured in the intial scan of the Pietà on site in Florence. Two merged patches are shown, in wire frame.

We have performed one scan of the geometry of the sculpture on site in Florence. Typical results for part of the sculpture are shown in Fig. 1b. A Virtuoso shape camera [14] was used to capture the geometry in approximately 620 overlapping patches, referred to as "shape photographs." The shape camera uses light striping and 6 black and white cameras to obtain three dimensional points, and one color camera to acquire a texture. The Virtuoso software processes the points into a triangle mesh with uv indices to a color image to be used as a texture map. The results shown in Fig. 1b are the result of merging two patches. Figure 2 (see color plate) shows the texture maps acquired for the two patches that were merged in Fig. 1b. The red dots are from lasers projected on the sculpture to facilitate the alignment of patches. The differences in overall lightness and hue in the texture maps are due to the automatic exposure control and color balance of the color camera used in the Virtuoso. The images clearly show highlights and shadows that are unique to the particular lighting conditions. Even without the red laser spots, these textures are not suitable for realistically rerendering the sculpture. The geometry is more than adequate for the questions of form and composition. We only need finer geometric detail for rendering the appearance at a smaller scale, and a method to capture a better approximation of the spectral reflectance of the surface.

2 Previous Work

Recently, more high-precision three dimensional scanners have become available [12]. Many scanners can produce very dense point clouds representing the object's geometry. The problem of efficiently finding a triangle mesh or other surface approximation to fit these points, and then reducing the number of vertices needed to define the surface is a very active area of research [1]. A reduced number of vertices is necessary for efficient storage and rendering.

Some researchers have considered the problem of retaining detail from the point cloud while simplifying the base geometric model. Krishnamurthy and Levoy [7] describe a method for deriving B-Spline patches with an accompanying height field or bump map from a dense triangle mesh. Although this provides primitives useful for rendering, it still requires an high precision initial scan, fitting a triangle mesh to a dense cloud, and significant processing.

Most scanning systems, like the Virtuoso, acquire color texture data by taking color images that are precisely aligned with the geometry. As illustrated by Fig. 2, this can produce poor results because the maps include the effects of the specific lighting conditions and camera characteristics. One approach to improving this has been developed by Sato et al. [11]. In their method, multiple color images are used to estimate the BRDF (bidirectional reflectance distribution function) of an object which has been digitized by a light striping range finder. Their method eliminates many problems with texture artifacts, although they do not discuss the issues of how to find absolute (rather than relative) reflectance, or how to correct for the alteration in color due to the light source and camera transfer function.

An alternative, more expensive, technology, for acquiring accurate maps of object color is to scan with three color lasers. The result is a dense, unorganized cloud of three dimensional points, each with a color attached. Soucy et al. [13] describe a method for deriving simplified meshes and texture maps from these dense three dimensional clouds.

In [10], the use of a version of photometric stereo is described for obtaining bump maps and maps of the approximate relative values of diffuse reflectance. Photometric stereo can produce much denser descriptions of an object for a given camera resolution than binocular (or multiview) stereo because an estimate can be made at each pixel, rather than at object edges or the edges of projected stripes. Photometric stereo is good for estimating surface normals, but poor for reconstructing the three dimensional surfaces. In the field of computer vision various researchers, e.g. Ikeuchi [5] and Wolff [15] have developed hybrid techniques combining photometric stereo and multiple camera views. In these methods multiple cameras take images for multiple lighting conditions in order to reconstruct a denser range image than could be obtained with multiple cameras alone.

In our application, we do not need additional three dimensional points. The Virtuoso output is robust, and is more than adequate for our geometric model. In our system we use photometric stereo to compute normals only, which will be aligned with the three dimensional output and used as bump maps.

3 Proposed Method

Because of the need for a relatively small and mobile system to capture areas in the cavities of the sculpture, and because of the constraints of time and budget, we are building our acquisition system around the Virtuoso. To augment the output, we are adding a photometric system, similar to that described in [10], to obtain bump and texture maps at a finer level of detail.The texture maps will be calibrated by spot measurements using a Colorton II Digital Color Ruler [8] to produce maps of approximate spectral diffuse reflectance.

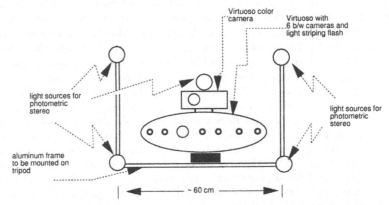

Figure 3: A diagram showing the Virtuoso system augmented by a photometric stereo system.

A diagram of the system currently being tested is shown in Fig. 3. Five lights are positioned around the Virtuoso. The Virtuoso color camera is used to capture images with each of the lights turned on in sequence. The light source positions must be known with respect to the Virtuoso's internal coordinate system. This is required so that the coordinate system of the normals computed from the photometric images is aligned with the coordinate system of the three dimensional points.

A Faro digitizing arm [2] is used to measure the light locations. In a one time only measurement, the Virtuoso coordinates of specified points on a simple target geometry were found by taking a shape photograph of the target. These coordinates were then entered as data as the Faro arm digitized the target points, thus aligning the Faro coordinate system with the Virtuoso system. With the Faro arm realigned, the coordinates of three points on the Virtuoso camera body were found in the Virtuoso's own coordinate system. The reference points on the camera body are being used in subsequent experiments to align the Faro arm to the Virtuoso coordinate system for measuring light source positions.

The format of the shape photograph produced by the Virtuoso development software is an Open Inventor mesh of 3-D points with uv texture coordinates, and a color image trimmed to just the rectangular area covered by the captured geometry. The photometric software takes as input the 5 colored images for the 5 light positions, after trimming. Normals for each pixel in these images are computed using the photometric stereo method described in [10].

The issues to consider in successfully applying photometric stereo to non-Lambertian, non-convex surfaces are discussed in [10]. Briefly summarizing the basic calculations, let ρ and $\hat{n}$ be the diffuse reflectance and surface normal for a point P on the object. The lights used in the photometric system each have radiance L_o and subtend solid angle $\Delta\omega$ from point P. The radiance $L_{r,j}$ is reflected from P when it is illuminated by light j. The vector $\hat{l}_j$ points from P in the direction of light j. Grey level values are computed for each color image by combining the red, green and blue values for each pixel. The highest and lowest of the 5 grey level values for each pixel are discarded to minimize the effects of specular highlights and shadows. In the image captured by the color camera, the 0 to 255 grey scale value of the pixel in which P is visible is equal to $\alpha L_{r,j}$, where α is the constant of proportionality accounting for the linear response of the CCD camera. With the remaining three grey level values the following equation is solved for $\rho L_o \Delta\omega \hat{n}/\alpha\pi$:

$$\rho L_o \Delta\omega/\pi \begin{bmatrix} l_{1,1} & l_{1,2} & l_{1,3} \\ l_{2,1} & l_{2,2} & l_{2,3} \\ l_{3,1} & l_{3,2} & l_{3,3} \end{bmatrix} \begin{bmatrix} n_1 \\ n_2 \\ n_3 \end{bmatrix} = \begin{bmatrix} \alpha L_{r,1} \\ \alpha L_{r,2} \\ \alpha L_{r,3} \end{bmatrix} \tag{1}$$

Since $\hat{n}$ has a magnitude of 1 we can obtain $\hat{n}$ and $\rho L_o \Delta\omega/\alpha\pi$ separately. Since the source radiance, solid angle subtended, and camera response is the same for all pixels, $\rho L_o \Delta\omega/\alpha\pi$ is equal to ρ_{rel}, the reflectance of the object point viewed through this pixel, relative to the points viewed through the other pixels in the image. As discussed in [10], approximate relative reflectances for red, green and blue (RGB) channels can be obtained by adjusting the ratio of red to green to blue found in the images outside of the shadow and highlight regions. For pixel n, the adjusted values of $R'G'B'$ then are:

$$[R', G', B']_n = \rho_{rel,n}[R, G, B]_n/greylevel([R, G, B]_n) \tag{2}$$

Equation 2 gives a crude relative estimate of reflectance for the imprecisely defined red, green and blue channels. To obtain more precise, absolute measurements, we make spot measurements of absolute diffuse reflectance using the Colortron II. The Colortron II measures absolute spectral reflectances in 32 wavelength bands in the visible spectrum. To use the measurement to adjust the relative reflectances, we select values at three wavelengths, 700, 580, and 450 nm to correspond roughly with the R, G and B channels. For the location of the spot measurement c, we have reflectance measurements $\rho_{700,c}$, $\rho_{580,c}$ and $\rho_{450,c}$, and we can obtain the adjusted $R'G'B'$ values computed for c from the processed photometric images. Estimates of reflectance at the three selected wavelengths are then made for all of the image pixels n using:

$$[\rho_{700}, \rho_{580}, \rho_{450}]_n = [\rho_{700,c}R'_n/R'_c, \rho_{580,c}G'_n/G'_c, \rho_{450,c}B'_n/B'_c)] \qquad (3)$$

Because we only obtain three values to represent color in our images, we only obtain three values estimating the spectral reflectance at each point. A series of spot measurements of the spectral reflectance on the actual Pietà showed that the spectrum for the marble, like many other common materials, is quite smooth. Using three values gives a reasonable representation for our application. Analogous to the bump map recording small variations between more sparsely measured three dimensional points, the processed color images are used to record small variations in spectral reflectance between spatially sparse diffuse reflectance measurements.

Using the uv coordinates from the Open Inventor model, the resultant map of surface normals computed using Eq.1 and map of reflectances computed using Eqs. 2 and 3 can be rendered with the three dimensional geometry.

Using this plan for scanning, we can adjust our system to acquire data at the resolution needed for rendering. The resolution of the geometric model needed determines the distance at which we want to focus the cameras for the light stripe image, since for the same stripe pattern the spacing between the points increases with the distance between the camera and target. We have some control over the relative resolution acquired by the photometric portion of the process by using a color imaging camera with the appropriate resolution relative to the light stripe frequency. Along with this specification for the color camera we require manual control of exposure, so that the exposure is the same for all of the images used for the photometric solution.

The scan of one patch of the sculpture proceeds as follows:

- Turn on laser dots for geometric alignment.
- Turn on ambient light for adequate illumination for stripes
- Take light stripe pictures
- Turn off laser dots and ambient light
- Take five photometric images

This scan procedure will be under software control, with most of the elapsed time required for each scan being the time to download data from the camera to a removable storage card. We are currently estimating the precise resolution needed for the geometry and normals maps based on our inital scan of the actual piece and photometric tests in our laboratory.

4 Results

Figures 4 to 9 show some results from initial tests of the combined Virtuoso/photometric system that we performed in our laboratory. As a test object we used a 1/10 scale model of the Pietà, which is an approximate reproduction sold at the museum gift shop. The detail on this scale model, such as the fingers on the hand of the rightmost figure, are the same absolute size, 1 mm, as the features we wish to capture on the full scale model.

Figure 4: Results directly from the Virtuoso system for a small model of the Pietà. Results are shown (a) with texture map on the left, (b) without texture map in the center, and (c) in wire frame on the right.

Figure 4 shows the output from the Virtouso camera as purchased. Figure 4a shows the texture mapped geometry from the direction that it was acquired. The lighting in this view looks correct since it is just a display of the image as captured. There are holes in the geometry because points can only be computed for object points that are visible unshadowed from the six black and white cameras used to image the light stripes. A full model is assembled by merging together geometry from multiple views. The captured geometry is shown shaded without texture map and in wire frame in Figs. 4b and 4c . From a distance of approximately 1 m, the Virtuoso has digitized points on the 22.5 cm high statue at a spatial frequency of approximately 2 mm. This level of detail is inadequate to capture features such as the fingers on the right figure. However, it is more than enough to capture the overall coarse scale geometry. To simulate a coarser sampling, we ran the model through a simplification program. The results of using the simplification method described in [4] to reduce the number of vertices from 2927 to 1086 are shown shaded and in wire-frame in Figures 5a and 5b. The algorithm used a tolerance of 1% of the diagonal of the bounding box to determine the allowable change in the shape due to the simplification. Little detail has been lost relative to the original captured geometry.

Figure 6 shows the result of the photometric calculations. Figures 6a and 6b were rendered with the photometric results under two different novel lighting conditions (i.e. light positions that were not used in computing the normals.) The spacing of the normals on the map is approximately 0.4 mm. The spatial density of the normals is higher than the three dimensional points in part because the color camera has a higher resolution and in part because it takes a width greater than one pixel to determine the edge of the light stripes.

Figure 7 contrasts the results using the Virtuoso output alone with the photometric output mapped onto a simplified geometry. In Fig. 7, the light source direction is different from the conditions under which any of the Virtuoso or pho-

88

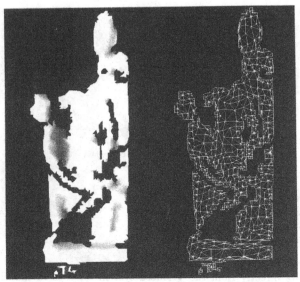

Figure 5: The geometry from Fig. 4 after simplification, (a) shown shaded on the left and (b) in wire frame on the right.

(a)   (b)

Figure 6: Images rendered using photometric results under two new lighting conditions.

tometric images were acquired. Figures 7a and 7c show the result from Virtouso under the new lighting conditions. Figure 7a shows the original view, and Fig. 7c shows a view from the right side. The Virtuoso results have the obvious advantage that correct silhouettes are obtained when a new view is selected, which cannot be done with just the photometric results. The texture on the Virtuoso looks very odd in 7a and 7c – particularly noticeable are the shadow lines from the original image that are not consistent with the new lighting.

Figure 7: Comparison of original Virtuoso geometry and texture map (in 7a and 7c, the leftmost image and image third from the left) with simplified geometry and photometric results (7b and 7d).

Figures 7b and 7d show the results of mapping the normals from the photometric process onto the simplified geometry obtained from the original Virtuoso mesh. In this initial application the lighting is computed using the photometric normals and a diffuse reflectance in software, and the resulting texture map is displayed on the simplified mesh using hardware texture mapping. Even though the model has fewer triangles than the original Virtuoso mesh, a correct silhouette is obtained when the model is viewed from the right. The bump map allows the display of the detailed hand of the right figure placed on the central figure. As shown in the detailed images in Fig. 8, the appearance of the geometric resolution has been increased by a factor of 5, without any explicit acquisition and processing of additional three dimensional points.

Figure 9 shows the results for a larger colored object – a ceramic cookie jar with a somewhat glossy finish similar to parts of the Pietà. The Virtuoso-only results with and without the acquired texture map are shown in 9a and 9b respectively. Figure 9b illustrates another problem common in scanning systems. Under new lighting conditions (from above and the right) slight diagonal artifacts appear on the squirrel's head and arm. The number of ridges on the squirrel's tail has been increased. The artifacts occur because even small sub-millimeter errors in computing the coordinates can result in noticeable artifacts in the slopes on the object surface. As shown in Fig. 10 even a 0.25 mm error (which is quite small by the standards of current non-contact digitizers) in samples spaced 2 mm apart can result in an angle of more than 28 degrees between normals that should be the same.

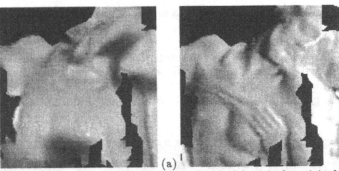

Figure 8: Details from Fig. 7, showing the rendering (a) with the original texture map, and (b) with photometric results.

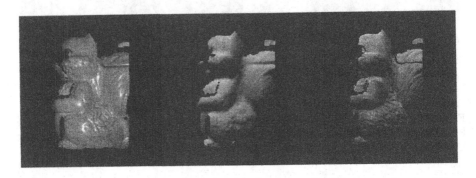

Figure 9: Results of scanning a squirrel-shaped cookie jar. The results directly from the Virtuoso are shown (a) texture mapped on the left and (b) without texture map in the center. The results of the photometric normals calculation are shown on the right (c) with the same lighting as used in (b.)

A typical remedy for small slope errors is to perform additional scans of the same area, producing even a denser cloud of points and averaging the results. However, without taking more points we already have a good description of the base geometry in one scan. Instead of taking more points, we use our photometric results, shown in Fig. 9c to replace the normals computed from the triangle mesh and eliminate the scanning artifacts and to more faithfully reproduce the portion of the jar representing the squirrel's fur. The results in Fig. 9c, while a better representation of the pattern on the tail and the fur, are somewhat noisy. The useful data for the photometric calculations is in the brown, darker areas in the images. We do not have a full 8 bits of image data for the pixels we use to compute the normals. Because our test camera has automatic exposure control, we were not able to adjust the exposure and use the full 8 bit range to eliminate the noise in this test.

The cookie jar is basically brown, but has subtle variations in the color of the diffuse portion of reflectance to enhance the impression of fur on the squirrel's body. The relative reflectance maps were computed from the photometric images, and then adjusted by a Colortron II spot measurement on the squirrel's arm.

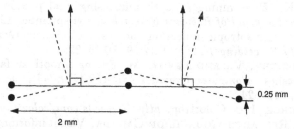

Figure 10: The solid line shows the actual solid surface. The dotted line shows the results of a scan with a small error that results in noticeable errors in the surface normals.

The resulting map of reflectances is compared with the original texture map in shown in Fig. 11 (see color plate), with the R, G and B channels used to display $\rho_{700}, \rho_{580}, \rho_{450}$.

5 Future Work

A number of issues need to be addressed to obtain high quality results from this system. For example, it is clear from the cookie jar example that a method is needed to estimate the specular component of reflectance. We can not measure the absolute value of the specular component with the limited dynamic range of our camera, however we can estimate the width of the specular peak using a modified form of Sato et al.'s method. This may need to be supplemented by spot measurements.

With our planned system, we can scan the statue at a coarser resolution than our original 600 patch scan. However, the result will still be over 100 overlapping maps of normals and reflectances. While careful calibration can ensure that these maps are consistent with one another, having numerous overlapping maps is inefficient for storage and rendering. A method for merging and reducing the number of surface maps without loss of resolution is needed.

In this paper we have presented a step forward in acquiring input at different levels of input directly, rather than starting with a very dense cloud of points and colors. Refining this idea by combining various modes of point and image acquisition has great potential for acquiring data for rendering complex objects efficiently and economically.

References

1. BERNARDINI, F., BAJAJ, C., SHEN, J., AND SCHIKORE, D. Automatic reconstruction of 3D CAD models from digital scans. *International Journal of Computational Geometry and Applications* (to appear).
2. FARO TECHNOLOGIES, INC. The FaroArm. *http://www.faro.com/*.
3. FRAUNHOFER INSTITUTE FOR COMPUTER GRAPHICS. SCULPTOR. *http://www.igd.fhg.de/www/igd-a7/Projects/model/examples.html*.
4. GUÉZIEC, A. Surface simplification inside a tolerance volume. *IBM Research Report*, RC20440 (1997).

5. IKEUCHI, K. Determining a depth map using dual photometric stereo. *International Journal of Robotics Research 6*, 1 (September 1987), 15–37.

6. KAJIYA, J. Anisotropic reflection models. *Computer Graphics (SIGGRAPH '85 Proceedings) 19*, 3 (July 1985), 15–22.

7. KRISHNAMURTHY, V., AND LEVOY, M. Fitting smooth surfaces to dense polygon meshes. *Computer Graphics (SIGGRAPH '96 Proceedings)* (August 1996), 313–324.

8. LIGHT SOURCE, INC. Colortron. *http://www.ls.com/colortron.html*.

9. NATIONAL RESEARCH COUNCIL OF CANADA. Visual information technology. *http://www.vit.iit.nrc.ca/image/masque.html*.

10. RUSHMEIER, H., TAUBIN, G., AND GUÉZIEC, A. Applying shape from lighting variation to bump map capture. *Proceedings of the Eighth Eurographics Rendering Workshop* (June 1997), 35–44.

11. SATO, Y., WHEELER, M., AND IKEUCHI, K. Object shape and reflectance modeling from observation. *Computer Graphics (SIGGRAPH '97 Proceedings)* (August 1997), 379–388.

12. SCHECHTER, J. 3D digitizers make their mark. *Computer Graphics World* (March 1998), 79–84.

13. SOUCY, M., GODIN, G., BARIBEAU, R., BALIS, F., AND RIOUX, M. Sensors and algorithms for the construction of digital 3-D colour models of real objects. *Proceedings of the International Conference on Image Processing* (September 1996), 409–412.

14. VISUAL INTERFACE, INC. Virtuoso. *http://www.visint.com/*.

15. WOLFF, L., AND ANGELOPOULOU, E. 3D stereo using photometric ratios. *Journal of the Optical Society of America – A*, 11 (November 1994), 3069–3078.

Editors' Note: see Appendix, p. 327 for colored figures of this paper

Interactively Modeling with Photogrammetry

Pierre Poulin   Mathieu Ouimet   Marie-Claude Frasson

Département d'informatique et de recherche opérationnelle
Université de Montréal

Abstract. We describe an *interactive* system to reconstruct 3D geometry and extract textures from a set of photographs taken with arbitrary camera parameters. The basic idea is to let the user draw 2D geometry on the images and set constraints using these drawings. Because the input comes directly from the user, he can more easily resolve most of the ambiguities and difficulties traditional computer vision algorithms must deal with.

A set of geometrical linear constraints formulated as a weighted least-squares problem is efficiently solved for the camera parameters, and then for the 3D geometry. Iterations between these two steps lead to improvements on both results. Once a satisfying 3D model is reconstructed, its color textures are extracted by sampling the projected texels in the corresponding images. All the textures associated with a polygon are then fitted to one another, and the corresponding colors are combined according to a set of criteria in order to form a unique texture.

The system produces 3D models and environments more suitable for realistic image synthesis and computer augmented reality.

1 Introduction

Realism in computer graphics has greatly evolved over the past decade. However very few synthetic images simulating real environments can fool an observer. A major difficulty lies with the 3D models; creating realistic models is an expensive and tedious process. Unfortunately the growing need for this level of accuracy is essential for realistic image synthesis, movie special effects, and computer augmented reality.

One attractive direction is to extract these models from real photographs. Although two decades of computer vision research has led to important fundamental results, a fully automated and reliable reconstruction algorithm in general situations has not yet been presented, at least for 3D models satisfying computer graphics general requirements. Misinformation in computer vision algorithms resulting from false correspondences, missed edge detections, noise, etc. can create severe difficulties in the extracted 3D models. We base our premise on the fact that the user *knows* what he wants to model, and within which accuracy. He can decide what must be modeled by geometry, and what could be simulated by a simpler geometry with a texture applied on it. To provide this functionality, we developed a fully interactive reconstruction system.

Getting an accurate 3D model requires the solution of several problems, which are all interrelated. We must first compute correct camera parameters, and then use the cameras and constraints to reconstruct the 3D geometry. After discussing some related work, we outline our geometry reconstruction system. A set of correspondences and incidences result in simple and efficient linear constraints. Although these constraints are not new, the improvements obtained in accuracy and speed demonstrate the importance of considering all of them together. User intervention at every step of this process, results in more satisfying general reconstructed 3D models.

Simple 3D geometry will be effective only with good quality textures. We focus in Section 3 on a more complete, view-independent, treatment of textures. Textures are extracted for each 3D geometry from all images it projects to. The *best* texels (2D texture elements) are then combined into a single texture according to various criteria including visibility, projected areas, color differences, and image quality.

By solving accurately each problem, we will better understand the robustness, stability, and precision of our techniques. It should become easier later on to extend our interactions with more automatic computer vision and image processing techniques in order to alleviate some of the more cumbersome and tedious tasks, while keeping user intervention where required. The results of our system should help us create more precise textured synthetic models from real 3D objects in less time than current 3D modelers, and more robustly than fully automated geometry extraction algorithms.

2 Extracting 3D Geometry

Twenty years of active research on 3D reconstruction from 2D images in computer vision and robotics have left a considerable legacy of important results [12, 6, 5].

The first problem to address concerns camera calibration, i.e. computing camera parameters. This is a difficult and unstable process often improved by the use of specific targets. By putting in correspondence points or lines between images, it becomes possible to calibrate cameras. Similarly with known camera parameters, one can reconstruct a 3D scene up to a scale factor [4]. In these classical approaches, segmentations and correspondences are automatically determined. One typical example of the results obtained by these approaches was recently presented by Sato *et al.* [19].

A few recent projects such as REALISE [3, 10], *Façade* [1], *PhotoModeler* [16], and AIDA [21] propose to integrate more user intervention into the reconstruction process. They are derived from projective geometry [13], and are applied to the reconstruction of man-made scenes from a set of photographs and correspondences.

REALISE [3, 10] integrates user intervention early in the correspondence process, letting the user specify the first correspondences, and then returning to a more classical approach to identify automatically most of the other correspondences. The more stable initial solution greatly helps to reduce the errors of subsequent iterations. Nevertheless the same errors of fully automatic systems can still occur, and the user must then detect and correct the origin of the errors, which is not a simple task as the number of automatic correspondences increases.

Façade [1] develops a series of parameterized block primitives. Each block encodes efficiently and hierarchically several constraints frequently present in architectural design. The user must first place the blocks with a 3D modeler, and then set correspondences between the images and these blocks. Non-linear optimization of an objective function is then used to solve for all these constraints. The system has proven to be quite efficient and provides precise 3D models with little effort. However it requires the user to build with the blocks the model he wants to reconstruct. We believe this might be more difficult when general 3D models cannot be as easily created with these blocks.

PhotoModeler [16] is a commercial software for performing photogrammetric measurements on models built from photographs. Once the camera is calibrated, the user has to indicate features and correspondences on the images, and the system computes the 3D scene. The models obtained appear quite good, although it seems to be a lengthy process (they reported a week for a model of 200 3D points) which uses images of very high resolution (around 15 MB each). We also noticed many long thin triangles and gaps in some of their models. The system can apply textures coming from the photographs but does not seem to perform any particular treatment since the shadows, highlights, etc. are still present. No details are provided on the algorithms used.

The AIDA system [21] is a fully automatic reconstruction system that combines surface reconstruction techniques with object recognition for the generation of 3D models for computer graphics applications. The system possesses a knowledge database of constraints, and selects the constraints to apply to the surface under reconstruction after performing a scene interpretation phase. We believe it might be safer, less cumbersome, and more general to let the user choose which constraints he wants to apply to its 3D primitives rather than letting the system pick some constraints from a knowledge database created specifically for the type of scene to reconstruct.

For these reasons, we introduce a system essentially based on user interaction. The user is responsible for (almost) *everything*, but also has the control on (almost) *everything*. This should provide a comprehensive tool to improve on the modeling from real 3D objects and on the computer graphics quality of these 3D models, while offering the opportunity to focus on the details important to the designer.

2.1 Our Reconstruction System

System Overview. We have developed an *interactive* reconstruction system from images. The images define the canvas on which all interaction is based. They can come from any type of cameras (even a virtual synthetic camera) with any settings and position. The user *draws* points, lines, and polygons on the images which form our basic 2D primitives. The user interactively specifies correspondences between the 2D primitives on different images. He can also assign other constraints between reconstructed 3D primitives simply by clicking on one of their respective 2D primitives. These additional constraints include parallelism, perpendicularity, planarity, and co-planarity.

At any time, the user can ask the system to reconstruct all *computable* cameras and 3D primitives. The reconstructed 3D primitives can be reprojected on the images to estimate the quality of each recovered camera and the 3D primitives. The user then has the choice to iterate a few times to improve on the mathematical solution, or to add new 2D primitives, correspondences, and constraints to refine the 3D models. This process, illustrated in Fig. 1, demonstrates the flexibility and power of our technique. The 3D model is reconstructed incrementally, refined where and when necessary. Each error from the user can also be immediately detected using reprojection. Contrarily to Debevec *et al.* [1], the user does not create a synthetic model of the geometry he wants to recover, although the reconstructed 3D model can as easily be used to establish new constraints between 2D and reconstructed 3D primitives.

Each image thus contains a set of 2D primitives drawn on it, and a camera computed when the set of resolved constraints is sufficient. To bootstrap the reconstruction process, the user assigns a sufficient number of 3D coordinates to 3D primitives via one of their corresponding 2D primitives. For instance, six 3D points in one image allow the computation of the corresponding camera. Once two cameras are computed, all 3D geometry that can be computed by resolving the constraints is reconstructed. With the assigned and the newly computed 3D values, the constraints are resolved again to improve the reconstructed cameras. This process iterates until no more constraints can be resolved, and the 3D geometry and cameras are computed to a satisfactory precision. Typically, a convergence iteration solving the equation systems for computing all the cameras and 3D positions takes between 0.05 and 2 seconds,[1] depending on the complexity of the scene (50 to 200 3D points) and the constraints used.

---

[1] on a 195 MHz R10000 SGI Impact with unoptimized code

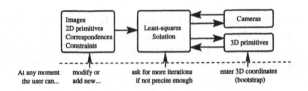

Fig. 1. Geometry reconstruction process

All our constraints are expressed as linear equations, typically forming an over-determined set of equations. A least-squares solution to this system is computed by singular value decomposition [17]. We use this solution for the unknown camera parameters, and the unknown 3D coordinates of points and lines. As an *exact* correspondence is hardly achievable by drawing 2D points on the image plane, we compute the *best* camera parameters in the least-squares sense. Hartley [7] demonstrate simple conditions under which linear systems of equations used to determine the camera parameters are as precise as their non-linear counterparts. Moreover this technique is simple to implement, efficient, general, always provides a solution, and we observed that it is more robust than the non-linear systems.

Geometrical Considerations. We express the geometrical constraints in our system as a set of incidences onto 3D planes. We model the perspective projection associated with each image with a 4×4 transformation matrix $\mathbf{T}$ [18]. This matrix is sufficient for our purposes because we do not intend to extract real camera parameters. A 3D point P expressed in homogeneous coordinates is transformed by $\mathbf{T}$ into a normalized homogeneous 2D point p on the image plane as

$$
\begin{bmatrix} p_u \\ p_v \\ 0 \\ 1 \end{bmatrix} = \begin{bmatrix} T_0 & T_1 & T_2 & T_3 \\ T_4 & T_5 & T_6 & T_7 \\ 0 & 0 & 0 & 0 \\ T_8 & T_9 & T_{10} & 1 \end{bmatrix} \begin{bmatrix} P_x \\ P_y \\ P_z \\ 1 \end{bmatrix}.
$$

In projective geometry, a 2D point p on an image defines a line in 3D. Any point P on this 3D line projects onto p. A 3D line can be defined as the intersection of two 3D planes. We define two orthogonal planes A_h and A_v such that they respectively project as horizontal and vertical 2D lines l_h and l_v intersecting at point p on the image plane. This is illustrated in Fig. 2 (left). We compute the plane A that contains the 2D line (a, b, c) and its 3D corresponding line as

$$
A = \begin{bmatrix} a & b & \star & c \end{bmatrix} \mathbf{T} = \begin{bmatrix} aT_0 + bT_4 + cT_8 \\ aT_1 + bT_5 + cT_9 \\ aT_2 + bT_6 + cT_{10} \\ aT_3 + bT_7 + c \end{bmatrix}^T.
$$

A 3D point P lies on a plane $A = \begin{bmatrix} a & b & c & d \end{bmatrix}^T$ when $A \cdot P = 0$. For the plane A_h defined by the 2D line l_h as $v = p_v$ in Fig. 2 (left), the constraint such that the 3D point P lies on this plane is satisfied when

$$
A_h \cdot P = \begin{bmatrix} 0 & 1 & \star & -p_v \end{bmatrix} \mathbf{T} \cdot P = 0
$$

that can be rewritten as

$$
P_x(T_4 - p_v T_8) + P_y(T_5 - p_v T_9) + P_z(T_6 - p_v T_{10}) = p_v - T_7. \tag{1}
$$

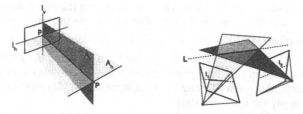

Fig. 2. Constraints for point position (left) and line correspondences (right)

Similarly the 3D point P lies on the vertical plane A_v defined by the 2D line $u = p_u$ if

$$P_x(T_0 - p_u T_8) + P_y(T_1 - p_u T_9) + P_z(T_2 - p_u T_{10}) = p_u - T_3. \qquad (2)$$

Each 2D point with a known 3D position thus provides two equations. If we want to solve for the eleven unknowns T_i of our projection matrix, we need a minimum of six such points. Although less correspondences could resolve a non-linear system of equations [11], we consider finding six points in one image into which we want to reconstruct geometry not to be an overwhelming request. Once the cameras for two images are recovered in this way, it becomes possible to compute the position of a 3D point given its two projected 2D points in the images using Eq. (1) and (2).

In the case of 3D lines, any pair of 3D points P_1 and P_2 ($P_1 \neq P_2$) satisfying the condition $A \cdot P = 0$ defines a 3D line on A, the plane going through the 3D line and its projection l. We represent a 3D line L by using the *point-form* matrix L, a 4×4 antisymmetric matrix containing the six *Plücker coordinates* [20] of the line. With this representation, L lies on A when

$$AL = 0 \Rightarrow \begin{bmatrix} aT_0 + bT_4 + cT_8 \\ aT_1 + bT_5 + cT_9 \\ aT_2 + bT_6 + cT_{10} \\ aT_3 + bT_7 + c \end{bmatrix}^T \begin{bmatrix} 0 & P_{xy} & P_{xz} & P_x \\ -P_{xy} & 0 & P_{yz} & P_y \\ -P_{xz} & -P_{yz} & 0 & P_z \\ -P_x & -P_y & -P_z & 0 \end{bmatrix} = \begin{bmatrix} 0 \\ 0 \\ 0 \\ 0 \end{bmatrix}$$

where for any two points P_1 and P_2, we have $P_{xy} = x_1 y_2 - x_2 y_1$, $P_x = x_1 - x_2$, etc.

By isolating the six Plücker coordinates, we obtain four constraints for each plane going through the 3D line. With more than one plane, we can determine the line in a least-squares sense as illustrated in Fig. 2 (right).

Additional 3D Constraints. Most man-made scenes exhibit some form of planarity, parallelism, perpendicularity, symmetry, etc. Using correspondences only, reconstructed geometry often does not respect these properties, which can lead to objectionable artifacts in the reconstructed 3D models. It is therefore very important to integrate this type of constraint in a reconstruction system. They are unfortunately difficult to detect automatically, as perspective projection does not preserve them in the image. The user can however very easily indicate each such constraint directly onto the images. They are integrated into our convergence process by adding equations to the system of linear equations used to compute 3D coordinates. Although they are not strictly enforced because they are simply part of a least-squares solution, they often result in more satisfying 3D models especially with respect to the needs of computer graphics models.

Coplanarity: A planar polygon with more than three vertices should have all its vertices on the same plane A. For each polygon with 3D vertices P_i, we add a *planarity constraint* of the form $A \cdot P_i = 0$ that will be used during the computation of each P_i. Polygons, points, and lines can also be constrained to lie on the same supporting plane. To compute this plane $A = [\ a\ \ b\ \ c\ \ d\ ]^T$, the user can specify a normal direction $N = [\ a\ \ b\ \ c\ ]^T$. We compute the *best* value for d by using the known 3D points. If there is no information about the plane orientation, we compute the *best* plane in the least-squares sense, that passes through at least three known 3D points of the coplanar primitives, and apply the same constraint.

Parallelism: Similarly, several polygons and lines can be parallel to each other, providing additional constraints. For each polygon, we get its orientation N_i from its plane equation, if available. These orientations allow us to calculate an average orientation N_{avg} that will be attributed to all the parallel polygons, even those for which no orientation could be first calculated. A *planarity constraint* is added to the computation of the polygons 3D points. For parallel lines, we compute their average direction.

Perpendicularity: If the normals N_{perp} to perpendicular polygons are known, we can add other constraints for the computation of N_{avg}. Two perpendicular polygons have perpendicular normals, thus for every polygon orthogonal to a set of parallel polygons, we have $N_{perp} \cdot N_{avg} = 0$.

Many other constraints should be exploited. *Symmetry* could constrain characteristics such as lengths or angles. *Similarity* between models could specify two identical elements at different positions. *Incidence* of points and lines can be extended to different primitives. These are only a few of the constraints we observe in 3D scenes.

Each basic constraint described above can be used as a building block for more elaborate primitives. A cube for instance becomes a set of planar faces, with perpendicularity and parallelism between its faces and segments, and constrained length between its 3D vertices. Rather than letting the user specify all these constraints, a new primitive for which all of these are already handled represents a much more efficient tool for the user. These new primitives can be described in a library of primitives organized hierarchically. Debevec *et al.* [1] shows how this representation can also reduce significantly the number of constraints to resolve.

We can also weight the contributions of the constraints depending on their importance in the current reconstruction. The default weights assigned to each type of constraint can be edited by the user. The resolution of our equation systems is simply extended to a weighted least-squares.

2.2 Results of Geometry Reconstruction

To evaluate the precision and the convergence of our iterative process, we constructed a simple synthetic scene made of seven boxes. Five images of resolution 500×400 where rendered from camera positions indicated by the grey cones in Fig. 3 (left). 2D polygons were manually drawn and put in correspondences within 60 minutes on a 195 MHz R10000 SGI Impact. The 3D coordinates of six points of the central cube on the floor were entered to bootstrap the system. The three curves in Fig. 3 (right) represent the distance in world coordinates between the real 3D position of three points in the scene ((-2,3,0),(0,2,-2),(1,2,-2)) and their reconstructed correspondents.

200 iterations without any constraints other than the point correspondences took about 5 minutes. We then applied successively the constraints of planarity, coplanarity, and parallelism between all the 3D polygons. Calculating all these constraints typically adds a few tenths of a second per iteration depending on the complexity of the 3D scene.

The three curves reach a plateau after a certain number of iterations. This does not mean that the 3D model is then perfectly reconstructed, but rather that the solution is

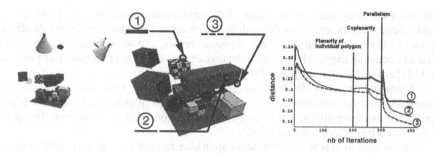

Fig. 3. Reconstruction of a synthetic scene

stable and should not change significantly with more iterations. When we introduce the planarity and then the coplanarity constraints for individual polygons, the points move slightly. In this scene where parallelism is preponderant, the addition of this last constraint improves significantly the reconstruction for all three points, which is shown by the drop of all three curves after iteration 300. The introduction of a new constraint can sometimes perturb the whole system, affecting more the less-constrained elements as demonstrated by the sudden spike in curve 2. In most observed cases, the system quickly returns to an improved and more stable state.

In Fig. 3 (center), we reproject in wireframe mode the reconstructed model using the computed camera from one of the original images. Distances between the 2D drawn points and the reprojected reconstructed points all lie within less than one pixel from each other. When 3D constraints are used to improve the model, this distance can reach up to two pixels. The 3D scene then corresponds more to reality but does not fit exactly the drawn primitives when reprojected with the computed projection matrix. The constraints thus compensate for the inaccuracy introduced by the user interaction or by the primitives far from the entered coordinates. Because the user draws 2D primitives at the resolution of the image, getting a maximum of two pixels is considered satisfactory. Sub-pixels accuracy is obtained if these primitives are drawn at sub-pixel precision, but this lengthen the user interaction time.

3 Extracting Texture

Textures have been introduced in computer graphics to increase the realism of synthetic surfaces. They encode via a surface parameterization the color for each point on the surface. While the contribution of textures to realism is obvious, it is not always easy to extract a texture from real images. One must correct for perspective foreshortening, surface curvature, hidden portions of the texture, reflections, shading, etc. All these limitations have restricted the type of extracted real textures. However our reconstructed geometry and cameras provide a great context within which we can extract these textures.

Most current approaches are based on view-dependent textures. Havaldar *et al.* [8] use the projection of the 3D primitive in all the images to determine the *best* source image for the texture. Then we apply to the texture the 2D transformation from the projected polygon in this best image, to the projected polygon in the image from a new viewpoint. Unfortunately, this 2D deformation of the texture is invalid for a perspective projection, and prone to visibility errors.

Debevec *et al.* [1] reprojects each extracted texture for a given primitive as a weighted function based on the viewing angle of the new camera position. The technique provides better results with view-dependent information. However, neglecting the distance factor in the weights can introduce important errors, and aliasing can appear from the use of occlusion maps. Moreover all the textures must be kept in memory as potentially all of them might be reprojected for any new viewpoint.

Niem and Broszio [14] identifies the best image for an entire polygon (according to angle and distance criteria), and samples the texture from this image. Because adjacent polygons can have different best images, they then proceed to smooth out the adjacent texels, possibly altering the textures.

In our system, a texture is extracted upon user request for a given primitive projecting in a number of images. The texture is the result of recombining the estimated *best* colors for each point of the surface projected in each image. Even though extracting a single texture is prone to errors as will be discussed later, it is more suitable for general image synthesis applications such as applying it to different primitives, filtering, and use of graphics hardware.

3.1 Sampling Colors from Images

Assume a one-to-one mapping between every point on a primitive and its surface parameterization. We reconstruct the texture as a regular grid of 2D texels from a sampling of this parameterization.

Each texture point (s, t) on the surface maps to a 3D point. For each image with a computed projection matrix, we can easily find the 2D point corresponding to the projection of the 3D point. Therefore each 3D point on the surface goes through the extraction pipeline of Fig. 4 to determine its color as viewed from each computed camera. These colors are then recombined according to a set of criteria to determine the final color associated with this point.

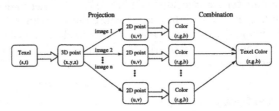

Fig. 4. Texture-to-color extraction pipeline

3D primitives and cameras are reconstructed as least-squares solutions from drawn 2D primitives. The precision is considered acceptable for 3D models, as demonstrated by the results of the previous section. However the small variations from one image to the other, or assuming a surface is perfectly planar while it is not can slightly affect the alignment of the corresponding extracted texture in each image. Matching directly each texel in the set of extracted textures thus tends to produce aliasing artifacts, or blurry textures if a wider filter is used. User intervention again allows the correction of these errors. For each extracted texture, the user selects (draw) a few (three to six) 2D feature points in all the corresponding extracted textures. They create a set of 2D-to-2D correspondences. A linear transformation of the form

$$\begin{bmatrix} M_0 & M_1 & M_2 \\ M_3 & M_4 & M_5 \\ 0 & 0 & 1 \end{bmatrix}$$

is solved again using a least-squares method in order to find a transformation to better fit two textures together.

We reconstruct the color of a 3D point by applying a filter kernel on the neighboring pixel colors at the 2D point it projects to. We can also improve on the texture definition by supersampling each texel, i.e. by using more samples to represent the texel color. Each such sample is weighted by the filter kernel on the texel. As the texture space can be very different than the 3D surface space, we adaptively supersample the texels for which adjacent samples project to more distant (> 2) pixels. This adaptive sampling scheme can also be used to determine the finest texture resolution required according to the projection of the 3D primitive in all the images.

3.2 Occlusion

For a given 3D point, its projection in one image will most likely be a visible point because the user drew its supporting 3D primitive as a corresponding 2D primitive on the image. However some portion of the 3D primitive might be blocked by another 3D primitive closer to the image plane, thus leading to an incoherent color.

A simple test determines the zone with an occlusion risk by intersecting the 2D primitive with all 2D primitives on this image. If there is intersection, we must determine the 3D intersection between the corresponding 2D point on the image (two planes) and the potentially occluding 3D polygon (a third plane). If there is intersection and the depth is smaller than the one of the 3D point, we simply mark this occluded color sample as invalid.

3.3 Weighting the Contributions for the Final Color

We extract a color for each texel in each image. The final color for this texel must be computed from these colors.

The size in pixels of the texel projected in the image is a good indication of the quality of the color extracted for this texel. The larger the projected area of a texel, the more precise the texture should be. Therefore, for all the valid colors of a given texel, we weight its color contribution as a relative function of the projected areas of the texel in all selected images.

Ofek *et al.* [15] stores each pixel in each image into a *mipmap* [22] pyramidal structure for the texture. Color information is propagated up and down the pyramid, with some indication of *certainty* according to color variations. The structure fits each texel to the image pixel resolution, thus adapting its processing accordingly. However unless very high resolution textures are required, we believe the extra cost of propagating the information in the pyramid and the inevitable loss of information due to the filtering between levels might not be worth the savings.

Our solution is simple, but might require sampling many texels projecting within a small fraction of a pixel area. However all information is kept at the user-specified texel resolution, and as such, we get much flexibility in ways of interpreting and filtering the information. It is also fairly simple to integrate various new criteria to improve on the process of combining the extracted texels.

3.4 Color Differences

Unfortunately we must be aware that several situations might invalidate any such texture extraction algorithm. Any view-dependent feature that changes the aspect (color) of a 3D point as the camera moves might be a source of errors. These include specular reflections (highlights), mirrors, transparencies, refractions, ignored surface deformations (sharp grooves and peaks), participating media, etc. Without user intervention, a combination of these *artifacts* can hardly be handled automatically.

When one color at one texel is very different than the others for different images, we simply reject its contribution, assuming it was caused by view-dependent features or noise in the image. When all the colors are very different from each other, we simply mark this texel as invalid. We will discuss in the conclusion how these differences could be used to extract such view-dependent information.

The color of a pixel at silhouettes and edges of polygons includes not only the texel color, but also background and other geometry colors. Camera registration errors also introduce slight misalignments between the reconstructed 3D primitive and each 2D primitive it should project to. We again simply mark such a texel as invalid.

We therefore end out with an extracted texture with some undefined texel colors. Fortunately, these typically represent a small portion of the entire texture. We currently fill in these texels by applying a simple filter, although some *filling* algorithms have proven to be quite efficient [9]. They should be even more effective for gaps as narrow as those observed so far in our tests.

3.5 Results of Extracted Textures

The quality of extracted textures is really important for computer graphics applications. To demonstrate the quality of our sampling approach, we reconstructed a simple scene from four photographs of a real scene displayed in Fig. 5 (left). Each photo is digitized at a 640×480 resolution. 2D polygons were drawn by the user for five faces of the Rubik cube, the cover of the magazine, and the top and one face of the book. The 3D positions of 8 corners of the cube were entered to *bootstrap* the projection reconstruction, and the cover and book 3D geometry have been reconstructed from point correspondences and polygons planarity. The entire process took less than 10 minutes.

The four corresponding extracted textures are displayed in Fig. 5 (center). Each texture was sampled at a higher 320×400 resolution, one sample per texel, without filtering or supersampling. The black cut to the right of the top right texture is due to the book occluding this part of the texture in the corresponding photograph.

Fig. 5. Four images and the extracted textures

A zoom on the **GR** in Fig. 5 (top right) shows the ghosting of the default reconstruction and the improvements after texture alignments by the user using four feature points. Although one texture was partly occluded, its valid texels were used to reconstruct the combined final texture. The final recombined texture with samples from all extracted textures (except for occluded parts) is displayed as the larger front cover in Fig. 5. One can observe the quality of the reconstruction on the title of the magazine.

Color plates of two reconstructed scenes with geometry and textures appear in the Color Section. All photos have a 640 × 480 resolution. The tower took about 6 hours to reconstruct from a set of ten photos. 400 convergence iterations led to the model which contains over a hundred 3D polygons. The desk scene was made from a set of six photos in about 3 hours. It was greatly improved by the use of constraints which corrected the slanted surfaces due to the distance from the small reference primitive (entered coordinates of the Rubik cube corners). The space constraints and image reproduction limitations do not demonstrate well the quality and problems of reconstruction algorithms. We invite the reader to visit the site associated with this paper from http://www.iro.umontreal.ca/labs/infographie/papers to better appreciate additional images, 3D models, and textures.

4 Conclusion

We have presented a simple algorithm to register cameras and reconstruct 3D geometry from a set of 2D primitives drawn directly by a user on 2D images. Various correspondences and linear constraints are introduced to improve on the 3D model accuracy. Our system permits user interaction at every stage of the iterative process. Because the user specifies all geometry, correspondences, and constraints, the reconstructed 3D models are better adapted to general graphics requirements.

We presented an algorithm to extract the textures for a polygon and to combine their colors into a unified texture. When the color differences are too large, these colors are simply rejected. Missing information can be replaced by a filling algorithm.

While the results are very encouraging and already satisfying for many scenes, we have merely scratched the surface of all the possibilities. Our interactive system is rather basic, and it would greatly benefit from a better interface, standard 2D interactive editing tools, and efficient vision tools. These improvements combined with a tool to detect potential correspondence errors would dramatically reduce modeling time.

We are currently investigating the extraction of surface reflection properties from the radiance [2] differences at each texel. As the number of images necessary to extract a precise reflection model can be fairly large, we must extend our user-driven scheme to model-tracking in video sequences. With known geometry and emission of light sources, we can easily determine the regions in shadow from one light source. This direct illumination is very important to extract reflection properties. Once a first estimate of this information is computed, we will again use an iterative process to try to extend this solution to interreflections in the scene.

Our extracted geometrical models are fairly simple and mostly polygonal. We are also investigating a more automatic reconstruction of complex geometry in a form of displacement textures applied on simpler enclosing primitives.

We finally expect to develop bounds on the precision of the extracted 3D points, surface normals, colors, reflection coefficients, illumination, etc. These bounds would then provide a measure of the accuracy of the extracted information.

Acknowledgments. Pat Hanrahan provided very important ideas at the early stages of this project while the first author was a postdoc at Princeton. Craig Kolb also helped with the first implementation. We acknowledge financial support from NSERC. The third author also acknowledges a CFUW fellowship.

References

1. P.E. Debevec, C.J. Taylor, and J. Malik. Modeling and rendering architecture from photographs: A hybrid geometry- and image-based approach. In *SIGGRAPH 96 Conference Proceedings*, pages 11–20. August 1996.

2. P.E. Debevec and J. Malik. Recovering high dynamic range radiance maps from photographs. In *SIGGRAPH 97 Conference Proceedings*, pages 369–378. August 1997.

3. O. Faugeras, S. Laveau, L. Robert, G. Csurka, and C. Zeller. 3D reconstruction of urban scenes from sequences of images. Tech. Report 2572, INRIA Sophia-Antipolis, May 1995.

4. O.D. Faugeras. What can be seen in three dimensions with an uncalibrated stereo rig? In *Proceedings of Computer Vision (ECCV '92)*, pages 563–578, May 1992.

5. O. Faugeras. *Three-dimensional Computer Vision: A Geometric Viewpoint*. MIT Press, 1993.

6. W.E.L. Grimson. *From Images to Surfaces: A Computational Study of the Human Early Vision System*. MIT Press, 1981.

7. R.I. Hartley. In defense of the eight-point algorithm. *IEEE Trans. on Pattern Analysis Machine Intelligence*, 19(6):580–593, June 1997.

8. P. Havaldar, M.-S. Lee, and G. Medioni. View synthesis from unregistered 2-D images. In *Proceedings of Graphics Interface '96*, pages 61–69, May 1996.

9. D.J. Heeger and J.R. Bergen. Pyramid-Based texture analysis/synthesis. In *SIGGRAPH 95 Conference Proceedings*, pages 229–238. August 1995.

10. F. Leymarie *et al.* Realise: Reconstruction of reality from image sequences. In *International Conference on Image Processing*, pages 651–654. September 1996. http://www.inria.fr/robotvis/projects/Realise

11. Y. Liu, T.S. Huang, and O. Faugeras. Determination of camera location from 2-D to 3-D line and point correspondences. *IEEE Trans. on Pattern Analysis Machine Intelligence*, 12(1):28–37, January 1990.

12. H.C. Longuet-Higgins. A computer algorithm for reconstructing a scene from two projections. *Nature*, 293:133–135, 1981.

13. J. Mundy and A. Zisserman, editors. *Geometric Invariance in Computer Vision*. MIT Press, Cambridge, Massachusetts, 1992.

14. W. Niem and H. Broszio. Mapping texture from multiple camera views onto 3D-object models for computer animation. In *Proceedings of the International Workshop on Stereoscopic and Three Dimensional Imaging*, September 1995.

15. E. Ofek, E. Shilat, A. Rappoport, and M. Werman. Multiresolution textures from image sequences. *IEEE Computer Graphics and Applications*, 17(2):18–29, March 1997.

16. http://www.photomodeler.com

17. W.H. Press, S.A. Teukolsky, W.T. Vetterling, and B.P. Flannery. *Numerical Recipes in C: The Art of Scientific Computing (2nd ed.)*. Cambridge University Press, Cambridge, 1992.

18. D.F. Rogers and J.A. Adams. *Mathematical Elements for Computer Graphics*. McGraw-Hill, 1976.

19. Y. Sato, M.D. Wheeler, and K. Ikeuchi. Object shape and reflectance modeling from observation. In *SIGGRAPH 97 Conference Proceedings*, pages 379–387. August 1997.

20. J. Semple and G. Kneebone. *Algebraic projective geometry*. Oxford University Press, 1952.

21. S. Weik and O. Grau. Recovering 3-D object geometry using a generic constraint description. In *ISPRS96 - 18th Congress of the International Society for Photogrammetry and Remote Sensing*, July 1996.

22. L. Williams. Pyramidal parametrics. In *Computer Graphics (SIGGRAPH '83 Proceedings)*, pages 1–11, July 1983.

Editors' Note: see Appendix, p. 328 for colored figure of this paper

Efficient View-Dependent Image-Based Rendering with Projective Texture-Mapping

Paul Debevec, Yizhou Yu, and George Borshukov

Univeristy of California at Berkeley

debevec@cs.berkeley.edu

Abstract. This paper presents how the image-based rendering technique of view-dependent texture-mapping (VDTM) can be efficiently implemented using projective texture mapping, a feature commonly available in polygon graphics hardware. VDTM is a technique for generating novel views of a scene with approximately known geometry making maximal use of a sparse set of original views. The original presentation of VDTM by Debevec, Taylor, and Malik required significant per-pixel computation and did not scale well with the number of original images. In our technique, we precompute for each polygon the set of original images in which it is visibile and create a "view map" datastructure that encodes the best texture map to use for a regularly sampled set of possible viewing directions. To generate a novel view, the view map for each polygon is queried to determine a set of no more than three original images to blend together in order to render the polygon with projective texture-mapping. Invisible triangles are shaded using an object-space hole-filling method. We show how the rendering process can be streamlined for implementation on standard polygon graphics hardware. We present results of using the method to render a large-scale model of the Berkeley bell tower and its surrounding campus enironment.

1 Introduction

A clear application of image-based modeling and rendering techniques will be in the creation and display of realistic virtual environments of real places. Acquiring geometric models of environments has been the subject of research in interactive image-based modeling techniques, and is now becoming possible to perform with time-of-flight laser range scanners. Photographs can be taken from a variety of viewpoints using digital camera technology. The challenge, then, is to use the recovered geometry and the available real views to generate novel views of the scene quickly and realistically.

A desirable quality of such a rendering algorithm is to make judicious use of all the available views, including when a particular surface is seen from different directions in several images. This problem was addressed in [2], which presented view-dependent texture mapping as a means to render each pixel in the novel view as a blend of its corresponding pixels in the original views. However, the technique presented did not guarantee smooth blending between images as the viewpoint changed and did not scale well with the number of available images.

In this paper we adapt view-dependent texture mapping to guarantee smooth blending between images, to scale well with the number of images, and to make efficient use of polygon texture-mapping hardware. The result is an effective and efficient technique for generating virtual views of a scene when:

- A geometric model of the scene is available

- A set of calibrated photographs (with known locations and known imaging geometry) is available
- The photographs are taken in the same lighting conditions
- The photographs generally observe each surface of the scene from a few different angles
- Surfaces in the scene are not extremely specular

2 Previous Work

Early image-based modeling and rendering work [16, 5, 8], presented methods of using image depth or image correspondences to reproject the pixels from one camera position to the viewpoint of another. However, the work did not concentrate on how to combining appearance information from multiple images to optimally produce novel views.

View-Dependent Texture Mapping (VDTM) was presented in [2] as a method of rendering interactively constructed 3D architectural scenes using images of the scene taken from several locations. The method attempted to make full use of the available imagery in novel view generation using the following principle: to generate a novel view of a particular surface patch in the scene, the best original image from which to sample reflectance information is the image that observed the patch from as close a direction as possible as the desired novel view. As an example, suppose that a particular surface of a building is seen in three original images from the left, front, and right. If one is generating a novel view from the left, one would want to use the surface's appearance in the left view as the texture map. Similarly, for a view in front of the surface one would most naturally use the frontal view. For an animation of moving from the left to the front, it would make sense to smoothly blend, as in morphing, between the left and front texture maps during the animation in order to prevent the texture map suddenly changing from one frame to the next. As a result, the view-dependent texture mapping approach allowed renderings to be considerably more realistic than static texture-mapping allowed, since it better represented non-diffuse reflectance and can simulate the appearance of unmodeled geometry.

Other image-based modeling and rendering work has addressed the problem of blending between available views of the scene in order to produce renderings. In [6], blending is performed amongst a dense regular sampling of images in order to generate novel views. Since scene geometry is not used, a very large number of images is necessary in order to produce relatively low-resolution renderings. [4] is similar to [6] but uses approximate scene geometry derived from object silhouettes. Both of these methods restrict the viewpoint to be outside the convex hull of an object or inside a convex empty region of space. The number of images necessary and the restrictions on navigation make it difficult to use these methods for acquiring and rendering a large environment. The work in this paper leverages the results of these methods to render each surface of a model as a light field constructed from a sparse set of views; since the model is assumed to conform well to the scene and the scene is assumed to be predominantly diffuse, far fewer images are necessary to achieve good results.

The implementation of VDTM in [2] computed texture weighting on a per-pixel basis, required visibility calculations to be performed at rendering time, examined every original view to produce every novel view, and only blended between the two closest viewpoints available. As a result, it was computationally expensive[1] and did not always guarantee the image blending to vary smoothly as the viewpoint changed. Subse-

---

[1]Approximately ten minutes per frame on a 200 MHz SGI Indigo2 for a twelve-image model

quent work [9, 7] presented more efficient methods for optically compositing multiple re-rendered views of a scene. In this work we associate appearance information with surfaces, rather than with viewpoints, in order to better interpolate between widely spaced viewpoints in which each sees only a part of the scene. We use visibility preprocessing, polygon view maps, and projective texture mapping to implement our technique.

3 Overview of the Method

Our method for VDTM first preprocesses the scene to determine which images observe which polygons from which directions. This preprocessing occurs as follows:

1. **Compute Visibility**: For each polygon, determine in which images it is seen. Split polygons that are partially seen in one of the images. (Section 5).
2. **Fill Holes**: For each polygon not seen in any view, choose appropriate vertex colors for performing Gouraud shading. (Section 5).
3. **Construct View Maps**: For each polygon, store the index of the image closest in viewing angle for each direction of a regularly sampled viewing hemisphere. (Section 7).

The rendering algorithm (Section 8) runs as follows:

1. Draw all polygons seen in none of the original views using the vertex colors determined during hole filling.
2. Draw all polygons which are seen in just one view.
3. For polygon seen in more than one view, calculate its viewing direction for the desired novel view. Calculate where the novel view falls within the view map, and then determine the three closest viewing directions and their relative weights. Render the polygon using alpha-blending of the three textures with projective texture mapping.

4 Image-Based Rendering with Projective Texture Mapping

Projective texture mapping was introduced in [10] and is now part of the OpenGL graphics standard. Although the original paper used it only for shadows and lighting effects, it is directly applicable to image-based rendering because it can simulate the inverse projection of taking photographs with a camera. In order to perform projective texture mapping, the user specifies a virtual camera position and orientation, and a virtual image plane with the texture. The texture is then cast onto a geometric model using the camera position as the center of projection. The focus of this paper is to adapt projective texture-mapping to take advantage of multiple images of the scene via View-Dependent Texture-Mapping.

Of course, we should only map a particular image onto the portions of the scene that are visible from its original camera viewpoint. The OpenGL implementation of projective texture mapping does not automatically perform such visibility checks; instead a texture map will project through any quantity of geometry and be mapped onto occluded polygons as seen in Fig. 1. Thus, we need to explicity compute visibility information before performing projective texture-mapping.

We could solve the visibility problem in image-space using ray tracing, an item buffer, or a shadow buffer (as in [2]). However, such methods would require us to compute visibility in image-space for each novel view, which is computationally expensive

108

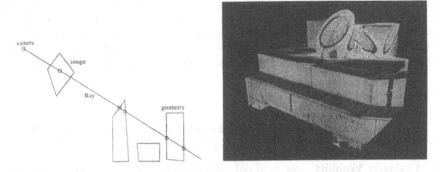

Fig. 1. The current hardware implementation of projective texture mapping in OpenGL lets the texture go through the geometry and get mapped onto all backfacing and occluded polygons on the path of the ray, as can be seen in this rendering of the San Francisco Museum of Modern Art. Thus it is necessary to perform visibility pre-processing so that only properly visible polygons are texture-mapped with a particular image.

and not suited to interactive applications. Projective texture-mapping is extremely efficient if we know beforehand which polygons to texture-map, which suggests that we employ a visibility preprocessing step in object-space to determine which polygons are visibile to which cameras. The next section describes our visibility preprocessing method.

5 Determining Visibility

The purpose of our visibility algorithm is to determine for each polygon in the model in which images it is visible, and to split polygons as necessary so that each is fully visible or fully invisible to any particular camera. Polygons are clipped to the camera viewing frustums, to each other, and to user-specified clipping regions. This algorithm operates in both object space [3, 15] and image space and runs as follows:

1. Assign each original polygon an ID number. If a polygon is subdivided later, all the smaller polygons generated share the same original ID number.
2. If there are intersecting polygons, subdivide them along the line of intersection.
3. Clip the polygons against all image boundaries and any user-specified clipping regions so that all resulting polygons lie either totally inside or totally outside the view frustum and clipping regions.
4. For each camera position, rendering the original polygons of the scene with Z-buffering using the polygon ID numbers as their colors.
5. For each frontfacing polygon, uniformly sample points and project them onto the image plane. Retrieve the polygon ID at each projected point from the color buffer. If the retrieved ID is different from the current polygon ID, the potentially occluding polygon is tested in object-space to determine whether it is an occluder or coplanar.
6. Clip each polygon with each of its occluders in object-space.
7. Associate with each polygon a list of photographs to which it is totally visible.

Using identification numbers to retrieve objects from Z-buffer is similar to the item buffer technique introduced in [14]. The image-space steps in the algorithm can quickly obtain the list of occluders for each polygon. Errors due to image-space sampling are largely avoided by checking the pixels in a neighborhood of each projection in addition to the pixels at the projected sample points.

Our technique also allows the user the flexibility to specify that only a particular region of an image be used in texture mapping. This is accomplished with an additional clipping region in step 3 of the algorithm.

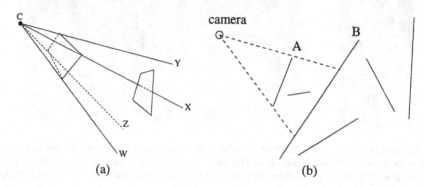

Fig. 2. (a) To clip a polygon against an occluder, we need to form a pyramid for the occluder with the apex at the camera position, and then clip the polygon with the bounding faces of the pyramid. (b) Our algorithm does *shallow clipping* in the sense that if polygon A occludes polygon B, we only use A to clip B, and any polygons behind B are unaffected.

The method of clipping a polygon against image boundaries is the same as that of clipping a polygon against an occluding polygon. In either case, we form a pyramid for the occluding polygon or image frame with the apex at the camera position (Fig. 2(a)), and then clip the polygon with the bounding faces of the pyramid. Our algorithm does *shallow clipping* in the sense that if polygon A occludes polygon B, we only use A to clip B, and any polygons behind B are unaffected(Fig. 2(b)). Only partially visible polygons are clipped; invisible ones are left intact. This greatly reduces the number of resulting polygons.

If a polygon P has a list of occluders $O = \{p_1, p_2, ..., p_m\}$, we use a recursive approach to do the clipping: First, we obtain the overlapping area on the image plane between each member of O and polygon P; we then choose the polygon p in O with maximum overlapping area to clip P into two parts P' and S where P' is the part of P that is occluded by p, and S is a set of convex polygons which make up the part of P not occluded by p. We recursively apply the algorithm on each member of S, first detecting its occluders and then performing the clipping.

To further reduce the number of resulting polygons, we set a lower threshold on the size of polygons. If the object-space area of a polygon is below the threshold, it is assigned a constant color based on the textures of its surrounding polygons. If a polygon is very small, it is not noticeable whether it is textured or just a constant color. Fig. 3 shows visibility processing results for two geometric models.

6   Object-Space Hole Filling

No matter how many photographs we have, there may still be some polygons invisible to all cameras. Unless some sort of coloring is assigned to them, they will appear as undefined regions in novel views in which they are visible.

Fig. 3. Visibility results for a bell tower model with 24 camera positions and for the university campus model with 10 camera positions. The shade of each polygon encodes the number of camera positions from which it is visible; the white regions in the overhead view of the second image are "holes" invisible to all cameras.

Instead of relying on photographic data for these regions, we instead assign colors to the unseen surfaces based on the appearance of their surrounding surfaces, a processed called *hole filling*. Previous hole-filling algorithms [16, 2, 7] have operated in image space, which can cause flickering in animations since the manner in which a hole is filled will not necessarily be consistent from frame to frame. Object-space hole-filling can guarantee that the derived appearance of the each invisible polygon is consistent between viewpoints. By filling these regions with colors close to the colors of the surrounding visible polygons, the holes can be made difficult to notice.

Fig. 4. The image on the left exhibits black regions which were invisible to all the original cameras but not to the current viewpoint. The image on the right shows the rendering result with all the holes filled. See also Fig. 8.

The steps in hole filling are:

1. **Determine polygon connectivity**. At each shared vertex, set up a linked list for those polygons sharing that vertex. In this way, from a polygon, we can access all

its neighboring polygons.

2. **Determine colors of visibile polygons.** Compute an "average" color for each visible polygon by projecting its centroid onto the image planes of each image in which appears and sample the colors from there.

3. **Iteratively assign colors to the holes.** For each invisible polygon, if it has not yet been assigned a color, assign to each of its vertices the color of the closest polygon which is visible or that has been color-filled in a previous iteration.

The reason to have an iterative step is that an invisible polygon may not have a visible polygon in its neighborhood. In this way its filling color can be after its neighboring invisible polygons are assigned colors.

Due to slight misalignments between the geometry and the original photographs, the textures of the edges of some objects may be projected onto the background. For example, a sliver of the edge of a building may project onto the ground nearby. In order to avoid filling the invisible areas with these incorrect textures, we do not sample polygon colors at regions directly adjacent to occlusion boundaries.

Fig. 4 shows the result for holefilling. The invisible polygons, filled will Gouraud-shaded low-frequency image content, are largely unnoticeable in animations. Because we assume that the holes will be relatively small and that the scene is mostly largely diffuse, we do not use the view-dependent information to render the holes.

7 Constructing and Querying Polygon View Maps

The goal of view-dependent texture-mapping is to always use surface appearance information sampled from the images which observed the scene closest in angle to the novel viewing angle. In this way, the errors in rendered appearance due to specular reflectance and incorrect model geometry will be minimized. Note than in any particular virtual novel view, different visible surfaces may have different "best views"; an obvious case of this is when the novel view encompasses an area not entirely observed in any one view.

In order to avoid the perceptually distracting effect of surfaces suddenly switching between different best views from frame to frame, we will blend between the available views as the angle of view changes. This section shows how for each polygon we will create a *view map* that encodes how to blend between at most three available views for any given novel viewpoint, with guaranteed smooth image weight transitions as the viewpoint changes. The view map for each polygon takes little storage and is simple to compute as a preprocessing step. A polygon's view map may be queried very efficiently: given a desired novel viewpoint, it quickly returns the set of images with which to texture-map the polygon and their relative weights.

To build a polygon's view map, we construct a local coordinate system for the polygon that represents the space of all viewing directions. We then regularly sample the set of viewing directions, and assign to each of these samples the closest original view in which the polygon is visible. The view maps are stored and used at rendering time to determine the three best original views and their blending factors by a quick look-up based on the current viewpoint.

The local polygon coordinate system is constructed as in Equation 1:

$$\mathbf{x} = \begin{cases} \mathbf{y}^W \times \mathbf{n} & \text{, if } \mathbf{y}^W \text{ and } \mathbf{n} \text{ are not collinear,} \\ \mathbf{x}^W & \text{otherwise} \end{cases}$$

$$\mathbf{y} = \mathbf{n} \times \mathbf{x} \tag{1}$$

where $\mathbf{x}^W$ and $\mathbf{y}^W$ are world coordinate system axes, and $\mathbf{n}$ is the triangle unit normal.

We transform viewing directions to the local coordinate system as in Fig. 5. We first obtain $\mathbf{v}$, the unit vector in the direction from the polygon centroid $\mathbf{c}$ to the original view position. We then rotate this vector into the $\mathbf{x} - \mathbf{y}$ plane of the local coordinate system for the polygon.

$$\mathbf{v}_r = (\mathbf{n} \times \mathbf{v}) \times \mathbf{n} \tag{2}$$

This vector is then scaled by the arc length $l = \cos^{-1}(\mathbf{n}^T\mathbf{v})$ and projected onto the $\mathbf{x}$ and $\mathbf{y}$ axes giving the desired view mapping.

$$
\begin{aligned}
x &= (l\mathbf{v}_r)^T\mathbf{x} \\
y &= (l\mathbf{v}_r)^T\mathbf{y}
\end{aligned} \tag{3}
$$

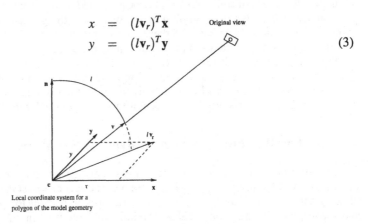

Local coordinate system for a
polygon of the model geometry

Fig. 5. The local polygon coordinate system for constructing view maps.

We pre-compute for each polygon of the model the mapping coordinates $\mathbf{p}_i = (x_i, y_i)$ for each original view i in which the polygon is visible. These points $\mathbf{p}_i$ represent a sparse sampling of view direction samples.

To extrapolate the sparse set of original viewpoints, we regularize the sampling of viewing directions as in Fig. 6. For every viewing direction on the grid, we assign to it the original view nearest to its location. This new regular configuration is what we store and use at rendering time. For the current virtual viewing direction we compute its mapping $\mathbf{p}_{virtual}$ in the local space of each polygon. Then based on this value we do a quick lookup into the regularly resampled view map. We find the grid triangle inside which $\mathbf{p}_{virtual}$ falls and use the original views associated with its vertices in the rendering (4, 5, and 7 in the example of Fig. 6). The blending weights are computed as the barycentric coordinates of $\mathbf{p}_{virtual}$ in the triangle in which it lies. In this manner the weights of the various viewing images are guaranteed to vary smoothly as the viewpoint changes.

8 Efficient 3-pass View-Dependent Texture-Mapping

This section explains details of the implementation of the view-dependent texture-mapping algorithm.

For each polygon visible in more than one original view we pre-compute and store the viewmaps described in Section 7. Before rendering begins for each polygon we find

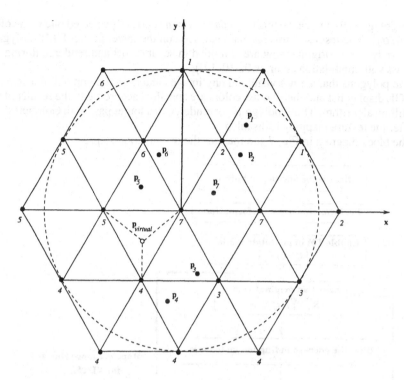

Fig. 6. The space of viewing directions for each polygon is regularly sampled, and the closest original view is stored for each sample. To determine the weightings of original views to be used in a new view, the barycentric coordinates of the novel view within its containing triangle are used. This guarantees smooth changes of the set of three original views used for texture mapping when moving the virtual viewpoint.

the coordinate mapping of the current viewpoint $\mathbf{p}_{virtual}$ and do a quick lookup to determine which triangle of the grid it lies inside of. As explained in Section 7 this returns the three best original views and their prescribed weights $\alpha_1, \alpha_2, \alpha_3$.

Since each VDTM polygon must be rendered with three texture maps, the rendering is performed in three passes. Texture mapping is enabled in modulate mode, where the new pixel color C is obtained by multiplying the existing pixel color C_f and the texture color C_t. The Z-buffer test is set to *less than or equal* (GL_LEQUAL) instead of the default *less than* (GL_LESS) to allow a polygon to blend with itself as it is drawn multiple times with different textures. The first pass proceeds by selecting an image camera, binding the corresponding texture, loading the corresponding texture matrix transformation $\mathbf{M}_{texture}$ in the texture matrix stack and rendering the part of the model geometry for which the first best camera is the selected one with modulation color $(\alpha_1, \alpha_1, \alpha_1)$. These steps are repeated for all image cameras. The results of this pass can seen on the tower in Fig. 8 (b). The first pass fills the depth buffer with correct depth values for the entire view. Before proceeding with the second pass we enable blending in the frame buffer, i.e. instead of replacing the existing pixel values with incoming values, we add those values together. The second pass then selects cameras and renders polygons for which the second best camera is the selected one with colors $(\alpha_2, \alpha_2, \alpha_2)$. The results of the second pass can seen on the tower in Fig. 8 (c). The third pass proceeds similarly ren-

114

dering polygons for which the third best camera is the currently selected one with colors $(\alpha_3, \alpha_3, \alpha_3)$. The results of this last pass can seen on the tower in Fig. 8 (d). Polygons visible only in one original view are compiled in separate list and rendered during the first pass with modulation color $(1.0, 1.0, 1.0)$.

The polygons that are not visible in any image cameras are compiled in a separate OpenGL display list and their vertex colors are specified according to the results of the hole-filling algorithm. Those polygons are rendered in another pass with Gouraud shading after the texture mapping is disabled.

The block diagram in Fig. 7 summarizes the display loop steps.

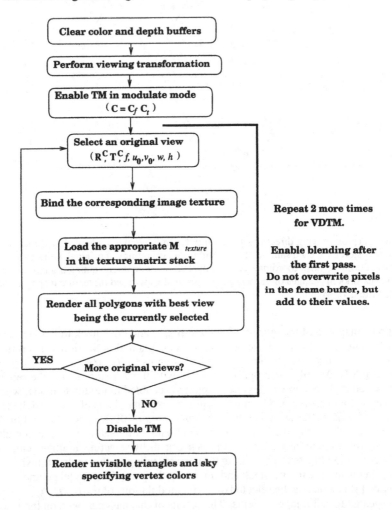

Fig. 7. The multi-pass view-dependent projective texture mapping rendering loop.

9 Discussion and Future Work

The method we have presented proved effective at taking a relatively large-scale image-based scene and rendering it realistically at interactive rates on standard graphics hardware. Using relatively unoptimized code, we were able to achieve 20 frames per second on the Silicon Graphics InfiniteReality hardware for the full tower and campus models. Nonetheless, many aspects of this work should be regarded as preliminary in nature. One problem with the technique is that it ignores the spatial resolution of the original images in its selection process – an image that shows a particular surface at very low resolution but at just the right angle would be given greater weighting than a high-resolution image from a slightly different angle. Having the algorithm blend between the images using a multiresolution image pyramid would allow low-resolution images to influence only the low-frequency content of the renderings. However, it is less clear how this could be implemented using standard graphics hardware.

While the algorithm guarantees smooth texture weight transitions as the viewpoint moves, it does not guarantee that the weights will transition smoothly across surfaces of the scene. As a result, seams can appear in the renderings where neighboring polygons are rendered with very different combinations of images. The problem is most likely to be noticeable near the frame boundaries of the original images, or near a shadow boundary of an image, where polygons lying on one side of the boundary include an image in their view maps but the polygons on the other side do not. [2] suggested feathering the influence of images in image-space toward their boundaries and near shadow boundaries to reduce the appearance of such seams; with some consideration this technique should be adaptable to the object-space method presented here.

The algorithm as we have presented it requires all the available images of the scene to fit within the main memory of the rendering computer. For a very large-scale environment, this is unreasonable to expect. To solve this problem, spatial partitioning schemes [13], image caching [11], and impostor manipulation [11, 12] technqiues could be adapted to the current framework.

As we have presented the algorithm, it is only appropriate for models that can be broken into polygonal patches. The algorithm can also work for curved surfaces (such as those acquired by laser scanning); these surfaces would be need to be broken down by the visibility algorithm until they are seen without self-occlusion by their set of cameras.

Lastly, it seems as if it would be more efficient to analyze the set of available views of each polygon and distill a unified view-dependent function of its appearance, rather than the raw set of original views. One such representation is the Bidirectional texture function, presented in [1], or a yet-to-be-presented form of compressed light field. Both techniques will require new rendering methods in order to render the distilled representations in real time. Extensions of techniques such as model-based stereo [2] might be able to perform a better job than linear blending of interpolating between the various views.

10 Images and Animations

Images and Animations of the Berkeley campus model may be found at:
                    http://www.cs.berkeley.edu/~debevec/Campanile

116

References

1. DANA, K. J., GINNEKEN, B., NAYAR, S. K., AND KOENDERINK, J. J. Reflectance and texture of real-world surfaces. In *Proc. IEEE Conf. on Comp. Vision and Patt. Recog.* (1997), pp. 151–157.
2. DEBEVEC, P. E., TAYLOR, C. J., AND MALIK, J. Modeling and rendering architecture from photographs: A hybrid geometry- and image-based approach. In *SIGGRAPH '96* (August 1996), pp. 11–20.
3. FOLEY, J. D., VAN DAM, A., FEINER, S. K., AND HUGHES, J. F. *Computer Graphics: principles and practice.* Addison-Wesley, Reading, Massachusetts, 1990.
4. GORTLER, S. J., GRZESZCZUK, R., SZELISKI, R., AND COHEN, M. F. The Lumigraph. In *SIGGRAPH '96* (1996), pp. 43–54.
5. LAVEAU, S., AND FAUGERAS, O. 3-D scene representation as a collection of images. In *Proceedings of 12th International Conference on Pattern Recognition* (1994), vol. 1, pp. 689–691.
6. LEVOY, M., AND HANRAHAN, P. Light field rendering. In *SIGGRAPH '96* (1996), pp. 31–42.
7. MARK, W. R., MCMILLAN, L., AND BISHOP, G. Post-rendering 3D warping. In *Proceedings of the Symposium on Interactive 3D Graphics* (New York, Apr.27–30 1997), ACM Press, pp. 7–16.
8. MCMILLAN, L., AND BISHOP, G. Plenoptic Modeling: An image-based rendering system. In *SIGGRAPH '95* (1995).
9. PULLI, K., COHEN, M., DUCHAMP, T., HOPPE, H., SHAPIRO, L., , AND STUETZLE, W. View-based rendering: Visualizing real objects from scanned range and color data. In *Proceedings of 8th Eurographics Workshop on Rendering, St. Etienne, France* (June 1997), pp. 23–34.
10. SEGAL, M., KOROBKIN, C., VAN WIDENFELT, R., FORAN, J., AND HAEBERLI, P. Fast shadows and lighting effects using texture mapping. In *SIGGRAPH '92* (July 1992), pp. 249–252.
11. SHADE, J., LISCHINSKI, D., SALESIN, D., DEROSE, T., AND SNYDER, J. Hierarchical image caching for accelerated walkthroughs of complex environments. In *SIGGRAPH 96 Conference Proceedings* (1996), H. Rushmeier, Ed., Annual Conference Series, ACM SIG-GRAPH, Addison Wesley, pp. 75–82.
12. SILLION, F., DRETTAKIS, G., AND BODELET, B. Efficient impostor manipulation for real-time visualization of urban scenery. *Computer Graphics Forum (Proc. Eurographics 97) 16*, 3 (Sept. 4–8 1997), C207–C218.
13. TELLER, S. J., AND SEQUIN, C. H. Visibility preprocessing for interactive walkthroughs. In *SIGGRAPH '91* (1991), pp. 61–69.
14. WEGHORST, H., HOOPER, G., AND GREENBERG, D. P. Improved computational methods for ray tracing. *ACM Transactions on Graphics 3*, 1 (January 1984), 52–69.
15. WEILER, K., AND ATHERTON, P. Hidden surface removal using polygon area sorting. In *SIGGRAPH '77* (1977), pp. 214–222.
16. WILLIAMS, L., AND CHEN, E. View interpolation for image synthesis. In *SIGGRAPH '93* (1993).

Editors' Note: see Appendix, p. 329 for colored figure of this paper

Uniformly Sampled Light Fields

Emilio Camahort

http://www.cs.utexas.edu/users/ecamahor/
Department of Computer Sciences
The University of Texas at Austin

Apostolos Lerios

http://www-graphics.stanford.edu/~tolis/
Align Technology, Inc.

Donald Fussell

http://www.cs.utexas.edu/users/fussell/
Department of Computer Sciences
The University of Texas at Austin

Abstract.
Image-based or light field rendering has received much recent attention as an alternative to traditional geometric methods for modeling and rendering complex objects. A light field represents the radiance flowing through all the points in a scene in all possible directions. We explore two new techniques for efficiently acquiring, storing, and reconstructing light fields that can produce uniformly high quality images of an object from any point of view exterior to it. Both techniques discretize the light field by sampling the set of lines that intersect a sphere tightly fit around the object.

1 Introduction

Image-based modeling and rendering techniques have recently received much attention as an alternative to traditional geometry-based techniques for image synthesis. Synthetic images are created from a prestored set of samples of the light field of an environment instead of a geometric description of the elements in the scene. While image-based techniques have long been used to augment geometric models in the form of texture maps or environment maps, for example, newer approaches completely replace geometric information with image data. This allows the construction of models directly from real-world image data without conversion to geometry, and it can also support faster rendering of visually complex or photorealistic objects than geometric techniques.

Image-based modeling and rendering relies on the concept of a *light field* [9]. A light field represents the radiance flowing through all the points in a scene in all possible directions. For a given wavelength, we can represent a static light field as a 5D scalar function $L(x, y, z, \theta, \phi)$ that gives radiance as a function of location (x, y, z) in 3D space and the direction (θ, ϕ) the light is traveling. Images are discretized 2D slices of this 5D function. Given a set of viewing parameters, we can render an image by evaluating the light field function at the (x, y, z) location of the eye for a discrete set of directions within the field of view.

Previous light field and image based models assume some limitations on the set of directions from which an object will be viewed. For instance, [16, 11] used the

118

two plane parameterization (2PP), where the set of lines in space is parameterized by the intersection points of each line with two planes. Levoy and Hanrahan [16] call such a pair of planes a *light slab*. They also propose the use of several slabs, or *slab arrangements*, to sample a wide enough variety of light field directions to allow images from arbitrary points of view. Gortler et al. [11] use a set of six slabs arranged as a cube, which they call a *lumigraph*.

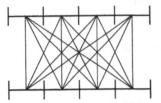

Fig. 1. 2D analogy of directional and positional bias in the 2PP: the thick lines represent two discretized planes, the thin lines represent the set of lines generated by joining discrete points on the planes. Note that the lines have seven possible orientations, but the number of lines for each orientation varies between 1 and 4. Also, note that the distance separating two neighboring lines varies with orientation.

Such arrangements show noticeable artifacts when the image being rendered uses samples from more than one slab. Even arrangements of 12 light slabs [16] do not suffice to avoid this *disparity problem* caused by a non-uniform parameterization of line space. The density of lines induced by the 2PP is biased towards certain directions and the spatial sampling induced by the representation is different for each direction (figure 1). The use of separate, individually parameterized slabs makes it difficult to orient filter kernels across abutting slabs.

In this paper we propose two related solutions to the disparity problem based on parameterizations with no directional biases. Both parameterizations are based on random processes which uniformly sample the space of oriented lines that intersect a tight bounding sphere of the object to be represented, that is, a process selects any line from this space with equal probability, given that its parameters are chosen uniformly. Thus our techniques do not inherit sampling biases induced by the parameterization, and provide uniform image quality from any angle of view.

(a)   (b)

Fig. 2. a 3D sphere: (a) select two random points P and Q uniformly distributed on the sphere's surface and join them with a line L, and (b) select a random great circle C with uniform probability, select a random point P uniformly distributed over C's surface, and choose the normal L to C passing through P.

Different choices of parameterizations can be found in the literature [27]. At least two of them support the selection of random lines following a uniform distribution. They are the *two-sphere parameterization* (2SP) [21] and the *sphere-plane parameterization* (SPP) as shown in figure 2. In the first case, choosing two points on a sphere with uniform probability implies that the line joining them follows a uniform distribution. In the second case, choosing an oriented great circle and a point on its plane following uniform distributions implies that the sampling of lines orthogonal to the great circle through the point follows a uniform distribution.

Given a parameterization, we develop a deterministic scheme to generate sets of parameter values regularly distributed throughout their domains. Samples are taken using these parameter sets and organized in memory in a way that they can be extracted efficiently for rendering and compressed efficiently. An on-line reconstruction algorithm exploits the data organization to provide high-quality real-time rendering from points of view external to the object.

We can generate samples using the 2SP by constructing a uniformly spaced set of points on the sphere, and taking all lines defined by pairs of these points. An SPP representation is created by using a similar set of points on the sphere to define oriented planes passing through the center of the sphere, and then, for each such plane, taking a uniformly spaced set of points on the interior of the great circle formed by the intersection of the plane and the sphere to define lines normal to the plane. Note that the sets of lines thus generated may not be uniformly spaced due to the difficulty of finding uniformly spaced points on a sphere. In the case of the SPP, the spacings of the positional and directional parameters can be chosen independently, allowing an extra degree of flexibility by allowing certain controlled violations of the uniformity of the samples.

Also, note that our deterministic sampling does not require a perfectly uniform set of light field samples, indeed, uniformity may not be desirable. Intentional departures from uniform sampling can give us more control over sample placement, for instance to divert samples from low-variance areas to detailed ones. In the case of the SPP, we can separately control the spatial resolution in each direction. For this reason, even though our methods start with a uniform sampling pattern, they both support hierarchical multiresolution so that groups of uniform samples in a low-variance area can be stored compactly. Furthermore, our methods provide many of the advantages of previous approaches, including interpolation, depth correction, and compression. Additionally, they allow progressive transmission and rendering of visually complex models at interactive rates.

2 Related Work

Early attempts at using images for scene representations are due to Chen and Williams, Laveau and Faugeras, and Faugeras et al. [3, 14, 7]. They use central planar projections to represent their samples of the light field. McMillan and Bishop [19], however, use cylindrical projections, as well as Chen's QuickTime© VR system [2]. All of these approaches use well-known image warping and manipulation techniques for rendering new images from a given set. Other researchers have concentrated on rendering by view interpolation [23, 4] and on improving geometry rendering by incorporating image-based techniques [17, 5, 24, 25].

Our work is more closely related to the work of Levoy and Hanrahan [16] and Gortler et al. [11] on the fundamental representation of a light field. More recent results have been obtained that improve the performance of the lumigraph [26], relate it to visibility events [12], and add illumination control to light fields [28]. Approaches similar to the SPP have been used before for different purposes, most notably for rendering of trees in [18]. Pulli et al. [20] use a similar approach based on perspective projections in developing a hybrid image/geometry based modeling scheme. Finally, error measures for image-based representations have been discussed by Maciel and Shirley [17], Shade et al. [24], and Lengyel and Snyder [15].

3 Building Light Field Models

Recall that a light field is represented by a scalar function $L(x, y, z, \theta, \phi)$ of five parameters. Since we are modeling single opaque objects, we assume that the light field remains constant in free space outside the object, so that an oriented line intersecting the object's bounding sphere corresponds to a light field sample. This will allow us to use four parameters instead of five, as is done by [16, 11].

3.1 Choosing Sample Directions

For the sphere-plane parameterization (SPP), we first choose an oriented plane P through the center of the sphere, represented by its normal direction (θ, ϕ). On this plane we define a 2D coordinate system (u, v). A light field sample is then parameterized by (θ, ϕ, u, v). For the two-sphere parameterization (2SP), we represent each sample of the light field by the coordinates of the two points of intersection of the oriented line with the sphere.

Now we need a method for selecting uniformly spaced samples given these parameterizations. We start by subdividing the surface of the sphere into (nearly) equilateral, (nearly) identical spherical triangles, called *patches*. The most popular subdivisions of the unit sphere's surface are based on polyhedra which approximate this surface: each point p on the polyhedron, the *generator*, is radially projected onto the sphere to yield a point p', in a 1-1 manner. Then, a patch of the subdivision comprises all points p' whose corresponding p's lie on the same polyhedron face.

The platonic solids are good generators: all their edges, faces, and vertices look identical; and, as a result, they induce a perfectly uniform subdivision. However, the most complex platonic solid, the icosahedron, has only 20 faces. We can refine this generator [8, 6, 10] using *quad subdivision*. We project the midpoint of each edge onto the circumsphere. Then we replace each face by the four triangles connecting the raised midpoints to the original vertices of the face. The result is a generator with 4 x 20 faces. Applying this process L times, we can get a solid with 4^L x 20 faces.

A uniform subdivision into k patches should produce patches whose area is $A = 4\pi/k$. Quad subdivision is not perfectly uniform, it produces patches with areas between 95% and 120% of A. If we project the centroid of each icosahedron face onto the circumsphere and replace the original face with the three triangles connecting its vertices to the new center point, we obtain a more uniform generator with 60 patches, area variation of only 97–107% of A, and triangles with a similar aspect ratio to those of the icosahedron. We use the former subdivision for the SPP and the latter for the 2SP.

3.2 Patches and Samples

To generate a light field model of an object using the 2SP we first choose some random rays crossing each ordered pair of patches of the representation. Then, for synthetic images, we use a ray tracer to determine the color of each ray. Finally, we combine the ray colors using some filter kernel (currently they are just averaged), and assign the result to the ray's line sample.

To generate a light field model of an object using the SPP, we let each patch generated represent a plane P_k through the bounding sphere's origin with normal defined by the direction vector $\vec{\omega}_k = \vec{P}_c - \vec{O}$, where P_c is the patch centroid and O is the bounding sphere origin. This patch will thus represent the pencil of directions defined by the set of all such vectors from the sphere origin through points on the patch.

Given a directional sample $\vec{\omega}_k$, we select a set of light field samples in a (u, v) coordinate system local to P_k. This is done for synthetic images by tracing a set of parallel rays orthogonal to P_k or by performing an orthographic projection of the object onto a plane parallel to P_k followed by scan conversion.

4 Storage and Rendering

4.1 2SP Based Models

2SP models are rendered using a fast sample retrieval algorithm. Given a ray to be used in rendering a view, we determine a color for that ray by retrieving prestored light field samples based on the ray parameters. Sample extraction is done by intersecting the ray with the unit sphere and determining which pair of patches are crossed by the ray, and in which order. We then return the light-field sample for the intersected pair. An obvious improvement to this approach uses reconstruction kernels which improve over our nearest-neighbor resampling scheme (piecewise-constant reconstruction).

Identifying the ordered patch pair that is intersected by a given ray is a two-step process. First, compute the two intersection points between the ray and the unit sphere. Second, retrieve the patches containing the two points. The first step is covered in most introductory textbooks on ray tracing, but it can be speeded up when a sequence of coherent rays are cast using incremental techniques similar to those in [13]. Rendering complexity is proportional to the projected image size of the bounding sphere.

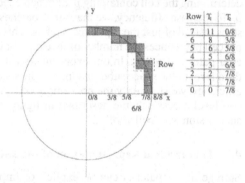

Fig. 3. 2D analogy of the compact storage for the 2SP: nonempty cells in the first quadrant are shown in gray, and cells sharing the same y form a *row*. Gray cells are stored in a linear array, row by row from bottom to top, and from left to right. We also construct two tables: T_1 maps each row to the number of gray cells below it, while T_2 contains the leftmost x coordinate of the row's cells. Given a point p, we use its y coordinate to find the row of cells containing p; then T_1's entry for that row tells us where in the linear array the row begins; finally, the distance between p's x coordinate and T_2's entry determines the array offset between the leftmost cell in the row and the one containing p. Due to symmetry we only need to maintain T_1 and T_2 for the first 3D octant.

Having a ray-sphere intersection, we must find the patch containing the point (we have to do this twice). We assume that the sphere subdivision is induced by a generator G which has been refined L times. We also assume that we know which projected face of G, i.e. which top-level patch, contains the point. The refinement algorithm generates four children per parent patch. We need only discover which of the four children contains the point. This classification is done by finding the placement of the point relative to the three planes defined by the sphere's center and the three great circle arcs outlining the children's patch boundaries within the parent patch. Since a constant number of tests are performed at each refinement level, the overall complexity of this search is linear in L.

To discover the top-level patch containing the given point p, we build a data structure dicing space into uniform cubical cells (see figure 3) and associate each cell with all

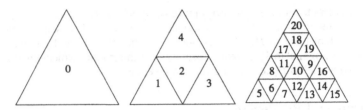

Fig. 4. The face indexing scheme. A top-level face, and the corresponding patch, are given the index 0. The next three indices, 1 to 4, are assigned to the four children in counter-clockwise fashion, starting at the bottom-left corner, as shown. At the next level, we assign indices 5 to 8 to the children of face 1, indices 9 to 12 to 2's children, etc. And, within each parent face, we follow the same counter-clockwise ordering, starting at the child face sharing its East-West and SW-NE edge with its parent. With this scheme, parent/child associations are easy, e.g. the child indices of face i are $4i + 1 \ldots 4i + 4$.

top-level patches which intersect it. The top-level patch containing p is found by first determining the cell containing p, and then checking which patch in the cell set contains p. To increase efficiency, we may store patches induced by some number of refinement steps instead of just the top-level patches. This increases the average number of patches per cell but reduces the number of levels we have to descend in order to reach the sought after patch at level L. In our experiments with $L = 5, 6$, we stored the patches of the third level of the space subdivision. For the sake of good cache performance and storage efficiency, we store only the non-empty cells, incurring an insignificant penalty via the two-level access scheme described in figure 3. In our experiments, we used a space subdivision with cell width 2^{-6}.

4.2 Hierarchical Representation of the 2SP

Storage of the patches can be explicit or implicit. An explicit scheme computes all the patch vertices and edges and stores their coordinates and connectivity. An implicit scheme is akin to dicing the plane into a rectangular grid: one need only store a few parameters (row and column count and size) and employ fast functions to compute all the other information, such as the top left corner of a grid cell. Fekete [8] presented such an implicit scheme for a subdivided sphere, and our work builds upon it. This scheme assigns indices to patches at every level as illustrated in figure 4 for a single top-level face and its descendants. Each such family of faces is distinguished from all others by means of some secondary tag. We extend Fekete's scheme to include an incremental numbering of edges and vertices that supports efficient extraction of geometric and topological feature information during patch hierarchy traversal. The details of these indexing schemes can be found in [1].

In practice, a mixture of explicit and implicit storage works best, aiming at the right trade-off between processor speed and memory system performance. For example, our system caches the equations of the patch boundaries to optimize point classification (discussed below) but stores no topological information.

The simplest way to store the overall 2SP model is a two-dimensional array, mapping each ordered patch pair (i.e., a pair of integer indices) to a light field sample. This array can be compressed using any of the standard image compression techniques; vector quantization (VQ) is probably best for fast reconstruction. However, vector quantization fails to compress the large empty portion of the light field efficiently. As a result, we opted for a common variant of VQ employing trees. Our data structure is

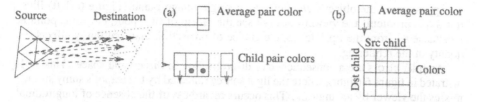

Fig. 5. The hierarchical storage scheme of the 2SP; for simplicity, we assume that the light field samples are ray colors. With each ordered pair of top-level patches — a source and a destination patch — we associate the average color of all rays crossing the pair, as well as a tree. This tree contains the colors for individual rays crossing the pair, as well as average colors for ray families that cross pairs of descendant patches at the same subdivision level. (a) Each tree grows only when a node — corresponding to a pair of patches at the same level — exhibits a variance across its children that exceeds some threshold; the children of a pair are the 16 pairs obtained by matching each of the four children of the source patch to each child of the destination. (b) If the children of a pair are all leaves — either because of low grandchild variances or because the children belong to the last subdivision level —, we use a 4x4 VQ block to represent the 16 children colors.

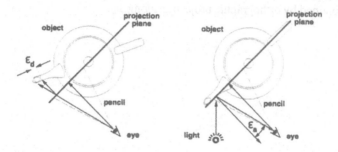

Fig. 6. Error measures. Left: ε_d is the longitudinal or translational error. Right: ε_a is the angular or rotational error.

summarized in figure 5.

4.3 SPP Based Models

Recall that an SPP based model is made of a collection of parallel projections. In its simplest form each projection is taken along a given direction $\vec{\omega}_k$ onto a projection plane orthogonal to $\vec{\omega}_k$ and located at the center of the object's bounding sphere. The locations of the projection planes containing the model's images are important because they approximate the geometry of the object as viewed along the patch direction. Such an approximation produces discontinuities or *seams* at rendering time. These are related to the errors incurred by multislab 2PP models.

Projection Planes and Error Control in the SPP. Errors in light field rendering should be measured using a radiance metric. While this may be impractical in general, we can define two geometric error measures related to the light field parameter approximation: longitudinal (translational) and angular (rotational) error (see figure 6) which can be used as correlates of radiance variations. Longitudinal errors occur when we approximate the object's geometry by a single image plane. Figure 6 (left) illustrates this problem: the viewer does not see the far tip of the spout, but some point at a distance ε_d from the tip. Hence, the choice of projection plane certainly affects the quality of the rendering.

Angular errors are a consequence of discretization in directional space. This is illustrated in figure 6 (right), where the light source reflected by the teapot's shiny surface misses the viewer by an angle ε_a. This occurs regardless of the absence of longitudinal error for that particular ray. Bounds for angular errors are given by the resolution of the (θ, ϕ) discretization of our representation. Angular error can not exceed the angle between a discrete direction $\vec{\omega}_k$ and the furthest direction within the patch it approximates.

Model construction should be done so as to minimize both types of error. We select an initial subdivision level, giving us a set of sample directions $\{\omega_k\}_{k=1}^{D}$. For each direction $\vec{\omega}_k$, we compute a parallel projection onto a plane P_k. P_k can be translated in the $\vec{\omega}_k$ direction and rotated with respect to this direction to minimize an error norm based on longitudinal errors. We can also determine the orientation of the (u, v) coordinate system to minimize the area of the projection of the object's bounding box along $\vec{\omega}_k$. Both optimizations can be accomplished by using depth information from the renderer's Z-buffer. Alternatively, we can just store the depth information together with the image obtained by orthographic projection along ω_k.

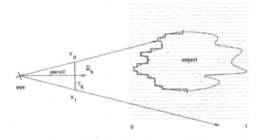

Fig. 7. 2D analogy of the depth map approximation of an object's surface. The blobby object is surrounded by a box representing an imaginary volumetric grid. Inside the grid the bold line represents the surface's approximation given by the depth map along direction $\vec{\omega}_k$. At rendering time rays are cast starting from the eye and passing through the vertices V_0 and V_1 of the flat triangle T_k. The top ray hits the surface's approximation after visiting 3 cells in the vertical direction. The bottom ray misses the object after visiting 5 cells. Note that the number of cells traversed decreases with the (solid) angle subtended by a pencil.

The set of directions can then be refined by adaptive subdivision of their associated pencils. Given an error analysis on the data associated with each pencil, we sort pencils by their error magnitudes normalized per unit solid angle. The pencil with the largest error is subdivided using the refinement procedure described for the 2SP above, error analyses are performed on the resulting pencils, these are inserted into the sorted list, and the process is continued until an error threshold or maximum number of pencils is reached. This allows us to concentrate the directional samples as needed to minimize errors. At rendering time we only use those pencils of the hierarchy that we can afford to render.

Rendering. The SPP rendering algorithm is an adapted version of the lumigraph algorithm [11]. Given the viewing parameters, we start by placing an imaginary sphere centered at the eye position. The sphere is subdivided according to the patch hierarchy of the SPP model. The algorithm determines which patches project onto the viewing frustum and, for each of those patches, renders an image on the portion of the frustum covered by the patch. Since patches form a hierarchy, the algorithm is a breadth first search. If a parent patch intersects the frustum, then its children are traversed. For each patch, intersection is determined by comparing the direction vectors through its vertices with the direction vector through the frustum vertices. If any of the patch's vectors falls within the frustum's then the patch is visible, and the search continues until a pencil is reached for which its patch direction vector and all three of its vertex direction vectors are within the frustum. For each such pencil the image I_k associated with its patch is retrieved. I_k approximates the object as seen along the directions through that patch. Therefore, if the object is visible through the patch, I_k is also visible through it and we display I_k in the area of the final image covered by the patch's projection.

In practice we approximate the patch by a flat triangle T_k, then we warp I_k and texture-map it onto T_k. Finally, we render the flat triangle T_k using standard graphics hardware. That way I_k ends up covering the portion of the final image that corresponds to its associated patch. T_k's texture coordinates determine which portion of I_k is visible through the patch. Since I_k is located on a known plane in world coordinates, it suffices to project the vertices of T_k onto that plane using a perspective projection centered at the eye position.

An alternative algorithm uses depth information for depth correction (see figure 7). For each vertex of the flat triangle T_k, a ray is constructed starting at the eye position and passing through that vertex. The ray is then cast into a volumetric grid, where the depth values represent a surface approximating the object's geometry. A Bresenham style incremental algorithm searches the grid for an intersection with the surface. Typically, less than four or five iterations are sufficient to find an intersection, depending on the size (solid angle) of the patch. However, it is often necessary to subdivide T_k, because either some of the rays miss the object, or the difference between the depths along two rays is too large.

The triangle-subdivision algorithm is steered by two threshold values, A_{min} and A_{max} as illus-

Fig. 8. The effect of the two threshold values A_{min} and A_{max} on the triangle subdivision of the image plane for a rendering of the bunny. All triangles have a maximum projected size of 400 pixels. At the boundaries of the bunny the triangles have subpixel size.

trated in figure 8. Any triangle T_k whose area in pixels is larger than A_{max} is subdivided, even if there is nothing but background behind it. This helps detection of small or skinny objects that otherwise may fall between vertices and can prevent artifacts at object boundaries. Of course, to render all such objects properly, A_{max} would need to be set to one pixel.

The threshold A_{min} ensures that the triangle subdivision process terminates in cases where a pencil triangle's vertices do not fall within the frustum (such as frequently happens at silhouette edges). Triangles are subdivided until their area in pixels falls below A_{min}. A_{min} is usually set to subpixel values in order to guarantee that object boundaries are properly sampled, although larger values can be chosen for speed. The depth corrected SPP images in this paper use $A_{min} = 0.5$.

Once the texture coordinates of the patch have been determined, its associated texture can be rendered using various reconstruction kernels. We have employed piecewise constant and piecewise linear kernels.

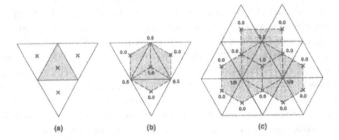

Fig. 9. Reconstruction kernels: (a) piecewise constant, (b) piecewise linear with small support, and (c) piecewise linear with large support. The crosses indicate $\vec{\omega}_k$ directions, the numbers α-value assignments. In (c) the support may reach anywhere between 9 and 12 neighboring patches due to the near-uniformity of our representation.

For a constant kernel we just render the triangle associated with a given patch (see figure 9(a)). For a linear kernel we use standard graphics hardware to perform bilinear interpolation in (u, v) space. In (θ, ϕ) space we use α-blending to implement spherical linear interpolation. To illustrate this procedure consider the center triangle in figure 9(b) and note that its associated image applies to all the pixels inside of the shaded hexagon. In order to implement linear interpolation between that image and its three neighbors', we define three quadrilaterals, one for each neighbor, and associate to their vertices the α-values shown in the figure. When rendering the quadrilaterals of two neighboring images, interpolation is achieved by α-blending the quadrilaterals. Figure 9(c) shows an alternative larger support for linear interpolation. We observed that the smaller support was likely to produce artifacts near the vertices with α-values of 0.5. Widening the support of the linear kernel eliminated these artifacts.

5 Implementation and Results

The results of our implementation of the 2SP indicate that our representation for the sphere subdivision and our techniques for mapping pixels to patches provide high performance: any function mapping one color to each patch, for up to $L = 6$ levels of subdivision, can be displayed at 30 frames a second, each measuring 300 x 300 pixels (where a $L = 6$ patch maps to a pixel), on a single 195MHz MIPS R10000 processor.

To obtain good image quality in the 2SP, we usually choose $L = 5$ or $L = 6$ yielding 61k and 245k faces, respectively. Why? Because if we were to render the sphere on a square display of 300 x 300 pixels, so that the sphere edge borders the display edge, each induced patch consumes at most 4.6 (for $L = 5$) or 1.15 ($L = 6$) pixels, hence producing a smooth image. For example, the 2SP model of the dragon rendered in color plate 1 (see appendix) used $L = 5$ subdivision levels and a 24-bit RGB representation for each sample.

Without compression, the 2SP dragon model consumes 11.3GB. With 4^2 VQ blocks and 16-bit codewords, pure VQ yields a 500MB data set. With hierarchical VQ, the figure drops to 150MB. In addition, disk storage can be reduced to 50MB using commercial Lempel-Ziv coding.

Results from our preliminary implementation of the SPP are shown in color plate 2 (see appendix). We can render images from an SPP model at interactive rates; for instance the Stanford Happy Buddha model can be rendered at 24 frames per second on an Onyx2 InfiniteReality, while geometric rendering to produce images of the same quality takes 1 second per frame on the same machine. Rendering a model with depth correction still takes more than a second per frame on an Onyx2. We expect substantial improvement after carefully optimizing our renderer, currently we do not use the fast retrieval techniques described for the 2SP. Also note that there is a time-quality tradeoff controlled by the two thresholds A_{min} and A_{max} of the triangle subdivision algorithm. The figures were produced by setting these parameters for optimal quality rather than performance.

Our data sets are very similar in size to those of the 2SP above. We used JPEG compression for the image data and Lempel-Ziv and Welch for lossless compression of different types of depth maps. Using these schemes has two advantages: we achieve reasonable compression rates, and data caching and decompression are fast due to hardware and systems support. Note that lossy compression is not acceptable for the most significant bit of a depth map since the outline of the object must be clearly defined to avoid seams at the boundaries of the object due to the loss of information.

Our images also illustrate several advantages accruing from the directional uniformity of our sampling. First, interpolation using linear reconstruction kernels is not critical. This allows avoidance of the blurring that can significantly degrade image quality as in current 2PP model implementations. Also, note that using uniform parallel projections provides bounds for the angular error. In fact the angular error is uniform and bounded and depends only on the directional sampling rate. This allows for fast ray-depth map intersections, as well as more accurate illumination effects like the highlights of the teapot in color plate 2 (see appendix).

Both of our representations are well-suited to supporting adaptive frame-rate control. Given a model made of D directions, the average number of directions n covered by a field of view fov is

$$n = D \frac{\text{fov}}{2\pi} \sin \frac{\text{fov}}{2}.$$

We can easily obtain an estimate of the rendering time for a frame and decide how far down the patch hierarchy we want to go in order to keep an animation running at a given frame rate. This can also be useful to decide whether to use an image-based or a geometry-based model in a hybrid rendering system.

6 Discussion

We have introduced two new, related schemes which differ from traditional approaches to light field or image based rendering in that they are designed to avoid preferred viewing directions. The schemes are based on parameterizations which support uniform random light field sampling, even though the deterministic algorithms we have developed intentionally allow for controlled non-uniform sampling. We have shown that we can achieve competitive rendering efficiency with high image quality, even though we haven not avoided the complexities of spherical parameterizations.

Our methods can guarantee error bounds in all dimensions of the light field, so provisions can be made to reduce or avoid interpolation in rendering. Image quality is largely independent of camera position or orientation. Furthermore, the hierarchical subdivision schemes presented in this paper allow for multiresolution support and adaptive non-uniform construction of light field models. Other features of our models are fast reconstruction algorithms, progressive transmission and progressive rendering, extraction of pre-filtered samples for distant views (much like texture mip-mapping), smooth transitions between LODs, and adaptive frame-rate control.

The 2SP and SPP are alternatives to the 2PP data structures described by Levoy and Hanrahan [16] and Gortler et al. [11], both of which rely on two-plane parameterizations. Rendering the 2SP takes approximately three times longer than a single-slab 2PP, for comparable image quality and piecewise-constant reconstruction. For piecewise-linear reconstruction, this performance gap narrows to a factor of two due the retrieval of fewer samples for the 2SP. Our results hint at two possible 2PP performance improvements: cull overlapping slabs (if any) and introduce barycentric interpolation by treating each grid rectangle as two abutting triangles. Rendering benefits from coherent access to the light field, which allows the 2PP and the SPP to showcase high performance via scan-conversion. The 2SP takes twice the time to access a single sample randomly.

The 2SP totally eliminates the seams that occur with multiple slab 2PPs, although at low sampling rates it does show artifacts that appear as "hair" over the entire image rather than localized seams. The SPP falls between the 2PP and 2SP approaches in terms of such artifacts. The SPP suffers from discrepancies across directional image boundaries, much like the 2PP's problems at slab boundaries. At low sampling rates seams are quite noticeable with the SPP, but at rates similar to those which produce good quality images with the 2SP, the SPP can be tuned to minimize both seams and hair. This is due to the fact that the SPP allows separate control of the directional and positional resolutions of the model, unlike the 2PP or the 2SP. For same number of samples represented by a single slab, many more directional images can be supported by the SPP.

For synthetic scenes, 2PPs and SPPs may be built using off-the-shelf, efficient renderers (assuming the latter support parallel as well as central projections). Unfortunately, the 2SP requires a ray tracer which can be instructed to shoot individual rays, joining pairs of points on the sphere. For acquired data, all the approaches require resampling: the usual camera motion is on a sphere (suiting the 2SP and SPP but requiring 2PP resampling), with each view being a central projection (suiting the 2PP, but requiring 2SP and SPP resampling). The three models can be built from one another by resampling.

As with the 2PP, we have achieved compression ratios approaching 200 using simple techniques for both the 2SP and SPP. We hope to employ (tensor products of) spherical wavelets [22] to compress full surround views of these models at once. It is unclear how one could similarly exploit correlations across slabs to achieve global compression for multiple slab 2PPs.

Since the sphere is the natural domain of several physical quantities, our techniques are relevant to applications which sample, store, and frequently retrieve (i) light intensity distributions, (ii) environment maps, (iii) topographical and geomagnetic data, (iv) 3D directions, (v) bidirectional reflectance distribution functions (BRDFs), etc.

Acknowledgments

The first and third authors would like to thank Nina Amenta, Steve Gortler, Al Bovik, Pat Hanrahan and Marc Levoy for useful discussions, and Makoto Sadahiro and Brian Curless for providing the data used for building their light field models. Their work was supported in part by NASA under grant NAGW-4247 and by the National Science Foundation under grant CDA-9624082.

The second author would like to thank Leo Guibas, Pat Hanrahan, and Marc Levoy for their guidance, Xing Chen and Lucas Pereira for many helpful discussions, and Matt Pharr for tweaking his ray tracer to generate 2SP models. Support by Interval Research is also gratefully acknowledged.

References

1. E. Camahort, A. Lerios, and D. Fussell. Uniformly sampled light fields. Technical Report TR-98-9, Department of Computer Sciences, The University of Texas at Austin, Austin, TX, March 1998.

2. S. E. Chen. QuickTime© VR - an image-based approach to virtual environment navigation. In *Proceedings of SIGGRAPH'95 (Los Angeles, CA, August 6–11, 1995). In Computer Graphics Proceedings, Annual Conference Series*, pages 29–38. ACM SIGGRAPH, 1995.

3. S. E. Chen and L. Williams. View interpolation for image synthesis. In *Proceedings of SIGGRAPH 93 (Anaheim, CA, August 1–6, 1993). In Computer Graphics Proceedings, Annual Conference Series*, pages 279–288. ACM SIGGRAPH, 1993.

4. L. Darsa, B. C. Silva, and A. Varshney. Navigating static environments using image-space simplification and morphing. In *1997 Symposium on Interactive 3D Graphics*, pages 25–34, New York, 1997. ACM.

5. P. E. Debevec, C. J. Taylor, and J. Malik. Modeling and rendering architecture from photographs: A hybrid geometry- and image-based approach. In *Proceedings of SIGGRAPH'96 (New Orleans, LA, August 4–9, 1996). In Computer Graphics Proceedings, Annual Conference Series*, pages 11–20. ACM SIGGRAPH, 1996.

6. G. Dutton. Locational properties of quaternary triangular meshes. In *Proceedings of the Fourth International Symposium on Spatial Data Handling*, pages 901–910, July 1990.

7. O. Faugeras, S. Laveau, L. Robert, G. Csurka, and C. Zeller. 3-d reconstruction of urban scenes from sequences of images. Technical Report RR-2572, INRIA, Sophia-Antipolis, France, June 1995.

8. G. Fekete. Rendering and managing spherical data with sphere quadtrees. In *Proceedings of Visualization'90*, pages 176–186, Los Alamitos, California, 1990. IEEE Computer Society Press.

9. A. Gershun. Svetovoe Pole (The Light Field, in English). *Journal of Mathematics and Physics*, XVIII:51–151, 1939.

10. J. S. Gondek, G. W. Meyer, and J. G. Newman. Wavelength dependent reflectance functions. In *Computer Graphics Proceedings*, Annual Conference Series, pages 213–220, New York, NY, July 1994. ACM SIGGRAPH. Conference Proceedings of SIGGRAPH '94.

11. S. J. Gortler, R. Grzeszczuk, R. Szeliski, and M. F. Cohen. The lumigraph. In *Proceedings of SIGGRAPH'96 (New Orleans, LA, August 4–9, 1996). In Computer Graphics Proceedings, Annual Conference Series*, pages 43–54. ACM SIGGRAPH, 1996.

12. X. Gu, S. J. Gortler, and M. F. Cohen. Polyhedral geometry and the two-plane parameterization. In *Eighth Eurographics Workshop on Rendering*, pages 1–12, Saint Etienne, France, June 1997. Eurographics.

13. H. Laporte, E. Nyiri, M. Froumentin, and C. Chaillou. A graphics system based on quadrics. *Computers & Graphics*, 19(2):251–260, 1995.

14. S. Laveau and O. Faugeras. 3-d scene representation as a collection of images and fundamental matrices. Technical Report RR-2205, INRIA, Sophia-Antipolis, France, February 1994.

15. J. Lengyel and J. Snyder. Rendering with coherent layers. In *Proceedings of SIGGRAPH'97 (Los Angeles, CA, August 3–8, 1997). In Computer Graphics Proceedings, Annual Conference Series*, pages 223–242. ACM SIGGRAPH, 1997.

16. M. Levoy and P. Hanrahan. Light field rendering. In *Proceedings of SIGGRAPH'96 (New Orleans, LA, August 4–9, 1996). In Computer Graphics Proceedings, Annual Conference Series*, pages 31–42. ACM SIGGRAPH, 1996.

17. P. W. C. Maciel and P. Shirley. Visual navigation of large environments using textured clusters. In *1995 Symposium on Interactive 3D Graphics*, pages 95–102. ACM SIGGRAPH, 1995.

18. N. Max. Hierarchical rendering of trees from precomputed multi-layer z-buffers. In *Seventh Eurographics Workshop on Rendering*, pages 166–175, Porto, Portugal, June 1996. Eurographics.

19. L. McMillan and G. Bishop. Plenoptic modeling: An image-based rendering system. In *Proceedings of SIGGRAPH'95 (Los Angeles, CA, August 6–11, 1995). In Computer Graphics Proceedings, Annual Conference Series*, pages 39–46. ACM SIGGRAPH, 1995.

20. K. Pulli, M. F. Cohen, T. Duchamp, H. Hoppe, L. Shapiro, and W. Stuetzle. View-based rendering: Visualizing real objects from scanned range and color data. In *Eighth Eurographics Workshop on Rendering*, pages 23–34, Saint Etienne, France, June 1997. Eurographics.

21. M. Sbert. An integral geometry based method for fast form-factor computation. In R. J. Hubbold and R. Juan, editors, *Computer Graphics Forum*, volume 12(3), pages C–409–C–420, C–538, Oxford, UK, Sept. 1993. Eurographics Association, Blackwell. Proceedings of Eurographics '93.

22. P. Schröder and W. Sweldens. Spherical wavelets: Efficiently representing functions on the sphere. In *Proceedings of SIGGRAPH'95 (Los Angeles, CA, August 6–11, 1995). In Computer Graphics Proceedings, Annual Conference Series*, pages 161–172. ACM SIGGRAPH, 1995.

23. S. M. Seitz and C. R. Dyer. View morphing. In *Proceedings of SIGGRAPH'96 (New Orleans, LA, August 4–9, 1996). In Computer Graphics Proceedings, Annual Conference Series*, pages 21–30. ACM SIGGRAPH, 1996.

24. J. Shade, D. Lischinski, D. H. Salesin, T. DeRose, and J. Snyder. Hierarchical image caching for accelerated walkthroughs of complex environments. In *Proceedings of SIGGRAPH'96 (New Orleans, LA, August 4–9, 1996). In Computer Graphics Proceedings, Annual Conference Series*, pages 75–82. ACM SIGGRAPH, 1996.

25. J. W. Shade, S. J. Gortler, L.-W. He, and R. Szeliski. Layered depth images. In *to appear in SIGGRAPH'98*. ACM SIGGRAPH, 1998.

26. P.-P. Sloan, M. F. Cohen, and S. J. Gortler. Time critical lumigraph rendering. In *1997 Symposium on Interactive 3D Graphics*, pages 17–23, New York, 1997. ACM.

27. H. Solomon. *Geometric Probability*, volume 28 of *CBMS–NSF Regional conference series in applied mathematics*. SIAM, Philadelphia, PA, 1978.

28. T.-T. Wong, P.-A. Heng, S.-H. Or, and W.-Y. Ng. Image based rendering with controllable illumination. In *Eighth Eurographics Workshop on Rendering*, pages 13–22, Saint Etienne, France, June 1997. Eurographics.

Editors' Note: see Appendix, p. 328 for colored figures of this paper

Forward Shadow Mapping

Hansong Zhang

Department of Computer Science
University of North Carolina at Chapel Hill, U. S. A.
zhangh@cs.unc.edu

Abstract. Forward shadow mapping is a new approach to real-time shadow generation. The traditional shadow map algorithm maps the pixels in the eye's view *backward* into the depth buffers of light sources (i.e. shadow maps), which is similiar to and often implemented as an extension to texture mapping. Our algorithm reverses this process by using 3-D image warping techniques to transform shadow map pixels *forward* into the eye's view to directly indicate which pixels are lit; it does not interfere with normal texture mapping and easily supports anti-aliased shadow edges and projective textures. Access to shadow maps and projective textures is in pixel-sequential order. This algorithm has advantages when speed of texture mapping becomes the performance bottleneck, which is often the case in visual simulation and game applications.

1   Introduction

Shadow generation has been a subject of active research in computer graphics. Surfaces in a scene not lit by the light source are in shadow. Shadows can significantly enhance the realism of synthesized images and often lead to better understanding of spatial relationships between objects.

The task of shadow determination is to tell whether portions of a scene visible from the eye are visible from the light sources. Areas seen by both the eye and a light are illuminated, while those visible to the eye but not the light lie in shadow with respect to that light. Multiple light sources can be handled individually in the same way. Obviously, shadow generation is visibility determination from multiple view points.

Depth buffering is a popular image-space approach to resolving visibility. A shadow algorithm can make use of the depth buffers of the light sources to determine whether a 3-D point is visible to the lights; these depth buffers are also called *depth maps* or most commonly *shadow maps*.

This paper presents a new algorithm for shadow generation. Based on the *forward* 3-D warping of shadow maps into the eye's view, the algorithm reverses the mapping process of traditional shadow map algorithms (discussed in the next section) which often share resources with texture mapping. Our algorithm shows its advantages when graphics performance is bound by texture mapping; it also easily supports anti-aliased shadow edges and integrates projective textures into the shadow generation process.

2   Related Work

Shadow map algorithms originated in [Wil78], which proposed an algorithm that renders the scene in two passes with depth-buffering. The first pass is from the light sources and generates the shadow maps. The scene is then rendered from the eye's view. 3-D points visible in the eye's view (corresponding to non-background pixels in the eye's image) are projected into the shadow maps; depth comparisons then show whether they are also visible to the lights. The image-space nature of this algorithm makes it simple to implement, applicable to any kind of primitives, easy to accelerate with hardware, and efficient enough to be used with complex scenes.

To solve the aliasing problem of the original shadow map algorithm, [HN85] proposed the use of per-pixel object identifiers and [RSC87] introduced percentage closer filtering. These im-

132

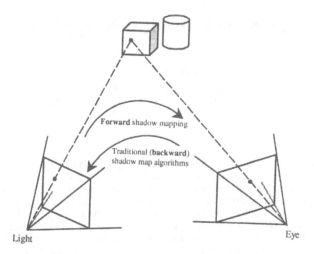

Fig. 1. The basic concept of forward shadow mapping. Pixels in the light's depth buffer (the shadow map) are transformed, using 3-D image warping technqiues, into the eye's view to directly show which pixels are illuminated. It reverses the traditional shadow map algorithm which is a projection into light's view followed by random look-ups into the shadow map.

provements alleviate aliasing and produce soft shadow edges. [SKW92] recognizes that projective textures and shadow maps are both a simple extension to perspective-correct texture mapping; as a result, graphics systems such as the RealityEngine[2] [Ake93] have sucessfully supported real-time shadows. The implementation of shadow maps on the PixelFlow [EMP97], a very different architecture, is also a variation of texture mapping. Extensive sharing of texture mapping hardware makes shadow maps easier and less expensive to implement. However, speed of texture mapping is often the limiting factor in modern graphics applications, and sharing of critical resources can lead to severe system imbalance.

Many other shadow algorithms have been developed. Compared to depth-buffer-based algorithms, they have limitations in generality, scalability and speed. Scanline shadow algorithms [App68, BK70] requires significant pre-processing and are thus not suitable for moving objects and lights. Polygon subdivision [AWG78, NN85], shadow volume algorithms [Cro77, BB84, TN85, Ber86, Max86, FGH85, Hei91], and the BSP-tree shadow volume method [CF89, CF92] are restricted to polygonal models and inefficient for scenes with a large number of polygons. Polygon subdivision algorithms algorithm is also difficult to implement robustly. For a survey of some of the algorithms see [WPF90]. Ray-tracing [Whi79, Kay79] and radiosity algorithms [GTGB84, CG85] produce high-quality shadows, but are still too expensive for real-time purposes.

Another area of research to which our work is related is image based rendering (IBR). The goal of IBR is to generate an image for a view (the *destination* image) directly from existing images for other views (*source* images), without an explicit geometric representation of the scene. The basic operation is the transformation of source image pixels to the destination image. [CW93] proposed a view interpolation method based on pre-computed per-pixel optical flow information. [MB95] points out that with per-pixel disparity values (which actually make pixels 3-D entities), source pixels can be re-projected directly into the destination image. This technique is also called 3-D image warping [MMB97], which is the term we use in this paper. An image with per-pixel disparity or Z value is often called a *depth image*. (It is often unclear whether a "depth" image has disparity or Z values. In this paper, depth means Z; this is consistent with the notion of a depth buffer.) [Zha97] discusses 3-D image warp using the OpenGL[WND85] graphics model.

Another way of generating a destination image from source images is to first reconstruct geometry from source images, and then render them for the destination view[DSV97, SDB97]. For

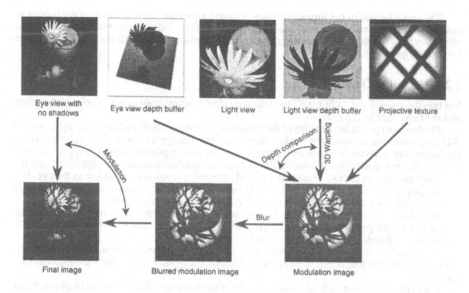

Fig. 2. The process of forward shadow mapping

scenes with large near-planar structures, this can be faster than per-pixel 3-D warping. On the other hand, it requires substantial computation in reconstruction, and thus is not applicable when source images change frequently.

3 Forward Shadow Mapping

The forward shadow mapping algorithm applies 3-D image warping techniques to the problem of shadow generation. This section describes in detail the basic algorithm and an architecture-specific variation. The next section gives a comparison between forward shadow mapping and traditional shadow map algorithms.

3.1 Motivation

Implementations of shadow maps as an extension of texture mapping are very natural, given the fact that shadow and texture mapping are similiar operations and can share similiar hardware resources. However, this makes the texture hardware often the bottleneck of system performance. In many applications, speed of texturing has become the bottleneck even without shadow maps. The use of a shadow map, which is equivalent to applying a texture map to all geometry in the scene (or all visible geometry, for deferred shading architectures[DWS88]) aggravates the situation. Shadow maps also compete with ordinary textures in the scene for texture memory. It is often desirable to use a *projective texture*[McR96] (an image projected by the light) to model a light's variation in intensities, in order to achieve effects like spotlights with gradual fall-off or lights shining through stained glass; this means more textures for the scene. Another problem is that applying more than one texture to the same polygon is a feature not commonly available in hardware and often has to be emulated using multi-pass rendering.

Still more shadow maps are required if we want to support shadow casting in wider angle than a single shadow map can handle. Real-world point lights or cylindrical lights can be modeled by multiple shadow maps covering the surface of a sphere or cylinder—or, in other words, modeled by multiple lights with limited fields of view. With so many shadow maps, the cost of multi-pass rendering soon gets out of control.

It is still true that we should maximize hardware sharing to minimize the amount of hardware (if any) specific to shadow generation. But for texture-rate bound applications, this sharing should not be of the texture mapping hardware, in order to maintain system balance.

The major benefit of the forward shadow mapping algorithm is the removal of the burden of shadow generation from the texture mapping resources. Our goal is to provide an algorithm that efficiently handles real-time, anti-aliased shadowing of textured scenes with wide-angle projective-texture-mapped lights.

3.2 Algorithm Overview

For brevity, we will use "a lit pixel" to mean a pixel that corresponds to a 3-D point in the scene that is lit and seen by the eye.

Shadow map pixels are samples of the portions of scene geometry visible to the light. We use 3-D image warping to transform shadow map pixels into the eye's view, so that we know where the illuminated portions fall in the eye's image and what their depths are in the eye space. Shadow map pixels are accessed in sequential order. A depth comparison between a warped pixel and the corresponding one in the eye's image decides whether the warped pixel is visible to the eye. If it is, the eye pixel is in the lit area of the eye's image. Note that this assumes the scene has already been rendered from the eye so that the eye's depth buffer is ready to be used.

If we have a projective texture we can access it in the same sequential order as shadow map pixels, and the warped pixels can then be assigned light intensities. Since memory access is all in pixel-sequential order, these effects are cheap to support.

The above is a high-level description of forward shadow mapping. Depending on what one does once a pixel is determined to be lit (or in shadow), many variations of the base algorithm are possible.

For deferred shading (DS) architectures, a light mask can be stored per-pixel to switch lights on/off, so warping can be scheduled before shading to set the masks. Actually, forward shadow mapping is DS-oriented since it requires that the eye's depth buffer be available before shadow computation (part of shading).

For architectures that shade before scan-conversion (SBSC), DS can be simulated to an extent with multi-pass rendering from the eye. This is inefficient for a large scene. Below we show a fast algorithm for SBSC architectures that has only one eye pass. It also produces anti-aliased shadow edges. 3-D warping (combined with optional projective textures) produces a modulation image, which is used to scale color values for each framebuffer pixel separately. This is an approximation to real shading, but in practice produces realistic-looking images. Modulation has been used for spotlight effects in [McR96].

It is also possible to draw modulation pixels directly into the framebuffer as soon as a shadow map pixel is warped. But, as will be discussed later, antialiasing can be easily done if we have the whole modulation image.

Steps of the algorithm are listed below. For brevity, the algorithm is given for one light source; extension to multiple lights is straightforward. Figure 2 illustrates how it works on a model by showing the intermediate and final images. Comments on the algorithm are given in the following subsections.

1. Render the scene from the light's view; get the resulting depth buffer as the shadow map.
2. Render the scene normally from the eye's view.
3. Clear the modulation image with the dimming factor for areas not lit by any light.
4. Each shadow map pixel is warped to the eye's view, carrying the corresponding intensities from the projective texture; for each pixel in the eye's image covered by the warped pixel, compare the transformed depth with the depth in the eye's depth buffer. If they are approximately equal, then the the light intensities are accumulated in the modulation image.
5. Blur the modulation image for antialiased shadow edges.
6. Apply the modulation image. It modulates the normally rendered, eye's image for shadows and lighting effects defined by the projective texture.

3.3 Transforming the Shadow Map

We use existing 3-D warping techniques [MB95, Zha97] to warp the shadow map pixels to the eye's view. These techniques take advantage of the incremental nature of pixel-by-pixel computation, which results in much less operations than full (non-incremental) 3-D warping per pixel. The method described in [Zha97] is especially suited to our purpose since it deals directly with z-values from the depth buffer and uses the same coordinate systems as in conventional 3-D graphics.

Due to the fact that pixels are not real points, and that the eye and the light view generally have different sampling rates, the area a warped shadow map pixel covers in the eye's image has to be computed. Detailed discussion of this computation is beyond the scope of this paper, but it should be noted that the area is often estimated since it is too expensive to get the exact answer; it is often overestimated to avoid cracks between warped pixels. For interactive applications, pixels in the area are often treated as having the same color and depth.

We do not use mesh-based IBR methods (see section 2) because we want to support dynamic scenes and moving lights, which potentially changes shadow maps for each frame. The time required for processing depth values makes these methods impractical for us.

Note that the forward shadow mapping algorithm is independent of the 3-D warping methods it uses. Any warping algorithm can be plugged in without changing the shadow algorithm.

3.4 Updating the Modulation Image

Below we will use *pixel* to refer to both the 2-D location on an image and the 3-D point(s) in the scene it corresponds to. They are easy to distinguish in context.

For each pixel in the eye's image covered by the warped pixel, we have to decide whether the corresponding shadow map pixel is visible to the eye. This is done by comparing the warped depth to depths from the eye's depth buffer. If pixels are infinitely small and there is no numerical error, the warped depth should be equal to the depth at an eye pixel (meaning the two pixels are the projections of the same point in the scene) for the shadow map pixel to be visible—which in turn means the eye pixel is lit.

In practice this exact relationship will not happen, requiring error tolerance, ϵ. We use an approximately-equal operator for depth comparison with an error bound. The error bound is an empirical value based on the resolution of the images and depth values. We could also use a *less-than* operator as in other shadow map algorithms, offsetting the warped depth by ϵ before the comparison. But the more strict *approximately-equal* operator also helps to trim away the outside part of the estimated coverage of a warped pixel that sits across the silhouette of a visible surface, if depth changes more than ϵ at the silhouette.

When an eye pixel is determined to be lit, the light intensities from the projective texture are added to the corresponding pixel in the modulation image. Warped pixels may overlap since their sizes are estimated, so we have to avoid adding more than once the intensities from the same light for an eye pixel. If there is only one light, the intensities can overwrite instead of being added to the modulation image. For multiple lights, the stencil buffer can be used to store light identifiers to avoid the problem.

3.5 Anti-aliasing

In order to produce smooth shadow edges, we simply blur the modulation image with some weighted-average filter. The bigger the filter kernel, the softer the edges appear, and the less precise the edge positions. It is interesting to note that the effect of blurring is analogous to the percentage closer sampling for antialiasing [RSC87] in the traditional shadow map algorithm, but enjoys better memory access behavior. Advanced filtering techniques and large filter kernels are easier to support in our algorithm.

4 Comparison

In pure-software implementations, forward shadow mapping (FSM) and traditional shadow map algorithms (or backward shadow mapping, BSM) differ mainly in strategies for resampling—in the eye space—the light-space samples of the scene geometry. Similar to texture mapping, BSM resamples by interpolating the nearby samples for a new sampling location. (Note that it is the results of depth comparison, not the depth values themselves, that are interpolated [RSC87].) FSM resembles splatting[Wes90], in which each light-space sample leaves a "footprint" (i.e. area covered by the warped pixel) on the eye's image. In this paper we have assumed simple footprints to reduce amount of computation.

In hardware implementations, BSM borrows the random look-up capacity built into texturing mapping hardware and adds depth comparisons as a shadow map-specific extension. FSM, on the other hand, shares hardware with an ordinary framebuffer (the color and Z buffers), which

Fig. 5. A walking robot in a spotlight. This is a frame from an interactive demostration of the forward mapping algorithm.

provides random memory access needed by the warped pixels for depth comparisons and accumulation of light intensities. No special-purpose extension is needed.

FSM is independent of texture mapping so that the latter can be fully dedicated to scene contents. It also makes it possible for shadow generation to run in parallel with texturing. It accesses shadow maps *and* projective textures in pixel-sequential order.

5 Implementation

Our algorithm has been implemented on Silicon Graphics workstations using OpenGL. Ideally, the algorithm should be implemented inside graphics hardware in order for it to work at its full potential, and to conduct direct performance comparisons with hardware implementations of traditional shadow map algorithms. Without support for such low-level programming, our current implementation nontheless serves as a proof of concept and also runs at interactive frame rates.

3-D warping is done on the host, and runs in parallel if multiple CPUs are available. The depth buffers from the lights and the eye are read into the main memory to support warping. The modulation image, once constructed, is drawn onto the framebuffer with the proper blending function which treats incoming colors as modulation factors for already existing colors in the framebuffer.

Figures 3 and 4 (see Appendix) show the effects of shadow map resolutions and filtering of the modulation image. The filter uses a 3×3 averaging kernel. As expected, filtering helps to enhance image quality, especially when the shadow map is low resolution. On a 4 processor SGI Onyx2 InfiniteReality, frame rate is 7 frames per second with filtering and 9 without.

Figure 5 is a frame taken from an interactive demonstration of our algorithm. A robot walks in a soft-edge spotlight; with the robot moving, the shadow map has to be updated every frame. It runs at 640×480 VGA resolution. On the same machine as mentioned above, the frame rate is on average 6 frames per second. Table 1 shows the allocation of frame time, averaged over 100 frames.

6 Conclusion and Future Work

In this paper we have introduced a new shadow algorithm based on forward 3-D warping of shadow maps. This algorithm provides new design alternatives for both software and hardware shadow determination. It does not interfere with ordinary texture mapping. It accesses shadow maps and projective textures in sequential order and easily supports anti-aliased shadow edges. The algorithm has been implemented and demonstrates shadowing at interactive rates.

Table 1. Frame time allocation for the robot demo

Operations	Time (ms)
Render from light	26.2
Render from light	26.2
Read light depth	11.7
Render from eye	27.2
Read eye depth	12.3
3-D warping	67.1
Modulation	8.8
Misc. overhead	14.0
Total frame	167.3

Future plans include implementation of the algorithm inside graphics hardware, faster implementation of 3-D warping algorithms, and better methods for computing the pixel coverage of warped pixels.

References

Ake93. Kurt Akeley. RealityEngine graphics. In *Computer Graphics (SIGGRAPH '93 Proceedings)*, volume 27, pages 109–116, August 1993.

App68. A. Appel. Some techniques for shading machine renderings of solids. In *IFIP*, volume 32, pages 37–45, 1968.

AWG78. P. Atherton, K. Weiler, and D. Greenberg. Polygon shadow generation. In *Computer Graphics (SIGGRAPH '78 Proceedings)*, volume 12, pages 275–281, August 1978.

BB84. L. S. Brotman and N. I. Badler. Generating soft shadows with a depth buffer algorithm. *IEEE Computer Graphics and Applications*, 4(10):71–81, October 1984.

Ber86. P. Bergeron. A general version of crow'S shadow volumes. *IEEE Computer Graphics and Applications*, 6(9):17–28, 1986.

BK70. W. J. Bouknight and K. C. Kelly. An algorithm for producing half-tone computer graphics presentations with shadows and movable light sources. In *Proc. AFIPS JSCC*, volume 36, pages 1–10, 1970.

CF89. Norman Chin and Steven Feiner. Near real-time shadow generation using BSP trees. In *Computer Graphics (SIGGRAPH '89 Proceedings)*, volume 23, pages 99–106, July 1989.

CF92. Norman Chin and Steven Feiner. Fast object-precision shadow generation for areal light sources using BSP trees. In *Computer Graphics (1992 Symposium on Interactive 3D Graphics)*, volume 25, pages 21–30, March 1992.

CG85. Michael F. Cohen and Donald P. Greenberg. The Hemi-Cube: A radiosity solution for complex environments. In *Computer Graphics (SIGGRAPH '85 Proceedings)*, volume 19, pages 31–40, August 1985.

Cro77. Franklin C. Crow. Shadow algorithms for computer graphics. In *Computer Graphics (SIGGRAPH '77 Proceedings)*, volume 11, pages 242–248, July 1977.

CW93. Shenchang Eric Chen and Lance Williams. View interpolation for image synthesis. In *Computer Graphics (SIGGRAPH '93 Proceedings)*, volume 27, pages 279–288, August 1993.

DSV97. Lucia Darsa, Bruno C. Silva, and Amitabh Varshney. Navigating static environments using image-space simplification and morphing. In *1997 Symposium on Interactive 3D Graphics*, pages 7–16, April 1997.

DWS88. Michael Deering, Stephanie Winner, Bic Schediwy, Chris Duffy, and Neil Hunt. The triangle processor and normal vector shader: A VLSI system for high performance graphics. In *Computer Graphics (SIGGRAPH '88 Proceedings)*, volume 22, pages 21–30, August 1988.

EMP97. John Eyles, Steven Molnar, John Poulton, Trey Greer, Anselmo Lastra, Nick England, and Lee Westover. PixelFlow: The realization. In *1997 SIGGRAPH / Eurographics*

138

Workshop on Graphics Hardware, pages 57–68. ACM SIGGRAPH / Eurographics, ACM Press, August 1997.

FGH85.   Henry Fuchs, Jack Goldfeather, Jeff P. Hultquist, Susan Spach, John D. Austin, Frederick P. Brooks, Jr., John G. Eyles, and John Poulton. Fast spheres, shadows, textures, transparencies, and image enhancements in Pixel-Planes. In *Computer Graphics (SIGGRAPH '85 Proceedings)*, volume 19, pages 111–120, July 1985.

GHC97.   Steven Gortler, Liwei He, and Michael F. Cohen. Rendering layered depth images. Technical Report MSR-TR-97-09, Microsoft Research, 1997.

GTGB84.   Cindy M. Goral, Kenneth E. Torrance, Donald P. Greenberg, and Bennett Battaile. Modelling the interaction of light between diffuse surfaces. In *Computer Graphics (SIGGRAPH '84 Proceedings)*, volume 18, pages 212–22, July 1984.

Hei91.   Tim Heidmann. Real shadows, real time. *Iris Universe*, 18:28–31, 1991.

HH97.   Paul S. Heckbert and Michael Herf. Simulating soft shadows with graphics hardware. Technical Report CMU-CS-97-104, School of Computer Sciene, Carnegie Mellon University, 1997.

HN85.   J. C. Hourcade and A. Nicolas. Algorithms for antialiased cast shadows. *Computers and Graphics*, 9(3):259–265, 1985.

Kay79.   Douglas S. Kay. Transparency, refraction, and ray tracing for computer synthesized images. Master's thesis, Cornell U., January 1979.

Max86.   Nelson L. Max. Atmospheric illumination and shadows. In *Computer Graphics (SIGGRAPH '86 Proceedings)*, volume 20, pages 117–24, August 1986.

MB95.   Leonard McMillan and Gary Bishop. Plenoptic modeling: An image-based rendering system. In *Computer Graphics (SIGGRAPH '95 Proceedings)*, pages 39–46. ACM SIGGRAPH, August 1995.

McR96.   Organizer: Tom McReynolds. Programming with opengl: Advanced rendering. *SIGGRAPH'96 Course Notes*, pages 27–28, August 1996.

MMB97.   William R. Mark, Leonard McMillan, and Gary Bishop. Post-rendering 3d warping. In *1997 Symposium on Interactive 3D Graphics*, pages 7–16, April 1997.

NN85.   Tomoyuki Nishita and Eihachiro Nakamae. Continuous tone representation of three-dimensional objects taking account of shadows and interreflection. In *Computer Graphics (SIGGRAPH '85 Proceedings)*, volume 19, pages 23–30, July 1985.

RSC87.   William T. Reeves, David H. Salesin, and Robert L. Cook. Rendering antialiased shadows with depth maps. In *Computer Graphics (SIGGRAPH '87 Proceedings)*, volume 21, pages 283–291, July 1987.

SDB97.   François Sillion, George Drettakis, and Benoit Bodelet. Efficient impostor manipulation for real-time visualization of urban scenery. In *Computer Graphics Forum (Proc. of Eurographics '97)*, volume 16, pages 207–218, Budapest, Hungary, September 1997.

SKW92.   Mark Segal, Carl Korobkin, Rolf van Widenfelt, Jim Foran, and Paul E. Haeberli. Fast shadows and lighting effects using texture mapping. In *Computer Graphics (SIGGRAPH '92 Proceedings)*, volume 26, pages 249–252, July 1992.

TN85.   E. Nakamae T. Nishita, I. Okamura. Continuous tone representation of three-dimensional objects taking account of shadows and interreflection. *ACM Transaction on Graphics*, 20(2):123–146, April 1985.

Wes90.   Lee Westover. Footprint evaluation for volume rendering. In *Computer Graphics (SIGGRAPH '90 Proceedings)*, volume 24, pages 367–376, August 1990.

Whi79.   T. Whitted. An improved illumination model for shaded display. In *Computer Graphics (Special SIGGRAPH '79 Issue)*, volume 13, pages 1–14, August 1979.

Wil78.   Lance Williams. Casting curved shadows on curved surfaces. In *Computer Graphics (SIGGRAPH '78 Proceedings)*, volume 12, pages 270–274, August 1978.

WND85.   Mason Woo, Jackie Neider, and Tom Davis. *OpenGL Programming Guide, 2nd Edition*. Addison-Wesley, 1985.

WPF90.   Andrew Woo, Pierre Poulin, and Alain Fournier. A survey of shadow algorithms. *IEEE Computer Graphics and Applications*, 10(6):13–32, November 1990.

Zha97.   Hansong Zhang. A traditionalist view of 3-d image warping. Technical Report 97-043, Department of Computer Sciene, UNC-Chapel Hill, September 1997.

Editors' Note: see Appendix, p. 330 for colored figures of this paper

Efficient Image Generation for Multiprojector and Multisurface Displays

Ramesh Raskar, Matt Cutts, Greg Welch, Wolfgang Stürzlinger

Department of Computer Science, University of North Carolina at Chapel Hill,
Chapel Hill, NC 27599, U.S.A.

Abstract. We describe an efficient approach to rendering a perspectively correct image on a potentially irregular display surface that may be illuminated with one or more distinct devices. The first pass of the technique generates an image of the desired graphics model using conventional rendering. The second pass projects that image as a texture onto a model of the display surface, then re-renders the textured display surface model from the viewpoint of each display device. The algorithm scales with the complexity of the display surface, and is constant with respect to the complexity of the graphics model.

1. Introduction

Along with ongoing increases in rendering power comes renewed hope for wide-field-of-view and high-resolution displays for an increased sense of immersion and improved visualization. Two opportunities for improved immersion include

a. *Multiprojector* case: the images formed on the visible display surface originate from more than one display device; and/or

b. *Multisurface* case: the visible illuminated display surface is irregular or non-planar.

Examples of these cases are spatially immersive display (SID) systems such as the Cave Automated Virtual Environment (CAVE™) [Cruz-Neira93], the Office of the Future system [Raskar98b], and Alternate Realities' VisionDome [Bennett98]. Other cases include head-mounted displays (HMDs), e.g. 6 tiles per eye or 15 tiles per eye wide field of view HMDs developed by Kaiser Electro-Optics, Inc. [Kaiser98]. Figure 1 (see Appendix) has examples of multiprojector and multisurface displays.

Since the proposed method requires only one graphics model scene traversal to generate a texture and then relies on relatively inexpensive rendering of the display surface, it effectively scales with the complexity of the display surface model.

2. Previous Work

For multisurface displays with a single projector, Dorsey et al. provided a useful framework in the context of theater set design [Dorsey91]. A projector is used to display a regular grid onto the backdrop, which is then seen as a distorted grid from the spectator's viewpoint. Applying the inverse transformation (using image warping and interpolation) to the slide creates a predistorted image. The pre-distorted image then appears correct when projected onto the curved backdrop. Nelson Max described a dome-based system in [Max91]. Given a dome, a 3D point to be imaged, and the viewer's eye location, the method extends a 'projecting ray' from the eye through the 3D point until the ray intersects the dome. A new ray is drawn from the dome intersection point to the center of a fisheye lens. That ray is traced through the lens to compute a point on a film frame. Raytracing provides a solution for arbitrary (and even implicit) surfaces, but may be time-consuming. Equipe Ltd. implements real-time distortion correction of a single projector in a dome system for a fixed viewpoint, but the texture mapping is static [Jarvis97]. The Luminous Room system at MIT uses a single projector coupled with an optical-mechanical design to allow projection in any direction in a room [Underkoffler97]. Pre-warping is performed on the image sprites to provide an undistorted view.

3. Image Generation Using Projective Texture Rendering

The inputs to the algorithm are the viewer's location, the location and orientation of each projector, a *graphics model* to be rendered, and a geometric model of the display surface—a *display surface model*. (See Figure 2.) The goal of the algorithm is to create a correct image E of the graphics model regardless of the display surface. That is, we wish to find an image P for each projector such that when the projector shows that image, the user will see the correct image of the graphics model.

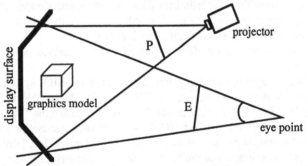

Fig. 2: Example of single-projector system with a non-planar display surface.

The first pass of the algorithm renders E, the correct or *desired image* of the graphics model. The cost of this pass is simply the standard cost of rendering the graphics model. For the second pass, imagine that the desired image is a slide and the user's eye is a light source, so that the desired image is projected onto the display surface. If we render this scenario from the viewpoint of a projector, we get the final image P. This final image, when displayed by the projector, will form the desired image E for the user. In our implementation, the slide projection and rendering is achieved using OpenGL *projective textures* from the user's eye location. The textured display surface is rendered from the viewpoint of each projector. The cost of the second pass for each projector is simply the cost of rendering the textured display surface model.

As Segal et al. [Segal92] note, "Projecting a texture image onto a scene from some light source is no more expensive to compute than simple texture mapping in which texture coordinates are assigned to polygon vertices. Both require a single division per-pixel for each texture coordinate; accounting for the texture projection simply modifies the divisor." Our method makes the *desired image* be the texture image, the user's eye be the light source, the display surface model be the scene, and each projector (in turn) be the viewpoint. Because projective textures use the hardware-accelerated texture stack of OpenGL, they are no more expensive than traditional texture mapping. Neither texture coordinates nor warp functions are explicitly computed because OpenGL generates the correct texture coordinates. However, projectors that can be modeled using standard OpenGL transformation matrices, such as digital micromirror device (DMD) projectors, are needed. The traditional graphics pipeline handles clipping of the display surface model. Provided the display surface model accurately represents the actual surfaces in the room, rendering from the projector's viewpoint will handle visibility of real-world surfaces correctly.

4. Comparison with Conventional Rendering

In [Raskar98a], we have analyzed the relative advantages of the method and addressed issues in synchronization, latency, and networking. Our technique is not advantageous for the trivial case of a single projector with a simple flat surface. However if the surface is not flat, our approach excels because it requires only one traversal of the graphics model—the desired image is computed only once. In contrast, one conventional technique would tessellate the surface then re-render for each planar portion. Another method would project the display surface model down into the image plane to explicitly recover texture coordinates. If the user's viewpoint changes, the texture coordinates have to be calculated again. When multiple projectors are used, our algorithm still requires only one traversal of the graphics model, as long as one can draw a plane perpendicular to the user that covers the user's field of view or that encompasses the projectors' display areas, whichever is smaller. The gains increase as the complexity of graphics model or the display surface model increase. Our method may require more than one scene traversal with very wide field-of-view systems (e.g. 360°); several smaller field-of-view images may be needed.

5. Implementation and Results

In all of the following systems, we use the texture stack of OpenGL for projective textures. On many types of Silicon Graphics (SGI) machines, the texture stack is hardware-accelerated as mentioned in [Segal92].

5.1 Protein Interactive Theater (PIT) System

The University of North Carolina's (UNC) PIT system is similar to a CAVE™. Instead of several walls and a floor, the PIT has two screens, with one projector per screen (see Figure 1). The PIT screens can be adjusted to meet at 90° or 120°. One use of the PIT is for performing walkthroughs of extremely large architectural databases. For example, the Walkthrough research group at UNC is working on a model of a power plant with 13 million triangles [Aliaga98]. Even after optimizing as much as possible, rendering such a large model can be prohibitively expensive.

For a comparison benchmark, a 454-frame path through the power plant model was recorded and then rendered on an Onyx. The program measured the average time to compute each frame in milliseconds/frame. Rendering twice at (1280x1024) required 231 milliseconds/frame. Using our method, rendering once at (1024x1024) and then texture-mapping two quadrilateral (1280x1024) needed only 176 milliseconds/frame. Differences between the images were difficult to see.

5.2 Kaiser Head-Mounted Display (KHMD) System

Our method was implemented and optimized for use with a 12-LCD wide field of view (153x48 degrees) KHMD [Kaiser98]. The HMD has 3x2 (horizontal x vertical) LCD's per eyes placed in circular arcs. Each LCD has 267x225 pixels. Special optics ensure that no seams are visible between the images. Figure 3 (see Appendix) shows the composite view for one eye (left image). The 6 quadrilaterals depict the viewing frustum for each of the 6 LCD's (right image).

The straightforward method of rendering images for the KHMD involves rendering each of the 12 views separately. This implies that the scene model has to be processed 12 times by the graphics hardware. The KHMD experiment was performed on an Onyx with R4400 processors and a two-pipe Infinite Reality (IR). Conventional rendering (rendering 12 views on one processor) can display a scene with 23940 polygons at 3 Hz. Our method (rendering two views, projective mapping 12 times, also on one processor) runs at 12.2 Hz with a 512x512 texture.

5.3 "Office of the Future" (OOTF) System

The OOTF system demonstrates multisurface rendering for circumstances described in [Raskar98b]. The portable nature of OpenGL allows the rendering to be done on different classes of machines, from an SGI Infinite Reality2 (IR2) down to SGI O2's and PCs. Currently we use an IR2 that is capable of simultaneously driving 3 projectors at 800x600 resolution. The display surface model has a desk located in the corner of a room. We have demonstrated interactive rates of 25-30 frames per second for a display model of about 100 polygons, without any optimization.

6. Issues in Multisurface/Multiprojector Display

The Projective Texture Rendering technique provides speed-up, but several issues must be addressed: the size and accuracy of the display surface model, aliasing, and effective resolution. Representing the display surface model with a large number of triangles to reduce the modeling error will degrade performance in the second pass. On the other hand, inaccuracies in the display surface model can cause incorrect views. Consequently a simplified model of the display surface that maintains minimum error is important.

Regarding aliasing, the system attempts to generate images with uniform sampling from the observer's viewpoint. This yields non-uniform resolution for rendered primitives in projector image space, which may increase the traditional aliasing problems. A related question is how large to compute the desired image. The perceived resolution depends on relative distance and angle of the viewer and projector from the display surface. One solution is to render the first pass at a higher resolution than desired. If the display has a wide extent in the user's field of view, e.g., a dome where images are projected in front of and behind the user, multiple desired images may be needed.

7. Conclusions and Future Work

We believe that Projective Texture Rendering provides a useful generalization of the typical 3D graphics pipeline to include multiple projectors and non-planar surfaces. The approach is simple, and nicely parameterizes the desired image in terms of the viewer position, projector positions, and display surface models. We have described analytical results in [Raskar98a] and also presented empirical results showing significant speedup on two systems (PIT and Kaiser HMD), and have demonstrated the technique in an "office of the future" application which would not have been possible with conventional rendering.

There are several research tasks that remain to be pursued in the future. For example, we plan to formally address the issue of blending between multiple projectors. We would like to characterize and quantify the sampling/resolution and aliasing issues encountered during the multiple passes in our approach. We also intend to study the approach across multiple distinct image generators, including a mix of high and low-power machines performing the first and second passes of our approach.

Acknowledgments

The authors wish to thank Henry Fuchs, Adam Lake, and Lev Stesin in the STC research group. Kevin Arthur and Rui Bastos collected data using the KHMD, Andy Wilson from the Walkthrough research group benchmarked performance on the PIT system, and Todd Gaul and David Harrison assisted with the video. This work was partially supported by the National Science Foundation Cooperative Agreement no. ASC-8920219: "Science and Technology Center for Computer Graphics and Scientific Visualization," and by the National Tele-Immersion Initiative. The KHMD was developed by Kaiser Electro-Optics, Inc., under contract to DARPA ETO.

References

[Bennett98] David T. Bennett. Chairman and Co-Founder of Alternate Realities Corporation, 215 Southport Drive Suite 1300, Morrisville, NC 27560, USA. Cited 29 March 1998, available at http://www.virtual-reality.com.

[Cruz-Neira93] Carolina Cruz-Neira, Daniel J. Sandin, and Thomas A. DeFanti. 1993. "Surround-Screen Projection-Based Virtual Reality: The Design and Implementation of the CAVE," SIGGRAPH 93 Conference Proceedings, Annual Conference Series, ACM SIGGRAPH, Addison Wesley.

[Dorsey91] Julie O'B. Dorsey, François X. Sillion, Donald P. Greenberg. 1991. "Design and Simulation of Opera Lighting and Projection Effects," SIGGRAPH 91 Conference Proceedings, Annual Conference Series, Addison-Wesley, pp 41-50.

[Jarvis97] Kevin Jarvis, "Real Time 60Hz Distortion Correction on a Silicon Graphics IG," in Real Time Graphics, Vol. 5, No. 7, pp. 6-7, February 1997.

[Kaiser98] Kaiser Electro-Optics, Inc., "Full Immersion Head Mounted Display," cited 29 March 1998, located at http://www.keo.com/Product_Displays_ARPA.html.

[Max91] Nelson Max. 1991. "Computer animation of photosynthesis," Proceedings of the Second Eurographics Workshop on Animation and Simulation, Vienna, pp. 25-39.

[Raskar98a] Ramesh Raskar, Matt Cutts, Greg Welch, and Wolfgang Stürzlinger. 1998. "Efficient Image Generation for Multiprojector and Multisurface Displays," UNC Computer Science Technical Report TR98-016, University of North Carolina at Chapel Hill, March 1998, located at ftp://ftp.cs.unc.edu/pub/technical-reports/

[Raskar98b] Ramesh Raskar, Greg Welch, Matt Cutts, Adam Lake, Lev Stesin, and Henry Fuchs. 1998. "The Office of the Future: A Unified Approach to Image-Based Modeling and Spatially Immersive Displays," to appear in SIGGRAPH 98 Conference Proceedings, Annual Conference Series, Addison-Wesley, July 1998.

[Segal92] Mark Segal, Carl Korobkin, Rolf van Widenfelt, Jim Foran, and Paul E. Haeberli. 1992. "Fast shadows and lighting effects using texture mapping," SIGGRAPH 92 Conference Proceedings, Annual Conference Series, Addison Wesley, volume 26, pp. 249-252, July 1992.

[UnderKoffler97] John Underkoffler. "A View From the *Luminous Room*," Springer-Verlag London Ltd., Personal Technologies (1997) 1:49-59.

Editors' Note: see Appendix, p. 330 for colored figures of this paper

Per-Object Image Warping with Layered Impostors

Gernot Schaufler, GUP, Altenbergerstr. 69, A-4040 Linz, AUSTRIA
gs@gup.uni-linz.ac.at

Abstract. Image warping is desirable in the context of image-based rendering because it increases the set of viewpoints for which a single image can be used. This paper proposes a method for image warping with adaptive accuracy compatible with current texture-mapping hardware. It is based on the observation that pixels at similar depth move in a similar way during warping. The method also generates approximate depth values at each pixel so that polygonal and image-based rendering can be applied in a mixed fashion.

1 Introduction and Motivation

In interactive computer graphics hardware-accelerated rendering of polygonal models is a well-established visualization technique. However, the ever growing complexity of polygonal models continues to outperform the advances made in hardware technology.

In recent years an alternative approach called image-based rendering (IBR) was proposed concerned with generating new images from existing ones. It decouples scene complexity from rendering complexity by employing work proportional to the number of pixels in the final image instead of work proportional to the number of modeling primitives in the scene.

Research on IBR has focused on how to best use the available image data and has progressed to warping images augmented with depth information for new viewpoints (view interpolation and extrapolation). Different approaches for this warp have been taken and will be reviewed in the related work section. In particular surfaces pose problems which were undersampled or occluded in the available images but are visible in the image to be generated causing objectionable holes and tears in the final image.

Pure image-based representations require overwhelming amounts of image data to represent a complex scene for every arbitrary point of view. Hybrid rendering systems apply image-based methods mostly for distant scene elements to exploit the coherence in the images of those scene parts. Image data generated for recent frames is reused in the next frames to avoid the expensive re-rendering of the complete scene.

The approach to be presented uses hardware-accelerated texture mapping to warp the image data for new viewpoints and is based on the observation that portions of an image at a certain distance from the viewer move in a uniform way: the warp is approximated by drawing several layers of texture-mapped quadrilaterals. This is similar in spirit to using several impostors per object like in the hierarchical image cache [20][22]. However, this is prohibitive both in texture memory consumption and texture generation cost. The presented method avoids these problems by caching only one image per object and warping it for new viewpoints. The warping accuracy is adaptive to the desired change to the image. Little changes to the image are computationally less expensive.

After an overview of the relevant work in this field the following section introduces layered impostors and describes the ideas behind them. Implementation details are given in section 4 followed by possible optimizations in section 5. Results are summarized in section 6 and conclusions drawn in section 7.

2 Previous Work

A useful classification of IBR methods is by complexity of the warping technique. It is most straightforward to re-use whole images or parts thereof to save rendering costs or to achieve greater temporal fidelity [7][8][15].

The next complex class of image reuse is as (possibly partially transparent) texture maps which are mapped onto simple geometry. This coarse type of reuse avoids tears in the images resulting from previously occluded surfaces. Maciel et al [11] pre-generate textures for objects and clusters of objects to achieve a uniform frame-rate in their walkthrough system. Schaufler et al [20] and Shade et al [22] dynamically update the textures to always match the object's or scene's appearance from the current point of view. Aliaga et al [1] warp the geometry to minimize the popping effect when switching between textures and geometry. The Talisman architecture [24] approximates the 3D warp with a hardware implementation of affine transformations. Schaufler [21] augments textures with depth to avoid visibility errors caused by the "flatness" of impostors.

In an attempt to better approximate the required changes to the reused images triangular meshes have been generated from the images and their depth map. While Mark et al [12] have found pixel-grained meshes to be expensive, meshes solve the problem of holes in the images. However, they introduce the problem of "rubber sheets" in place of the holes. Subsampling [23] and mesh-decimation [5] give meshes of manageable size. Pulli et al [17] use a "soft" z-buffering approach to blend together the textured meshes obtained from several range-images.

Instead of triangle meshes individual pixels have been warped as well. Both forward and backward mappings have been tried. With forward mapping [2][6][9][14][16] holes and tears in undersampled areas are avoided by splatting.

Backward mapping produces no holes but such a mapping is only simple to calculate if the current point of view is the same as the viewpoint from which the input images were taken [3][18]. Otherwise a search must be performed in the source image to identify the pixel which maps to the current output pixel [10].

Only one object-centered approach to image-based rendering has been published so far, also based on McMillan's warping algorithm, namely delta trees [4]. They are very efficient in storage requirements as redundancies among a number of input images are avoided.

In several publications hardware implementations of IBR-methods are being called for and the need for research on applicable hardware architectures for IBR has been pointed out [12][13][24].

3 Layered Impostors

For generating an image of an object a possibly approximate representation of the object's shape is needed. A polygonal model is one such representation, usually considered view-independent (it can be used for any viewpoint, only in extreme close-ups the polygonal approximation becomes apparent). An image of an object mapped onto a polygon facing the viewer is another approximation of the object's shape, but a view-dependent one. Such a textured polygon (or impostor [11]) can only be used for points of view close to the point from where the image was taken.

Layered impostors are a generalization of dynamically generated impostors [19] in the following sense: an impostor replaces an object by one transparent polygon onto which the opaque image of the object is mapped. Thereby the depth of the object is discarded and the polygon's flatness becomes apparent when the viewpoint moves.

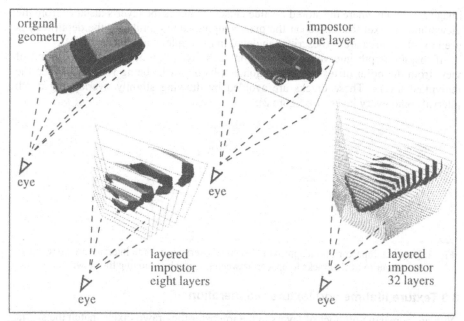

Fig. 1: Replacing an object (top left) with an impostor (top right) or a layered impostor (bottom)

A layered impostor consists of many such transparent polygons. On every polygon all drawn texels show those parts of the object's surface which are at a similar distance to the viewer as the polygon. In other words, every layer in the impostor depicts a slice of the object at some distance to the viewer (see figure 1). The more slices are used the better the approximation.

3.1 Texture generation and storage

The color information in the texture is generated in exactly the same way as for impostors: a tight viewing frustum is placed around the object and a view of the object is rendered from the current point of view. The object's surface can be textured as well. In this paper, however, object image and texture interchangeably refer to the image of the object which is mapped onto the layers. Texel refers to one pixel in such a texture. Final image refers to what is eventually displayed on the screen.

When the object's image is generated, instead of discarding the depth-buffer contents (the information describing the object's three-dimensional shape) the depth buffer contents are retained and stored together with the color information for each pixel. From this RGBz information a texture is defined in the format RGBα (with z in the α component) available on today's graphics workstations. (This is in contrast to McMillan's plenoptic modeling approach [16] for which disparity values must be derived from depth values).

3.2 Using the depth information in the texture

Instead of rendering one transparent polygon as with impostors several polygonal layers are rendered as a pyramidal stack having the point of view as it's apex (see figure 1). The object's image was generated from the current viewpoint and is mapped onto the layers as a partially transparent texture. In each layer only those areas of the image are drawn where the stored z-value closely matches the layer's distance from the viewpoint. A pixel test based on the pixel's alpha value (similar to the depth test) is used to select these areas (described in detail in the implementation section 4).

If disjoint depth intervals are drawn on each layer, any deviation of the point of view from the point of texture generation results in cracks being visible between the individual layers. These cracks are avoided by drawing slightly overlapping depth intervals onto every layer (see figure 2).

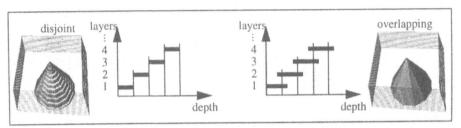

Fig. 2: Layered impostor for an approximated cone: overlaying depth intervals are assigned to each layer to avoid cracks to appear between layers when moving the viewpoint.

3.3 Texture lifetime and texture regeneration

With an increasing number of layers in the impostor the drawn pixels match the spatial location of the object's surface closer than the pixels on a single impostor polygon. As a result, in response to changes of the point of view the drawn pixels move in much the same way as the points on the surface of the object. Consequently, the layered impostor approximates the object's appearance over a much larger set of viewpoints than a single polygon. Compared to impostors the incurred error due to a mismatch of the spatial location of texels and object's surface points is halved by doubling the layers in the impostor.

The error introduced by impostors can be controlled by observing error angles [19]. Their derivation is repeated here briefly for comparison. When an impostor is used instead of an object, points on the object are drawn into the final image as texels on a textured polygon. When viewing the impostor from a point of view different from the one for which the texture was generated, a discrepancy between the point on the object's surface and the point on the texture will appear. The angle under which these two points are observed by the viewer can be calculated and used to limit the amount of error introduced. Whenever a given error threshold is exceeded (say the angle under which a pixel is observed on screen) the impostor's texture is regenerated.

A conservative estimation for the maximum error angle for all points on the object's surface can be calculated from a bounding volume. In figure 3 the error angles for extreme point's on the object's bounding box under orthogonal movements of the viewpoint are shaded in grey. These error angles are the result of the maximal distance of the object's points from the impostor. With layered impostors this distance is reduced with each additional layer. Error angles diminish accordingly as is evident when comparing figure 4 to figure 3.

3.4 Warping with variable accuracy

When drawing a layered impostor into the final image, the number of used layers can be selected. By not considering all the bits of the depth component in the texture, less

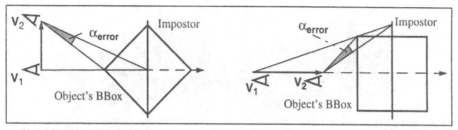

Fig. 3: Impostors: error angles on extreme points of the bbox for orthogonal motion cases.

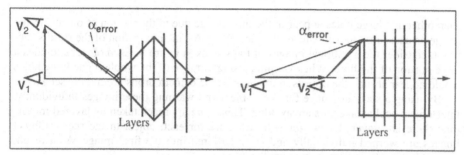

Fig. 4: Layered impostors: error angles on extreme points of the bbox for motion cases as above.

layers are obtained. As a result, the image warp is performed with less and less accuracy because texels from larger and larger depth intervals are warped in a uniform way (see figure 1).

Adapting the accuracy of the image warp is desirable because the warping accuracy required depends on the amount by which the viewpoint has moved from the point of view V_1 for which the texture was generated. As long as the viewpoint is coincident with the point of texture generation, one layer in the impostor is a sufficient approximation to the object's appearance.

The more the point of view moves away from the point of texture generation, the worse the approximation of the object's appearance by a single textured polygon. A warp of the object's image is required. However, for small deviations, a crude approximation to the image warp is sufficient. Layered impostors support different approximation accuracies by varying the number of layers used to draw the impostor into the final image.

3.5 Occlusion errors

When warping an image depicting a three-dimensional scene, occlusion errors can result. Consider the simple scene depicted on the top of figure 5. The cone, cube and cylinder hide portions of the wall behind them. When moving away from this point of view, the previously hidden areas of the wall become visible. When warping the image obtained from the initial point of view these areas are missing and holes in the final image will result (see figure 5 bottom). These holes are the bigger, the further the objects are separated in depth. Errors caused by objects occluding each other are called inter-object occlusion errors in this paper.

As the distance between two objects in a scene is potentially unbounded, inter-object occlusion errors are unbounded when warping a single image per scene and cause objectionable warping artifacts in the final image. The objects mutually occlude

Fig. 5: Inter-object occlusion errors: surface parts previously hidden by another object become visible. In one image for all the objects information on these surface parts is missing.

each other and large holes appear in the final image due to the motion of objects in the image as caused by movements of the viewpoint. As an anticipation of the results section 6 consider the additional images of the scenes in figure 8. Each object occludes a large portion of the floor. When warping a single image of the whole scene large holes would appear around the objects in the floor as the camera orbits around the scene.

By using a separate image for each object and warping these images individually, inter-object occlusion errors are avoided. This is the approach taken by layered impostors: there is a layered impostor with a texture for each object in the scene. Object images are warped individually and are z-buffered into the final image. As a result, inter-object occlusion errors are avoided.

Another class of errors remains, namely intra-object occlusion errors (see figure 6). Concave objects possibly occlude portions of their own surface, resulting in similar holes to appear in individually warped object images. In contrast to inter-object occlusion errors, the size of an object is bounded and so are the artifacts introduced by intra-object occlusion errors. The error is proportional to the maximum depth discontinuity in the image which - unfortunately - is not reported by the graphics hardware. Having it allows to bound both inter- and intra-object occlusion errors.

Fig. 6: Intra-object occlusion errors: a concave object hides parts of its surface. Even if one image is used per object, information on these surface parts is not available for warping.

4 Implementation

An implementation of layered impostors was done using the OpenGL graphics library which runs on a variety of platforms. As a prerequisite blending of RGB colors based on alpha values is disabled, so that every drawn pixel is always opaque.

The OpenGL standard provides an alpha test function to restrict the drawing of pixels to certain areas of a textured polygon. The alpha component stored in the texture is compared to a reference alpha value and pixels can selectively be drawn based on the outcome of the comparison. Drawing pixels with equal, smaller or greater alpha values than the reference value is possible.

For layered impostors depth values are stored in the alpha component of the texture. Ideally one would like to select pixels for a certain interval of depth values, namely the interval around the depth of each layer. As a test for depth intervals is not available in OpenGL, the implementation tests for equality of depth values and relies on the finite accuracy of depth values to actually result in an interval of depth values to be selected.

Scaling depth values by a factor smaller than one maps several depth values onto one (e.g. scaling by 0.5 removes one bit of accuracy and two values are mapped onto one). Assume four bits of accuracy in the following example. When the most significant bit of the depth value is set[1] and the scaling factor is increased slightly to the next representable number larger than 0.5 (with 4 bits 0.5 is represented as 1000_2, the next larger representable value is 1001_2 or 0.5625) the behavior shown in table 1 is achieved. Requiring the most significant bit to equal one insures that incrementing the scale factor actually influences the multiplication result.

Table 1: Fixed point value multiplication using 4 bits of accuracy. Values from 0.5 (1000_2) to 1.0 (1111_2) are scaled by 0.5 (middle column) and the next larger value 0.5625 (1001_2, right).

b	b * 1000_2	b * 1001_2
1000	0100	0100
1001	0100	0101*
1010	0101*	0101*
1011	0101*	0110
1100	0110	0110
1101	0110	0111*
1110	0111*	0111*
1111	0111*	1000

The regular pattern (highlighted by * in the above table) is used to select overlapping depth intervals. For successive layers depth values are multiplied alternately by 1000_2 or 1001_2 and the desired interval is selected by testing for equality with the multiplication result given in the table. This depth interval selection is depicted in figure 7. The same behavior is also obtained for more bits of accuracy and can be used to distinguish a larger number of overlapping depth intervals. This implementation has been tested on SGI Reality Engine, SGI O2 and SGI INDIGO graphics.

5 Optimizations

The layered impostor method provides a number of tunable parameters which can be used to optimize the drawing performance. As was already mentioned, the image warping accuracy can be adapted by varying the number of layers drawn per impostor. In the current implementation the number of layers is dynamically adapted to the error which must be compensated for. To achieve this, the scale factor is changed accordingly.

Another improvement of the basic algorithm is to provide a more uniform frame rate. As the number of textures to be regenerated varies from frame to frame, the frame rate varies as well. For walkthrough applications it has proven sufficient to update no more than one texture per frame. With a high frame rate this results in several texture

---

1.Requiring the highest bit of the depth values to be set means that the frustum surrounding the object must be made twice as deep with only the back half being occupied by the object.

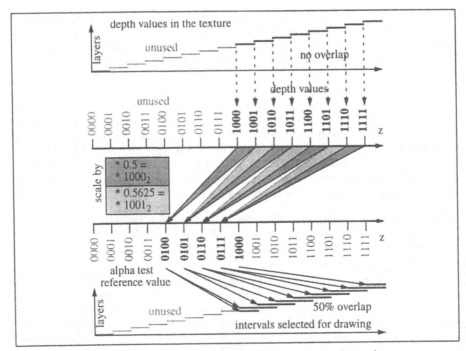

Fig. 7: Remapping depth values to obtain overlapping intervals.

updates per second. The warping accuracy of layered impostors is sufficient to hide the sparse updates of each texture even under object rotations.

While updating only one object image per frame guarantees that the frame-rate does not drop below a certain threshold, variations remain because in some frames no updates are necessary. These variations can be avoided with the saved time put to good use by updating exactly one object image per frame. This strategy has the advantage that any warping artifacts are kept to a minimum during slow movement of the point of view. When no object image would need to be regenerated under given error tolerance settings, the one with the highest error is updated. In particular intra-object occlusion errors are removed whenever possible.

6 Results and Discussion

The rendering performance of the layered impostor approach was tested on two scenes with different characteristics. The first one depicts a collection of cars where the complexity of the cars and other objects in the scene varies between 2816 and 21492 polygons. The second scene is made up from eighteen copies of one car model (the one with 21492 polygons).

Performance measurements were carried out on two machines, one from the low price range (an SGI O2 with 128MB memory and 175MHz R10000 CPU) and one high end graphics workstation (an SGI Onyx with RealityEngine2 graphics, 128MB memory and 195MHz R10000 CPU).

In the frame sequence measured the camera orbits in a circle around one of the scenes. This motion has been chosen because under rotation of the scene in front of the viewer the appearance of individual objects changes fastest. In all the previously pub-

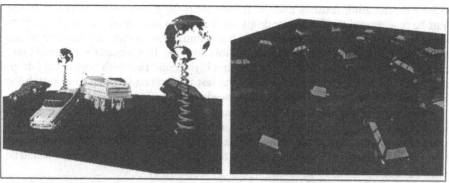

Fig. 8: Rendering with layered impostors: one texture update per frame removes most artifacts. **Left:** cars: 319847 vertices, 149488 polygons **Right:** jeeps: 1096096 vertices, 365368 polygons

lished image-based acceleration methods mentioned in the related work section this motion has been avoided. With layered impostors and their ability to warp object images this motion of the camera becomes feasible.

The window size was set to 768 by 576 pixels, the texture size was limited to 128 by 128 pixels which resulted in about one megabyte of texture memory being used in all cases. The maximal tolerable error angle was set to the angle under which one pixel is observed. As up to 32 layers are used per impostor this still results in an increase of the object image's lifetime by more than an order of magnitude in comparison to impostors.

For obtaining a more uniform frame rate with layered impostors the regeneration of exactly one object image per frame was forced by limiting the number of regenerations to one and using a very small error angle (much less than a pixel).

Average frame rates are given in table 2 for the two graphics platforms mentioned and the following three different rendering techniques:

- polygonal geometry only
- layered impostors with an error angle corresponding to viewing one pixel of the final image
- layered impostors with one texture update per frame (the one with the largest error angle)

Table 2: Average frame rates and frame rate variation for the different rendering methods.

rendering method	O2	RE2
Car scene		
polygonal geometry	1.0470 Hz $\sigma = 0.00413$	3.23215 Hz $\sigma = 0.0317553$
layered impostors	8.4143 Hz $\sigma = 1.69594$	47.1101 Hz $\sigma = 12.6673$
layered impostors (1 update/frame)	4.3317 Hz $\sigma = 0.995001$	20.8030 Hz $\sigma = 7.74516$
Jeep scene		
polygonal geometry	0.4853 Hz $\sigma = 0.00606$	1.42590 Hz $\sigma = 0.0154931$
layered impostors	3.9107 Hz $\sigma = 0.961839$	22.6625 Hz $\sigma = 9.22151$
layered impostors (1 update/frame)	2.4541 Hz $\sigma = 0.0433292$	11.2178 Hz $\sigma = 0.470177$

Rendering original geometry results in a very uniform frame rate because the same amount of geometry is visible in all frames. This rendering option is also slowest.

Layered impostors warp the object images and can be used for many frames even with the camera orbiting around the scene. As a result, most frames can be generated without regenerating textures and the frame rate is high. Whenever an object image must be regenerated the frame rate drops but usually no more than one image needs to be regenerated per frame. As described in the optimizations section a maximum of one update can be enforced without causing visible artifacts. It is a question of human factors whether it is a good thing to have such a high frame rate with occasional "dropouts". If a uniform frame rate is desired, the fast frames can be delayed to match the frame rate of those frames, when a texture needs regeneration.

Some variations in the frame rate is caused by the dynamic adaptation of the warping accuracy even if no object images need to be regenerated. These variations are mainly due to the fact that the current implementation only supports powers of two as the number of layers in an impostor. Finer control over the alpha test function would be required to remove this restriction.

Instead of delaying the fast frames it is better to regenerate one texture every frame as described in the optimizations section. Now the remaining frame rate variations are due to the varying complexity of the objects which need regeneration. Texture regeneration time dominates the whole rendering process due to the overhead of texture definitions in OpenGL.

In a scene where objects of similar complexity are replaced with layered impostors, a uniform frame rate is achieved (as in the jeep scene). Objects of uniform complexity can either be created when modeling the scene or by automatically clustering or partitioning too simple or too complex objects. Partitioning can be done arbitrarily as in the image caching approaches [20][22] as visibility is correctly resolved from the approximated depth values. In the current implementation objects of equal complexity must be created during scene modeling. If off-screen frame buffers were available, texture regeneration for complex objects could be distributed over several frames.

In general, the amount of experimentation possible with this OpenGL implementation of layered impostors is limited by the inflexibility of the alpha test functionality. Further evaluation of finer choices on the number of layers would be useful and different amounts of overlap among layers should be tried.

In addition more flexible rendering architectures have already been proposed from which layered impostors would benefit. For example in the talisman architecture [24] image layers are composited into the final image during video signal generation. This has the advantage that object-image regeneration and final image composition do not share one frame time but are pipelined over two frames.

As with the talisman architecture layered impostors allow to adapt the resolution on individual object images to the importance of the object. For example, distant background objects can be rendered into textures containing less texels than the pixels the object covers on screen. The talisman architecture offers high-quality filtering to decrease the so caused "pixelization" of object images. With OpenGL this is not possible because filtering RGBα textures would result in an interpolation of depth values of layered impostors. This interpolation disturbs the selection of depth intervals by testing for equality of finite accuracy depth values.

7 Conclusions and Future Work

This paper presented layered impostors, a novel object centered approach to render objects from depth augmented textures. Layered impostors unify image generation and warping into a single real-time rendering framework. They warp the images using the images' depth values to closely mimic the appearance of the depicted object from new points of view.

As with impostors only one texture is kept per object and reused as long as the errors introduced by the use of layered impostors remains below a given error threshold. The main advantage of layered impostors over dynamically generated impostors is the fact that they can be reused by an order of magnitude longer. Other advantages of layered impostors include:

- The drawing time for layered impostors is independent of the complexity of the represented object. Consequently, for sufficiently complex objects drawing time is saved by using layered impostors.
- As one layered impostor is used per object layered impostors work in dynamic scenes. Each layered impostor can be moved around individually.
- Layered impostors do not use more memory than impostors for storing the depth augmented texture. The usual eight bits of alpha are replaced by eight bits of depth.
- Approximate z-values are written into the depth buffer for correct visibility among layered impostors and objects rendered with polygonal primitives.
- The number of layers used in any layered impostor can be adapted to the required warping accuracy and the available fill-rate of the graphics hardware.
- Layered impostors cache the images of a scene on a per-object basis. Therefore, inter-object occlusion errors are avoided.
- Layered impostors are a hardware accelerated approach to 3D image warping.
- The warping quality is sufficient to accommodate object rotations.

Some disadvantages of layered impostors still need to be addressed:

- Large, highly overlapping polygons are used to draw the layers in the impostor. This results in high fill-rate requirements especially with many layers.
- Only one object image is warped. As a result, surface parts which were occluded when generating the image cause holes to appear in the warped image (intra-object occlusion errors). Always regenerating one texture per frame can partially compensate for this.

While the fillrate requirements may seem high polygonal scenes in real-time rendering typically cause a bottleneck in the vertex transformation stage of the rendering pipeline. Layered impostors provide a means to shift some of this work to the pixel processing stage and consequently better balance the work between the two stages.

One approach to deal with the fill rate requirements would be to detect and make use of the large amounts of transparent texels per layer. Intelligent pixel-fill hardware could detect these transparent areas and rapidly skip over them.

More flexible alpha test functions would allow further variations of the number of layers used and the amount of overlap between depth intervals.

Filling the holes in the warped image could be addressed by using more than one source image to calculate the warp. In particular, if a prediction of the user's motion is available, the new images could always be generated along the user's future trajectory. When the point of view for the image to be generated is within the convex hull of the viewpoints of available images, holes can be filled to a great extent.

References

[1] Aliaga, Daniel G., *"Visualization of Complex Models Using Dynamic Texture-based Simplification"*, IEEE Visualization '96, pp 101-106, Oct 28-Nov 1, 1996.

[2] Chen, Shenchang Eric and Lance Williams, *"View Interpolation for Image Synthesis"*, Computer Graphics (SIGGRAPH '93) 27 (August 1993) pp 279-288.

[3] Chen, Shenchang Eric, *"Quicktime VR - An Image-Based Approach to Virtual Environment Navigation"*, Computer Graphics (SIGGRAPH '95) (August 1995) pp 29-38.

156

[4] Dally, William J., Leonard McMillan, Gary Bishop, and Henry Fuchs, *"The Delta Tree: An Object-Centered Approach to Image-Based Rendering"*, MIT AI Lab Technical Memo 1604, May 1996.

[5] Darsa, Lucia, Bruno Costa and Amitabh Varshney, *"Navigating Static Environments Using Image-Space Simplification and Morphing"*, Proceedings of the Symposium on 3D Interactive Graphics, April 27 - 30, 1997, Providence, RI, pp 25 - 34.

[6] Debevec, Paul E., Camillo J. Taylor and Jitendra Malik, *"Modeling and Rendering Architecture from Photographs: A Hybrid Geometry- and Image-Based Approach"*, Computer Graphics (SIGGRAPH '96) (August 1996) pp 11-20.

[7] Dorsey, Julie, Jim Arvo, and Donald Greenberg, *"Interactive Design of Complex Time-Dependent Lighting"*, IEEE Computer Graphics and Applications. 15(2) (1995), 26-36.

[8] Duff, Tom, *"Compositing 3-D Rendered Images"*, Computer Graphics (SIGGRAPH '85) 19 3 (July 1985) pp 155-162.

[9] Gortler, Steven J., Li-wei He and Michael F. Cohen, "Rendering Layered Depth Images", Microsoft Technical Report MSTR-TR-97-09, March 1997.

[10] Laveau, Stéphane and Olivier Faugeras, *"3-D Scene Representation as a Collection of Images and Fundamental Matrices"*, INRIA Technical Report N°2205, February 1994.

[11] Maciel, Paulo W. and Peter Shirley, *"Visual Navigation of Large Environments Using Textured Clusters"*, Symposium on Interactive 3D Graphics (April 1995) pp 95-102.

[12] Mark, William R., Leonard McMillan and Gary Bishop, *"Post-Rendering 3D Warping"*, Proceedings of the 1997 Symposium on Interactive 3D Graphics (Providence, RI), April 27-30, 1997, pp 7-16.

[13] Mark, William R., Gary Bishop, *"Memory Access Patterns of Occlusion-Compatible 3D Image Warping"*, Proceedings of the 1997 SIGGRAPH / Eurographics Workshop on Graphics Hardware (Los Angeles, California), August 3-4 1997, pp 35-44.

[14] Max, Nelson, *"Hierarchical Rendering of Trees from Precomputed Multi-Layer Z-Buffers"*, Proceedings of the 7th Eurographics Workshop on Rendering '96, Porto, Portugal, pp165-174.

[15] Mazuryk, Thomasz and Michael Gervautz, *"Two-Step Prediction and Image Deflection for Exact Head Tracking in Virtual Environments"*, Computer Graphics Forum (EUROGRAPHICS '95) 14 3 pp 29-41.

[16] McMillan, Leonard and Gary Bishop, *"Plenoptic Modelling: An Image-Based Rendering System"*, Computer Graphics (SIGGRAPH '95), (August 1995) pp 39-46.

[17] Pulli, Kari, Michael Cohen, Tom Duchamp, Hugues Hoppe, Linda Shapiro, and Werner Stuetzle, *"View-based Rendering: Visualizing Real Objects from Scanned Range and Color Data"*, Proceedings of 8th Eurographics Workshop on Rendering, St. Etienne, France, June 1997, pp 23-34.

[18] Regan, Matthew and Ronald Post, *"Priority Rendering with a Virtual Reality Address Recalculation Pipeline"*, Computer Graphics (SIGGRAPH '94) (July 1994) pp 155-162.

[19] Schaufler, Gernot, *"Exploiting Frame to Frame Coherence in a Virtual Reality System"*, VRAIS '96, Santa Cruz, California (April 1996) pp 95-102.

[20] Schaufler, Gernot and Wolfgang Stürzlinger, *"A Three-Dimensional Image Cache for Virtual Reality"*, EUROGRAPHICS '96, (August 1996), 15 3, pp 227-236.

[21] Schaufler, Gernot, *"Nailboards: A Rendering Primitive for Image Caching in Dynamic Scenes"*, Proceedings of the 8th Eurographics Workshop on Rendering '97, St. Etienne, France, June 16-18, 1997, pp 151-162.

[22] Shade, Jonathan, Dani Lischinski, David H. Salesin, Tony DeRose and John Snyder, *"Hierarchical Image Caching for Accelerated Walkthroughs of Complex Environments"*, Computer Graphics (SIGGRAPH '96), (August 1996) pp 75-82.

[23] Sillion, François X., George Drettakis and Benoit Bodelet, *"Efficient Impostor Manipulation for Real-Time Visualization of Urban Scenery"*, Proceedings of Eurographics'97, September 4-8, 1997, pp 207-218.

[24] Torborg, Jay and James T. Kajiya, *"Talisman: Commodity Real-time 3D Graphics for the PC"*, Computer Graphics (SIGGRAPH '96), (August 1996) pp 353-363.

Interactive Volumetric Textures

Alexandre Meyer and Fabrice Neyret

iMAGIS*, laboratoire GRAVIR/IMAG-INRIA
{Alexandre.Meyer|Fabrice.Neyret}@imag.fr
http://w3imagis.imag.fr/Membres/Fabrice.Neyret/index.gb.html

Abstract:
This paper presents a method for interactively rendering complex repetitive scenes such as landscapes, fur, organic tissues, etc. It is an adaptation to Z-buffer of *volumetric textures*, a ray-traced method, in order to use the power of existing graphics hardware. Our approach consists in slicing a piece of 3D geometry (one repetitive detail of the complex data) into a series of thin layers. A layer is a rectangle containing the shaded geometry that falls in that slice. These layers are used as transparent textures, that are mapped onto the underlying surface (e.g. a hill or an animal skin) with an extrusion offset. We show some results obtained with our first implementation, such as a scene of 13 millions of virtual polygons animated at 2.5 frames per second on a SGI O₂.

1 Introduction

Visual complexity is part of the realism of a scene, especially for natural scenes like landscapes, fur, organic tissues, etc. When represented explicitly with facets, these complex -and often repetitive- details lead to very high rendering time and aliasing artifacts. In some cases these details are flat enough to be represented with flat textures. However in many case they are really three-dimensional, i.e. showing view-dependent appearance and parallax motion (e.g. trees on a hill). Moreover, mesh decimation algorithms are of no help on such complex objects. The situation is even worse in the scope of interactive rendering, where only very low complexity scenes can usually be dealt with in the available time.

The fact that a detail is not flat does not imply it has to be represented by a comprehensive - and costly - 3D representation such as a mesh. Indeed, the 3D impression is a progressive notion: it includes several properties, such as view-dependent contour, view-dependent apparent location, parallax motion, occlusion, shadowing, diffuse reflection and highlights, etc. Depending on the size of the object (or the detail) on the screen, some of these properties can be sufficient to convey a 3D impression. A means to do efficient rendering with few aliasing artifacts is thus to use a representation that refers to the minimum amount of information that is sufficient to reproduce what can be seen.

1.1 Related work

Volumetric textures, introduced in 1989 by Kajiya and Kay [4] and extended by us [8, 9], consider three different embedded scales to represent the information (see figure 1):
- large shape variations such as the surface of a hill or an animal skin are encoded using a regular surface mesh,
- the medium scale such as grass or skin, which is concentrated in the neighborhood of this surface, is encoded using a *reference volume* stored once and mapped several times in the spirit of textures (instances are named *texels*),

---

* *iMAGIS* is a joint research project of CNRS/INRIA/UJF/INPG.
Postal address: BP 53, F-38041 Grenoble cedex 09, France.

158

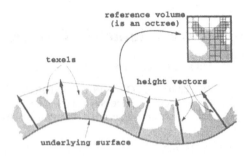

Fig. 1. Volumetric texture specification (cross-section).

the small scale, consisting of the microscopic shape of individual objects, is encoded by a reflection model stored in each voxel. In the multiscale extension of volumetric textures, this scale also corresponds to the pixel size.

To relate this to the progressive 3D impression mentioned before, one can see that explicit geometry is used only to specify the largest scale; volume data is sufficient[1] to reproduce occlusions and parallax effects at middle scale (i.e. a few pixels), while the illumination model stored in the voxels simulates the geometry below pixel size.

Texel rendering has some similarities with volume rendering, a previously costly ray-tracing method family. In 1994 Lacroute and Levoy introduced a new approach [5] adapted to graphics hardware, which makes volume rendering interactive. This approach consists of factoring the voxels by considering slices of the volume, that can be encoded by textured transparent faces. Volume rendering thus consists of superimposing these transparent slices. Since common 3D graphics hardware can deal with textures and transparency at no extra cost, rendering cost is now only proportional to the number of slices.

1.2 Overview

This paper presents a method for interactive rendering of complex repetitive geometry. The idea is to adapt the volumetric textures presented above [8, 9] to Z-buffer graphics hardware, using the same sort of approach that was used for volume rendering by Lacroute and Levoy [5]. We thus expect to obtain the same kind of complex scene as the first, with the same kind of interactiveness as the second.

Contrary to volume rendering the size of the volumes used in our method is small. We can thus render a relatively large number of such volumes interactively. For instance, a volume encoded with 64 slices can be rendered with a cost of 64 quadrilateral faces (i.e. 64 × 2 triangles), while the represented shape is built from a model that might have at least a thousand faces (and often ten or a hundred times more). Moreover, this cost is independent of the slice resolution.

On the other hand, using graphics hardware brings some limitations: with hardware rendering one cannot compute shading nor shadows for each individual texture pixel (i.e. for each volume cell), while ray-tracing can do this. Only color is stored in the volume, so the shading and shadows - if any - have to be captured inside the pattern at the creation stage, and will not be updated according to the main surface orientation and light position.

---

[1] Because of the concentration of the data complexity within the surface neighborhood, and the small volume resolution necessary to provide 3D location effect, in our context a volume is an efficient and compact way to store and render data.

The remainder of this paper is structured as follows. In section 2, we deal with the basic representation and rendering of interactive volumetric textures, that will be extended in section 4. In section 3 we describe how to encode the shape of a detail into a texel, in particular by converting existing representations. Animation approaches available for the ray-tracing method [7] are still usable for ours. We review these approaches in section 5. We discuss the results in section 6.

2 Basic representation and rendering

In this section, we present our method to encode a complex object made of repetitive details lying on a surface, and explain how to render the representation obtained. The modeling of the content is the object of the next section.

2.1 Data structure

In the same way as ray-traced volumetric textures [8], the specification of an object consists of a triangular mesh with (u, v) texture coordinates and a height vector at the vertices, plus a volumetric texture pattern. The height vectors control the direction and thickness of the third dimension of the texture (see figure 1).

The volumetric part of the model is different to the one used for the ray-traced version: it consists of a set of RGBA textures, representing infinitely thin horizontal slices of the volume. Empty parts have $A = 0$ (i.e. the slice is transparent there), and opaque parts have $A = 255$.

Fig. 2. A texel is drawn using extruded textured triangles.

2.2 Rendering

The rendering is done using a standard hardware-accelerated 3D graphics library (OpenGL [6], in our implementation), by drawing textured extruded facets above each geometric facet of a "volumetrically textured" surface. The three vertices of an extruded facet (corresponding to a slice) are obtained by linearly interpolating the position along the three height vectors at the three vertices of the surface facet (as illustrated in figure 2). Hardware MIP-mapping [13] can be used to deal with aliasing at grazing view angles and distant location. Notice that texture, transparency and MIP-mapping come at no extra rendering cost[2] on various 3D graphics cards.

It is known that transparency does not work well with a Z-buffer; correct transparency would require storing several Z, alpha and color values per pixel. As long as the alpha value A is 0 or 255, this is not a problem: transparent texture pixels are not drawn, and opaque texture pixels hide what is behind them. However a problem occurs when semi-transparent texture pixels exist, either because the content is smoothed or the MIP-mapping feature is on. To deal correctly with this problem, one has to draw the slices from back to front. This is easy to achieve within a single texel, but this would also require to sort the faces with Z, which is costly. Thus, we do not allow semi-transparent data in our implementation. However we do draw the slices from back to front, since this avoid the artifacts that may occur due to the lack of resolution in Z between slices.

---

[2] To a certain extent, beyond which hardware bottlenecks occur.

To choose the drawing order, it is sufficient to test the dot product of the normal to the surface facet and the view direction, assuming that the texel is not too distorted by the height vectors.

Each volume location is treated by the Z-buffer as a regular pixel fragment (i.e. it has its own Z-value), thus the intersection of two texels is dealt with correctly, which was not the case in the ray-tracing version. This important property is illustrated in figure 10(right) in the results section 6.

3 Modeling the pattern

In this section we describe how to encode in a texel one repetitive detail of a complex object. This detail is created using an existing modeling tool. However, using a textural approach to repeat the detail brings some constraints to the pattern shape: the 3D texture pattern, i.e. the reference volume of the volumetric texture, corresponds to the cubic box between $(u, v) = (0, 0)$ and $(u, v) = (1, 1)$ in the texture space. The mapping will

Fig. 3. *Left*: Pattern with torus topology. *Right*: Isolated pattern.

generate the (virtual) copies of the detail according to the (u, v) texture coordinates at the vertices. For the result of the mapping to appear continuous, the cubic pattern content has either to obey torus topology, or to consist of a disconnected shape that does not reach the borders of the reference volume, as shown in figure 3.

Once an appropriate shape has been chosen or modeled for the detail, it has to be encoded in the volume (i.e. the set of texture slices) in such a way that the texel reproduces the same visual effect. This leads to several stages in the encoding:
- slicing the 3D description,
- evaluating the shading at each location,
- filling the inside of the shape.

The last issue is a key point: if the description is a surface, and not a solid, each slice of it is a contour (see figure 4(left)), so gaps would appear between slices when the view direction is not orthogonal to the surface. Thus the shape has to be solid, or to be turned into solid if the description is a surface[3]. Some inside slice pixels are visible between two contours[4] as shown in figure 4(middle). We need to propagate the surface color toward the inside, in such a way that the image appears as continuous as possible.

This approach is not sufficient for grazing view angles, because the gap between slices appears, as illustrated in figure 4(right). This problem is solved in section 4.

Fig. 4. *Left*: Contours. *Middle*: Filled contours. *Right*: Gaps appearing between slices at grazing view angles.

---

[3] Of course, this does not concern shapes made of sparse polygons, e.g. foliage.

[4] Here, we only deal with opaque solids.

3.1 Slicing and shading the pattern shape

In the case of a standard surface description (e.g. an OpenInventor database), one can use a standard renderer (e.g. OpenGL) to do the slicing and shading at the same time. The view point is set at the top of the 3D pattern, a bounding box is defined by the user, then the front and back clipping planes are successively set around each slice (as illustrated in figure 5). Each resulting RGBA image is stored as a texture slice (including the alpha value, which is crucial), and the slices set is stored on disk.

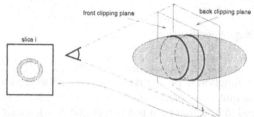

Fig. 5. Slices construction using a regular rendering tool.

A ray-tracer may be used as well for the rendering of the slices, which would allow for shadows. However, as for the shading, one has to keep in mind that the considered light directions will be fixed at this construction stage.

3.2 Filling the inside

For surface descriptions, the slices are empty contours, that need to be filled. Worse, these contours are incomplete. Thus the filling comprises three stages:
- closing the contours,
- marking the inside (i.e. where to propagate the color),
- performing the color propagation within a slice.

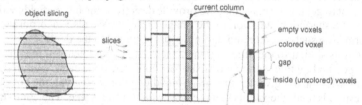

Fig. 6. Cross-section of the slices. The contours are bold. Note that there is sometimes a gap of several slices. This means that the intermediate slices have unclosed contours at this location.

The contours are generally not closed because the drawing of polygons viewed at a grazing angle (typically on the shape silhouette) generates consecutive pixels that can fall in very distant slices, i.e. their step in z is greater than the z interval between slices. This results in the contour not being drawn on the intermediate slices (see figure 6). To cope with this, we have to close the discretized shape surface by filling the gaps: let us call a *column* the set of pixels at a given location (x, y) through the set of slices. We now consider the volume as a set of columns. The segments in a column that fall inside the shape are made of uncolored pixels marked 'inside' bounded by two colored pixels, as shown in figure 6(right). A gap in the surface occurs when some of the uncolored column pixels directly neighbor the outside, because one or more of the four neighbor columns are empty at that position. Call *exposed* (i.e. to the outside) this part of the column. The problem of closing the contours is thus equivalent to filling these exposed voxels. Several algorithms are available for closing and the shape filling [10]. We have implemented a basic algorithm for our tests, which we present in appendix. Note that the contour completion process has also to get color values for the contour pixels added, to be obtained by interpolating the surrounding colors.

Once the inside is marked and the surface is closed (i.e. there are closed contours in every slice), we fill the contours by iteratively propagating the colors. If a pixel is marked as inside and uncolored, and at least one pixel in its neighborhood is marked as colored, then the pixel is painted. The used color is the average of the colored pixels of the neighborhood (it is drawn in a separate buffer to avoid bias due to the order of scanning).

3.3 Other kinds of shape specification

Implicit surfaces

Implicit surfaces [1] are easier, because they directly specify solid objects: a pixel is inside if the implicit function is greater than one, and the shading is obtained using the gradient of the function as a normal. Since the construction of the texture pattern does not especially need to be efficient (it is a precomputation), we simply evaluate the implicit function at each volume location along the slices, and thus no filling is required. Hypertextures [12] can be handled exactly the same way.

Height fields

Height fields are a popular way of specifying the details of a surface. We consider a 2D grey-level image with $[0, 255]$ range values. Each image (x, y) location corresponds to a column in the volume: we set the pixel of the slice that fits the z value encoded by the intensity of the image at that location. Thus, the slicing stage is trivial. The normals can be computed from the height gradient in the neighborhood and used with the Phong shading to get a color.

We have used Perlin noise [11] to produce grey-level images to be used as height fields (see figure 12 in section 6). Since this function is continuous, we prefer to directly use its gradient to get the normals, thus we provide directly to the height fields voxelizor an image of the shading in addition to the image of depth. We also incorporate in this image information such as color and darkness (proportional to the depth).

The filling is similar to the process done in section 3.2 (and detailed in appendix 7) for shapes specified by their surface. It is simpler however, since each column has only one segment, with the top given by the image and the bottom at the volume bottom. The inside corresponds to all the voxels that are below the given z value. The contours have to be closed as explained in section 3.2, with fewer special cases. The final color filling of the slices is exactly the same.

4 Dealing with grazing view angles

When coping with grazing view angles, typically on the silhouette of the underlying surface on which the texels are mapped, horizontal slices are no longer satisfactory because the gap between slices appears, as illustrated in figure 4(right). In this section, we introduce quality criteria to decide if the appearance of a texel is correct or not, and we additionally store alternate directions of slice sets to get a correct appearance for any view direction. This also provides some hints for optimizations.

4.1 Quality criteria

When the view direction has an large angle from the vertical (of a texel), one can may be able to see through the sliced shape, because the projections on screen of two consecutive slices are not superimposed (see figure 4(right)). This depends on the angle a, on the length h between slices, and on the (horizontal) thickness of the filled contours in the slices (call e the narrowest slice), as represented in figure 7(left). This provides a first quality criterion, $h\tan(a)/e \leq 1$, that indicates when such an artifact does not occur. This criterion can be used either to choose the number of slices to use to represent correctly a detail up to a given limit view angle, or to switch to an alternate slicing direction as describe in subsection 4.2.

However, as stated in section 3, the image may look degraded even for smaller view angles, because when the view direction is not orthogonal to slices one can see some pixels that are inside the contours. At some point an inside color differs from that of the surface. This provides for another quality criterion: if the user does not tolerate that the inside can be seen "deeper" than a constant d (see figure 7), the criterion is $h\tan(a)/d \leq 1$. A maximum value for d is $e/2$: since the shading of two opposite sides often has the opposite contrast, at some point (when closer to the other side of the contour) the deepest visible inside pixel color is closer to the opposite contrast than to the color of the border whose is should appear continuous (see figure 7(right)). The smallest contour thickness being e, this gives the limit of penetration $e/2$. Once again, such a criterion helps in choosing a correct number of slices. E.g. if one wants to allow view angles up to $a = \pi/3$ and the narrowest horizontal part of the shape is $e = 3$ texture pixels wide, then the distance between slices should be less than 0.9 times the length of a texture pixel.

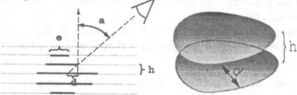

Fig. 7. *Left*: Slicing characteristics. *Right*: Limit for the second criterion: half of the inside of the narrowest slice is visible.

4.2 Alternate slice directions

To deal with grazing view angles (relative to the main slice set, called 'horizontal', i.e. parallel to the underlying surface), we also store the same volume as a set of slices in the two vertical slice directions (see figure 8).

Fig. 8. The three slicing directions.

At rendering time, the dot product c_i of the three directions and the view direction indicates which of the three sets to use (a classical solution when one has to draw objects organized along a 3D grid or an octree [3]). As suggested in the previous subsection, one should choose the slices direction that has the smaller quality criterion value. The tangent of the view angle is $\sqrt{1 - c_i^2}/c_i$, so that the direction to use is the i for which $(1/c_i^2 - 1) * h_i^2$ is minimum, with h_i the slice density for direction i, i.e. the size L_i of one texel in this direction divided by the number N_i of slices in this direction[5]. Thus, a

---

[5] However in our early implementation, we simply choose the direction for which the absolute value of c_i is maximal.

volume can be visualized correctly from any direction.[6]

4.3 Optimizations

Rendering a complex scene modeled with volumetric textures finally consists in drawing several thousands of textured polygons. There are two aspects in the rendering cost:
- the number of textured polygons that are drawn,
- the efficiency of the rendering process.

The number of polygons
There are two ways of decreasing the number of polygons to render: not drawing invisible texels (or slices), and using the minimum slice density for each texel.

The first issue is not easy to solve, since a texel lying on a back face can have some parts that are visible, so that culling is not trivial in general. A possible improvement would be to first draw (and temporarily) the bounding box of the texel, to check if at least one screen pixel was affected (e.g. using the stencil planes), and to proceed the rendering of this texel only in this case (quite like in [14] for visibility culling).

The second issue can be dealt with by deriving the minimum number of slices N_i to get a correct image from the quality criterion: $h_i \tan(a)/d \leq 1$, so $N_i \geq L_i \tan(a)/d$. This provides at the same time the direction and the number of slices to get a correct result with the lowest cost.

Another criterion can be used to decrease the number of slices with the distance: if one wants that the apparent distance between slices be less than p pixels on screen ($p \leq 1$), the criterion is $\frac{h \sin(a)}{z/f} \leq p$.

To avoid aliasing, we proceed quite similarly to MIP-mapping [13]: the number of slices in each set is a power of two, and we precompute several sets of slices. The criteria provide an optimal number of slices, which we round up to a power of two, that gives the set number to use. (None of these optimizations were used when running the tests presented in the result section.)

The efficiency of rendering
The effective rendering cost is strongly linked to the fact that the various graphics system bottlenecks can be avoided. In our case, a crucial one is the saturation of the texture cache. To minimize the potential texture cache faults, we use an alternate texel rendering method, that first draws all the occurrences of a given texel slice (in order to satisfy the back-to-front drawing requirement, the slices of front facing and back facing texels are drawn separately). Notice that this is valid only as long as there is no semi-transparent data, which would need to draw the back texels before the front texels.

5 Animating volumetric textures

Three ways of animating volumetric textures are mentioned in [7], that also correspond to three scales (illustrated on figure 9):
- deforming the underlying surface (e.g. for a flag or the skin of an animal),
- deforming the texture mapping, particularly the height vectors orientation (e.g. to simulate the wind on grass or fur),
- using several cycling volume contents along time, as for cartoons (e.g. for local oscillations).

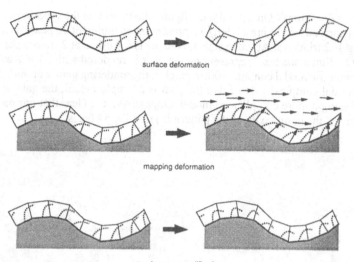

surface deformation

mapping deformation

texel content modification

Fig. 9. The three modes of animation, that also correspond to three scales.

These methods are still usable with our interactive rendering. Surface vertices or height vectors modifications need to recompute few items at each frame, which can be done using physical models (see [7]). Time constraints are the same as for any near real-time animation of simple surfaces. Cycling a volume set has some consequences on memory if different volumes of the set are visible in the same frame. Notice that the texture memory on SGI O_2 is the same as the main memory, so that this is not really a limitation on the machine we use for our tests. However this can be a problem on other platforms, thus the drawing of the instances of a given pattern has to be grouped together, so that the texture cache changes only once per kind of pattern.

6 Results

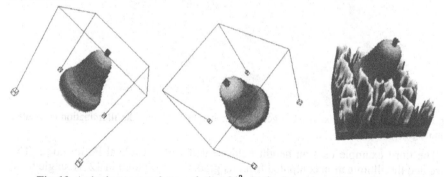

Fig. 10. A single pear texel at resolution 64^3. *Right*: Surperimposition of two texels.

The first example is an Inventor database of a pear, having about 1000 faces. The figure 10 shows a single texel at resolution $64 \times 64 \times 64$ from various viewpoints (using

---

[6] Notice that SGI has 3D texture facilities that also consist of a color volume, which can be indexed more easily (a single set of slices is sufficient). We have chosen not to use this feature because it is not available on most graphics hardware (in contrast to textures and Z-buffer), and because we want to control the texture memory usage to avoid swapping.

different slices directions). On the right, the figure illustrates that the superimposition of texels works correctly. In figure 11(left) we present the mapping of 96 pears on a sphere mesh having 192 triangles. At video size, this scene is refreshed at 2 frames per second on an SGI O_2. Since one texel representing the pear is rendered with 64×2 triangles, while the geometric model contains 1000 triangles, the rendering gain is about 7.5 times with equal visual complexity. Note that the pear is a simple model; the gain would be more when using a more complicated model. Oppositely, it is clear that our method is not interesting if the complexity of the pattern is less than 64 faces.

Fig. 11. Left: Mapping of 96 pear texels on a sphere. **Right:** Mapping of 16 bushes.

The second example is based on an AMAP [2] generated bush of 3,500 triangles. Since the data consists in sparse triangles, no filling is done. The texels have a $256 \times 256 \times 64$ resolution. The rendering of 16 instances shown in figure 11(right) is done at 6 frames per second. Note that because the number of instances is tuned at the mapping level, the cost would be the same even with many more bush instances.

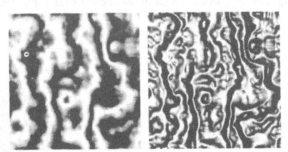

Fig. 12. Cyclical Perlin noise used to generate the height field, and the illumination computed from its gradient.

The third example uses an height field created with a cyclical Perlin noise. The noise and the illumination computed from its gradient are figured in 12. A single texel at resolution $256 \times 256 \times 64$ (i.e. with 64 slices) is shown on figure 13, with various deformations obtained (in real-time) by modifying the height vectors. Such an height field should be geometrically represented with $256 \times 256 \times 2 = 131,072$ triangles, while using this texel it is rendered with 64×2 triangles, with equivalent visual complexity. Here, the gain in polygon drawing is about 1000 times. Note that some artifacts occur on the top left of the deformed texel, where the quality criterion is not satisfied (by evaluating the criteria at the facet center, one assumes the deformation is small).

Image 15 (see color section) represents the mapping of 96 texels on a sphere mesh

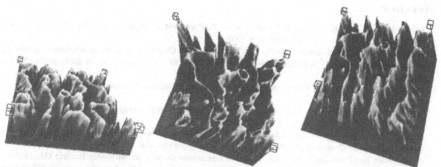

Fig. 13. *Left*: A texel at resolution $256 \times 256 \times 64$ created from the height field. *Middle and right*: Deformation of the texel by modifying the height vectors.

of 192 triangles. The frame rate is about 1.3 frames per second on an SGI O$_2$. The rendering of this scene could be optimized a lot, as suggested in section 4.3: no back-face culling is performed. Moreover, half of the texels appear on the sphere silhouette, thus 10% of the image represents 50% of the drawing, and 66% of the cost (because vertical slicing that is more dense due to the texture resolution is used near the silhouette). This is a waste, because on the silhouette keeping a lot of slices is useless considering the criteria seen in section 4.1. We thus expect to multiply the frame rate by about 5 by doing these optimizations. The figure 16 (see color section) illustrates the mapping of 100 texels on a jittered plane at 2.5 frames per second. This last scene has a visual complexity of 13 million of triangles.

7 Conclusion

We have presented a way to considerably increase the visual complexity of scenes displayed in the scope of interactive rendering, by adapting the ray-traced volumetric texture method [8, 9] to the graphics hardware features typically available on today's 3D graphics cards. Each texel mapped on a surface is rendered by drawing a set of extruded faces covered with transparent textures. The extrusion is controlled by height vectors located at the vertices (which can be animated). We propose several ways to build the texture content from various 3D descriptions (meshes, implicit surfaces, height fields).

Compared to the ray-tracing version, the rendering quality is of course lower (shadows and illumination are fixed). But compared to the low complexity of the scenes usually displayable at an interactive rate, our method brings a large improvement as shown by our results: the apparent complexity can be of 13 million polygons. The realism induced by the amount of visible details was previously totally unavailable for virtual reality applications. Among possible applications, we aim at introducing these apparent details in a surgery simulator we are working on. The main organ surfaces are reconstructed from scanner data, and only these surfaces are taken into account in the physical simulation of deformations. The 3D details added by the volumetric texture simply enrich the image by "dressing" these surfaces.

As future work, we want first to improve the frame rate by implementing the optimizations mentioned in section 4.3, and optimizing the OpenGL code, in order to get closer to real-time. We are currently working on the algorithm which uses an adaptive number of slices. Another issue is the development of a less naive filling algorithm to deal with more complicated patterns. We are also investigating ways of generating local illumination on the fly, possibly in the spirit of bump-mapping textures using high-end graphics capabilities.

168

References

1. J. Bloomenthal, C. Bajaj, J. Blinn, M.P. Cani-Gascuel, A. Rockwood, B. Wyvill, and G. Wyvill. *Introduction to Implicit Surfaces*. Morgan Kaufmann Publishers, 1997.
2. Phillippe de Reffye, Claude Edelin, Jean Françon, Marc Jaeger, and Claude Puech. Plant models faithful to botanical structure and development. In *Computer Graphics (SIGGRAPH '88 Proceedings)*, volume 22(4), pages 151–158, August 1988.
3. J. D. Foley, A. van Dam, S. K. Feiner, and J. F. Hughes. *Computer Graphics: Principles and Practices (2nd Edition)*. Addison Wesley, 1990.
4. James T. Kajiya and Timothy L. Kay. Rendering fur with three dimensional textures. In *Computer Graphics (SIGGRAPH '89 Proceedings)*, volume 23(3), pages 271–280, July 1989.
5. Philippe Lacroute and Marc Levoy. Fast volume rendering using a shear–warp factorization of the viewing transformation. In *Computer Graphics (SIGGRAPH '94 Proceedings)*, pages 451–458, July 1994.
6. Jackie Neider, Tom Davis, and Mason Woo. *OpenGL Programming Guide*. Addison-Wesley, Reading MA, 1993.
7. Fabrice Neyret. Animated texels. In *Eurographics Workshop on Animation and Simulation'95*, pages 97–103, September 1995.
8. Fabrice Neyret. Synthesizing verdant landscapes using volumetric textures. In *Eurographics Workshop on Rendering'96*, pages 215–224, June 1996.
9. Fabrice Neyret. Modeling animating and rendering complex scenes using volumetric textures. *IEEE Transactions on Visualization and Computer Graphics*, 4(1), January–March 1998. ISSN 1077-2626.
10. Theo Pavlidis. *Algorithms for Graphics and Image Processing*, pages 167–193. Springer-Verlag, 1982.
11. Ken Perlin. An image synthesizer. In *Computer Graphics (SIGGRAPH '85 Proceedings)*, volume 19(3), pages 287–296, July 1985.
12. Ken Perlin and Eric M. Hoffert. Hypertexture. In *Computer Graphics (SIGGRAPH '89 Proceedings)*, volume 23(3), pages 253–262, July 1989.
13. Lance Williams. Pyramidal parametrics. In *Computer Graphics (SIGGRAPH '83 Proceedings)*, volume 17(3), pages 1–11, July 1983.
14. Hansong Zhang, Dinesh Manocha, Thomas Hudson, and Kenneth E. Hoff III. Visibility culling using hierarchical occlusion maps. In *Computer Graphics (SIGGRAPH' 97 Proceedings)*, pages 77–88, August 1997.

Appendix: Filling the shapes

Marking the inside of the shape

This stage prepares the color propagation stage (and can also help the contour closing stage), by indicating where to propagate. It is thus a regular filling problem. We simply consider the parity of surface crossing between the current location and the top: for each (x, y) horizontal location, we traverse the 'volume' along z (i.e. the successive slices) from top to bottom, assuming that the top is outside the shape, and we flip a flag each time an opacity transition is found at a voxel (i.e. a texture pixel of a slice). Thus the inside area of each slice is marked. This method was easy to implement for our tests, but is known to fail for complicated shapes. A better filling method like non-recursive connectivity filling [10] should better be used in general.

Closing the contours

We have seen in 3.2 that this is equivalent to filling the exposed part of a column. We interpolate the color in the intervals defined by the top of the current column segment and the top of the neighbor columns segments that start below it (and we do the same for the bottom). We compute this interpolation for each direction in which the column is exposed to the outside (i.e. up to four), and we store the mean of these, thus coloring the missed contour pixels (see figure 14(left)). If the surface is vertical, the column has no neighbor in one direction. Then we interpolate the color from the top to the bottom of the segment (see figure 14(right)).

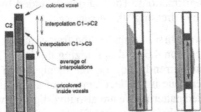

Fig. 14. Filling of the exposed part of the column. *Right*: Segment with one or two colored ends.

Editors' Note: see Appendix, p. 331 for colored figures of this paper

Efficiently Rendering Macro Geometric Surface Structures with Bi-Directional Texture Functions

Jean-Michel DISCHLER
Laboratoire MSI - Université de Limoges
I.U.T. Allée André Maurois
87065 Limoges Cedex (France)

Abstract. *Fast and realistic rendering of textures, characterised by a pronounced geometry, such as for example wickerwork, rattan or beach pebbles, is still a difficult and challenging problem in computer graphics. These effects can neither be simulated with 2D texture mapping nor with bump mapping. Direct geometric models or "volumetric" texturing approaches often become inevitable. All of them, however, imply excessive computational requirements. In this paper, the concept of bi-directional texture function (BTF) is introduced. A BTF permits a realistic simulation of the above mentioned effects at very low computational cost. Principles generally applied at microscale in the case of BRDFs are transposed at larger geometric scale. This allows us a direct generalisation of texture mapping and leads to many visual simulations beyond the possibilities of usual "fast" texturing techniques.*

Keywords: bi-directional texture function (BTF), bump mapping, 2D texture mapping, view dependent texture mapping, BRDF, rendering, realism.

1. Introduction

Because of the huge amount of different textures in nature, this domain has been topic of various researches in the field of image synthesis during the past two decades. Many techniques, such as 2D texture mapping, bump mapping [1], solid (3D) texturing [2-4], texel mapping [5], hypertexturing [6], view dependent texture mapping [7], etc. have been developed. Nevertheless, most complex coarse (pixel- or macro-) scale geometric surface structures still remain difficult to be rendered efficiently at low computational cost. Figure 1 illustrates some examples representing digitised photos of natural textures such as, from left to right and top to down, cordage, ground, a woven aluminium wire, a net, pebbles, a raffia weave, wickerwork and rattan (some of these textures were extracted from [8]).

Techniques reduced to a single colour modulation on the surface are not suited, even if view dependence is taken into account. The problem is due to the loss of some crucial parts of the underlying geometric information. Blinn showed that surface geometry cannot be rendered realistically with colour modulation, and, therefore, introduced his famous "bump mapping" (normal perturbation) technique [9]. Unfortunately, bump mapping, including its further extensions displacement mapping [10] and horizon mapping [11], are not sufficient to overcome the case of more complex geometric surface structures.

Because of the limitations of the mentioned "usual" methods, researchers introduced more sophisticated techniques, such as texel mapping [5] and hypertexturing [6]. In fact, most existing complex structures, including fur, fire, erosion, textiles, landscapes, rocks, etc. can be rendered with these kinds of "real 3D" (we may call them "volumetric") texturing techniques. This has been confirmed by some more recent works [12-15], based on the seminal works of [5] and [6]. Unfortunately, volumetric texturing remains in most practical cases avoided: ray marching, as required because of volume density rendering, turns out to be excessively time consuming. As noted in [16], this excludes any fast rendering application. In [12] and [17], we have investigated new methods for synthesising and applying hypertextures to divers objects and, likewise, experienced important computational requirements. In addition, it becomes necessary to provide an efficient antialiasing technique, if the viewpoint moves far away from the surface. But this is, in general, a difficult or extremely memory consuming task because of the explicit 3D nature of these kinds of textures (even in spite of compression techniques as presented in [13]). Our own confrontation with both problems (timings and aliasing) represents the main motivation behind this paper, and both encouraged us to develop a new and more efficient method.

Fig. 1: Some examples of natural textures (digitised photos), characterised by complex geometric structures, that are still difficult to be rendered efficiently with usual "fast" (2D texture mapping and bump mapping) texturing techniques.

Aside from "volumetric" texturing, the problem of complex geometric surface structures mainly has been investigated at a fine (sub-pixel or micro-) scale. This is known as bi-directional reflectance distribution function (BRDF). A significant work was done, for example in [18], in which general milliscale BRDFs were constructed from microscale scattering events. The pixel scale (or macroscale), however, generally called "bi-directional texture function" (BTF) has not been addressed, to our knowledge, by researchers in the field of image synthesis so far, though this notion is well-known in the field of vision [19] for example. Both notions (BRDF / BTF) are very closely related: at microscale, a BRDF relates incident irradiance to reflected radiance, and at macroscale, a BTF represents for example a certain colour (more generally BRDF), normal, transparency, etc. modulation, according to a certain viewpoint and position on the surface. Both determine surface details and a reflectance behaviour at different scales.

In many cases, a visual simulation of actual geometric surface complexity seems to be sufficient for rendering, which is well demonstrated by [7] and [16]. In fact, we believe that using "real 3D" (volumetric) textures often can be avoided, especially if the textures are not "deep" with respect to the whole size of the object. The notion of BTF, that we introduce in this paper, allows for an efficient visual simulation of complex surface structures, without using an explicit (and consequently often very time consuming) "volumetric" technique. In fact, bi-directional texture functions permit us to combine and include all of the principles of usual 2D texture mapping, bump mapping, etc., and provide a more general formulation of the texture mapping principle. Many visual effects beyond the possibilities of the usual "fast" texturing methods become realisable, still keeping the advantage of low computational cost. Colour plate 1 illustrates the limits of the conventional "fast" texturing methods: 2D texture mapping (even considering view dependence) and bump mapping, in the case of wickerwork. The part (c) represents the result that we obtained using a BTF. The simulation is visually convincing, though no real 3D depth was added. In fact, with BTFs the result is often visually so convincing, that it no longer looks like *texturing*, but rather does it look as if there existed indeed a real complex geometry wrapping up the surface.

In the following section, we first describe the basic principle of "mapping geometry" onto surfaces. We call this technique "virtual ray tracing". In section 3, we explain how to build a BTF from a geometry map, with similarity to some techniques used in the framework of BRDFs [18,20,21]. Note that we do not use digitised pictures of textures for synthesising the geometry maps as opposed to the mentioned image based rendering technique of [7] or as opposed to some analytical techniques that we have developed in the field of solid texturing [22-24] and hypertexturing [17]. In this case, the maps are synthesised by using directly 3D geometric sample models, as is done with some BRDF generation techniques. In section 4, the problem of antialiasing is discussed. Finally, graphical results are presented.

2. Virtual ray tracing

Virtual ray tracing consists of pasting a more or less complex 3D geometric landscape onto a certain "flat" surface. The landscape is called the map's world (see the right part of figure 2). It is composed of 3D objects that can be modelled using any kind of usual 3D modeller. In fact, it represents a user designed sample of the texture.

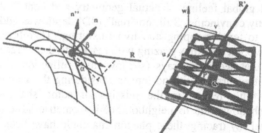

Fig. 2: The object's world (**left**) and the map's world (**right**). Rays that strike the object (left) to be rendered and textured, are also computed as "virtual" rays in the map's world (right) in order to find out the first visible point, its normal as well as its "material" (local BRDF).

Considering the left part of figure 2, whenever a ray R, with incoming direction (α,β) according to the normal $\mathbf{n}$, strikes the surface of the object on a point $P(x,y,z)$, a kind of "virtual" ray R' is shot in the map's world (right part). This "virtual" ray is defined by the same direction (α,β) with respect to the frame vector $\mathbf{w}$ of the map's world and the fact that it crosses a point $P'(u,v)$ on the origin's plane $w=0$. $P(x, y, z)$ and $P'(u, v, 0)$ are related according to a certain 2D texture mapping transformation M, i.e. $P' = M(P)$. Texture mapping transformations with particular properties, mainly with regard to distortions, have known much interest in computer graphics. We refer the reader to [25-27] for some examples. The virtual ray is used to compute the first visible point P'' of the map's world. If P'' exists (in the case of an interwoven iron-grid for example, the ray might pass throughout the map's world without striking any object), let be $\mathbf{n}''$ its normal and f_r its BRDF. Both are used to perform a usual shading computation on P. The original normal $\mathbf{n}$ is simply replaced by $\mathbf{n}''$ (relocated according to $\mathbf{w}$ and $\mathbf{n}$, which corresponds to a simple change of frame) and f_r is kept as BRDF on P. In the case of no intersection in the map's world, i.e. the point P'' does not exist, P is simply "ignored", just as if the object had a hole at this location.

Colour plate 2 illustrates an example of virtual ray tracing. The left part represents the map's world, and the right one, the visual result obtained by pasting the map onto spheres. The rendering was performed using ray tracing, but any other method such as for example Z-buffer, could have been used. All of the objects in the map's world are common 3D objects. The surfaces of these objects may be textured using "usual" techniques, such as 2D texture maps, solid textures or bump maps for example. As mapping transformation M, a -from Cartesian to polar transformation- was applied, which explains the distortions on the poles of the spheres (right part of the plate 2). With this virtual ray tracing technique, any geometry can, a priori, be mapped onto any kind of surface unless there exists no mapping transformation. As for bump mapping, this produces a certain "depth" feeling on the surface. And as for bump mapping, artefacts might become visible near to the borders, especially if the geometry in the map's world is high with respect to the size of the object. Note, however, that as opposed to usual bump mapping, some geometric structures might hide others, mainly according to the observer's position. This is an essential visual phenomenon. Normal vectors have not been computed by finite differentiation using basically a height map, but were directly extracted from the individual geometric structures. In addition, different objects in the map's world may have different materials (BRDFs, usual textures, etc.). All of these important supplements strongly reinforce the final global feeling of actual geometry and contribute to the illusion, making it so visually convincing. Still, no "real" (3D) depth was added to the surface, which is opposed to texel mapping and hypertexturing. Therefore, problems might arise when the viewpoint becomes grazing with respect to the surface.

We note that secondary rays (shadow rays, reflection, refraction, etc.) are directly shot from the surface with no longer considering the geometric structures in the map's world. This results in the missing of some shadows and reflection phenomena (self shadows due to neighbourhood geometric structures in the map's world). With virtual ray tracing, these phenomena could have been integrated, but in the case of BTFs, we did not consider them for simplicity and efficiency (see the next section).

3. Building a BTF from a geometry map

Virtual ray tracing is an efficient method for pasting repetitive geometric structures onto flat surfaces without adding real 3D depth. The principle is simple and avoids for example increasing further the geometric complexity of the scenes. However, it requires the shooting of rays and the computing of intersections. In this section, we introduce the notion of BTF, which is closely related to that of BRDF and to virtual ray tracing.

In the case of sub-pixel scale surface geometry, i.e. BRDF, the formulation is generally the following: given a certain incident direction (α,β) and a certain outgoing direction (α',β'), what is the ratio of leaving radiance according to the incident irradiance (some methods consider also wavelength). In the case of pixel-scale surface geometry, i.e. textures, a general formulation may be stated as follows: given a certain incident direction (α, β) and a certain map location (u, v), what is the normal **n**, the BRDF f_r, the colour c (both determine the "material"), and the transparency t. By transparency is meant, in fact, a Boolean value that indicates whether or not there is a hole.

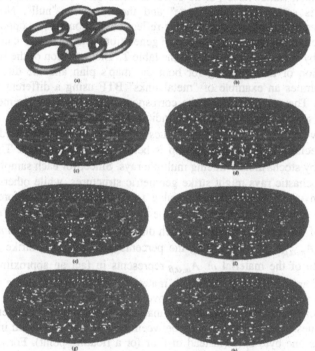

Fig. 3: An example of metal links. Rendering was performed using a BTF with a different precision of the sampling. (a) the map's world, (b) virtual ray tracing, (c) N=8, N_α=8, N_β=4, 0.0097Mb, (d) N=8, N_α=32, N_β=16, 0.156Mb, (e) N=32, N_α=8, N_β=4, 0.156Mb, (f) N=32, N_α=32, N_β=16, 2.5Mb, (g) N=128, N_α=8, N_β=4, 2.5Mb, (h) N=128, N_α=32, N_β=16, 40Mb.

Many BRDFs have been built by performing a direct computer simulation, i.e. by applying a deterministic or Monte Carlo simulation at microscale on a user modelled surface sample. In [20], the values of the simulation are stored in a table and then interpolated. In [21], the values of the simulation are analysed and then represented using spherical harmonics. In [28], the computed values are stored in a uniform subdivision of the hemisphere and in [29] in an adaptive subdivision of the geodesic sphere.

A similar approach, i.e. a computer simulation, may also be used in the case of textures and more particularly in the case of bi-directional texture functions (we call the textures bi-directional because of the bi-directionality of the underlying BRDF and because the terminology was already used in [19]). Therefore, the origin plane $w=0$ of the map's world is, first, sampled into discrete points $V(i,j)$, where $1 \leq (i,j) \leq N$. The direction space is also discretised into pyramidal shafts, each defined by two polar angles α and β. Second, a table $T(i, j, \alpha, \beta)$, corresponding to the BTF, is filled using a computer simulation based on ray tracing. The rays are shot towards all points V for all possible directions (α, β). Either a ray intersects an object O on a point P, in this case the normal on P (in fact two polar angles that determine the normal), its colour, as well as the corresponding BRDF are stored into $T(i, j, \alpha, \beta)$ and the transparency is set to the Boolean value "false", or no object is intersected. In the latter case, only the transparency is set to the value "true" and the others to "null". Note that this simulation does not take into account inter-reflection phenomena, as opposed to some techniques used in the framework of BRDF generation. The final BTF can be obtained and rendered by accessing the values of the table T. The precision of the BTF depends on the precision of the sampling for both the map's plan and the direction space. Figure 3 illustrates an example of "metal links" BTF using a different precision in several cases. The resolutions and the corresponding memory requirements without compression (see explanation below) are indicated in the caption.

Tracing only one ray to compute the values for $T(i, j, \alpha, \beta)$ might introduce aliasing artefacts, especially if the sampling step is below the Nyquist limit. Therefore, we oversampled by stochastically tracing multiple rays. Since, for each sample (i, j, α, β), only some stochastic rays might strike geometric structures, while others do not, we no longer can use a simple Boolean value for transparency. Therefore, we rather organised the table T as follows. First, the materials are separated. We call Γ the obtained set of different materials. For each direction (α, β) and material m, we obtain a 2D matrix $A_{m,\alpha,\beta}(i, j)$ representing the percentage of rays that strike a geometric structure made of the material m. $A_{m,\alpha,\beta}$ represents in fact an approximation of the percentage of area covered by a certain material m. The sum of A for all materials gives us the percentage of rays, that have hit objects in the map's world. The transparency can be directly derived from that value by taking its complement to one. All of the values of A, that actually lie between 0 and 1, are quantified in order to be stored each as one byte only (instead of four for a floating point). For all stochastic rays of one sample (i, j, α, β), the different normals and colours are simply averaged. Hence, we obtain for each direction (α, β) a 2D matrix $No_{\alpha,\beta}(i, j)$ of normals and $I_{\alpha,\beta}(i, j)$ of colours. The two polar angle values that define the normal are also quantified and stored each as two bytes, i.e. a short integer (one normal requires four bytes instead of eight for two floating points). The final uncompressed amount of memory storage required for the entire table T finally becomes $N^2 N_\alpha N_\beta (m+7)$ bytes, where N indicates the resolution on the map's plane, and N_α and N_β the number of shafts. The constant seven in the formula is due to the normal (four bytes) plus three bytes for the "rgb" colour. In the case of figure 3, there is only one material and no colour variation, hence the uncompressed memory requirement falls down to $5N^2 N_\alpha N_\beta$. Generally, a good approximation for the BTF is obtained with a resolution of $N=128$, $N_\alpha=32$ and $N_\beta=16$.

4. Antialiasing

The table T resulting from the previous computer simulation (section 3) may rapidly becomes huge for an increasing precision. In addition, as for usual 2D texture mapping, there may be problems of aliasing if the viewpoint moves too far away. Both problems, can be partly resolved using a multi-resolution scheme.

Antialiasing of 2D texture maps has been well investigated, and many approaches [30-34] can be found. Antialiasing of "bumpy textures", i.e. textures that are carrying a geometric information, however, raises generally many additional problems. In fact, we need to consider the surface at different levels of details: from the texture at the top level down to the corresponding BRDF at the lowest level. In [20], Kajiya mentioned the necessity of using hierarchical models. In [18], the levels of detail are called microscale and milliscale. A significant work was presented by Becker and Max [35], in which a complete smooth transition between all bump rendering techniques (displacement map, bump map and BRDF) was developed.

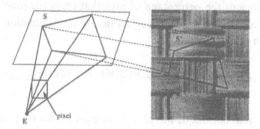

Fig. 4: Antialiasing.

Unfortunately, the principles of [35] cannot be applied in our case, since BTFs are generally holding a more complex information than a height variation (there are different geometric objects, materials, etc.). Figure 4 recalls the cause of aliasing. First, the usual approximation, that consists of projecting the concerned screen pixel onto the plane tangential to the object's surface, is computed. The obtained quadrilateral surface is called S. To this quadrilateral surface S corresponds also a quadrilateral surface S' in the texture space. An approximation of the global amount of the radiance received by the pixel is given by summing up over S:

$$L = \int_S L_{P \to E} ds \qquad (1)$$

where P represents a point on S and E the eye. $L_{P\text{-}>E}$ designates the radiance leaving the surface S towards the eye. $L_{P\text{-}>E}$ is obtained by summing up over the hemisphere [9] and the different materials of the BTF:

$$L_{P \to E} = \sum_{m \in \Gamma} A_m \int_\Omega L_{P' \to P} f_{r_m}(P', P, E, n') \cos(n', PP') d\omega + (1 - \sum_{m \in \Gamma} A_m) L_{behind} \qquad (2)$$

where A_m (see section 3) represents the area ratio covered by a material m according to the BTF (note that it is depending on the direction), $L_{P'\text{-}>P}$ the incoming radiance, P' a point on the hemisphere Ω, f_r and n' respectively the BRDF and the normal, both modulated according to the BTF (the original normal n of the surface is replaced by n', the BTF normal). L_{behind} represents the radiance coming from behind the surface, if there is a "hole" in the texture, as in the case of wickerwork for example. Let us first consider the two usual extreme cases:

• if the surface S' covers only one "pixel" of the BTF map, we do not need to sum up over S, since the normal, the BRDF, the colour and the transparency can, all of them, be supposed constant (this is the trivial case of no aliasing). The radiance of the pixel is then given by:

$$L = L_{P\text{->}E} \tag{3}$$

• if the surface S' covers the whole map, we need to compute a new BRDF, F_r, that matches the geometry of the map's world. This can be done for example by using the method of [WEST 92]. We just need to add to the basic model a global transparency term $\tau(\alpha, \beta)$, that indicates the percentage of rays that cross the map's world without striking any object. Note that τ is depending on the incident direction. The final pixel radiance is then given by:

$$L = (1 - \tau_{E\to P}) \int_{\Omega} L_{P'\to P} F_r(P', P, E, n) \cos(n, PP') d\omega + \tau_{E\to P} L_{behind} \tag{4}$$

We note that n, the real normal of the surface is used in this case, as opposed to n', the BTF normal. In all other cases, the computation is more complex, since we need to perform the sum of formula (1). We propose a simplified antialiasing solution, that consists of ignoring inter-reflections, arising because of the geometric structures in the map's world, as well as the specular effects. This choice is justified by the fact that we predominantly want to filter out aliasing artefacts, more than we want to obtain a high degree of physical accuracy. The simplified consideration allows us to apply directly a pyramidal-based prefiltering process on the transparency, the normals and the colours. Γ represents the set of different materials. We prefilter for each direction the colour $I_{\alpha,\beta}(i, j)$, the normal $No_{\alpha,\beta}(i, j)$ and the percentage of the area $A_m(i,j)$, that is covered by a material m, by computing pyramids as done with usual 2D texture maps [WILL 83]. The prefiltering process yields following pixel radiance formula, almost similar to formula (2):

$$L = \sum_{m\in\Gamma} A_{m,l} \int_{\Omega} L_{P'\to P} f_{r_m}(P', P, E, n'_l) \cos(n'_l, PP') d\omega + (1 - \sum_{m\in\Gamma} A_{m,l}) L_{behind} \tag{5}$$

where $A_{m,l}$ represents the area ratio covered by a material m at level l in the pyramid and n'_l the normal at the same level. The level l is determined according to the compression rate (figure 4). If S' tends to cover only one pixel in the texture space, then formula (5) converges on formula (3) which is obviously correct. Unfortunately, if S' tends to cover the whole texture map, we may notice a more or less important discrepancy between formula (5) and formula (4), since the BRDF model of formula (4) takes into account both neglected effects, namely inter-reflection and specularity. In order to avoid a too sharp transition, i.e. formula (5) is suddenly replaced by formula (4) though the viewpoint *continuously* moves away, we use, in fact, as final radiance value for the pixel, a weighted average between both formulas. The average is given by $k(5)+(1-k)(4)$, where the coefficient k guarantees a smooth transition and is empirically set according to the compression rate of the texture. Such a smooth transition technique, that is based on a compression rate coefficient, is frequently used in the framework of antialiasing.

Using a pyramidal prefiltering technique for the normals, the colours and the area ratios allows us in addition to compress the data using a Laplacian pyramid. For each direction, we have 2D fields $I_{\alpha,\beta}(i, j)$, $No_{\alpha,\beta}(i, j)$ and $A_{m,\alpha,\beta}(i,j)$, that all of them can be compressed using a JPEG-like scheme. For example, in the case of the metal links of figure 3, we obtained a global compression of about 18:1. We expect the compression rates to be even more important when using directly a 4D approach (as with the MPEG technique, that considers an additional third dimension). We note that our BTF concept has some relations with the light field problematic, and therefore all of the compression techniques and discussions developed in [36,37] can be directly applied. We also note that the entire texture not necessarily needs to be put into the memory as suggest in [38], but only relevant parts of the pyramid. This allows us to deal with multiple textures without overloading the memory capacity.

5. Results

One of the major motivation behind this paper was the efficient simulation of textures representing complex geometric structures (as illustrated by figure 1), that cannot be rendered correctly with usual fast texturing techniques. Colour plate 3 illustrates some examples of complex structures rendered using the new principle of BTF (including the bamboo-like texture in the back). The textures on the tori represent wickerwork (left), metal links (right), a kind of climber (back) and an iron wire (front). In all of these cases, the simulation is visually convincing and it looks as if there were a real complex geometry wrapping up the surface. The aspect of the texture changes according to the light source position and the observer's position. This is actually very important to guarantee a visually convincing result. In none of these cases, real depth was added to the surface. BTFs are, in fact, particularly suited for textures that behave at mid-way between a deep complex geometry such as fur (for which volumetric texturing is obviously more appropriated) and low height bumps, that can be rendered with usual bump mapping. We note that in this paper, we did not consider "texture self-shadowing". The problem is almost similar to that arising with usual bump mapping. A shadowing extension, as presented in [11] in the case of bump mapping, is conceivable.

We also compared the result with real geometry and measured the computational efficiency of our approach. Colour plate 4 illustrates a comparison in the case of rattan: (a) represents the result obtained with real geometry and (b) using a BTF. The first approach required about 28 minutes for computing on a Silicon Graphics O2 workstation, with a R5000 MIPS processor and 128Mb RAM at a resolution of 500×650 with 9 rays per pixel. Using a BTF the computation time was 2.33 minutes with one ray per pixel (with 9 rays the computation took 12.7 minutes, but oversampling for antialiasing is not necessary because the texture is already prefiltered). The BTF approach is faster for an approximately equivalent visual quality. Note that because direct timings are often difficult to be interpreted correctly, since the result is highly hardware and implementation depending, we suggest the reader to consider only relative timings. BTFs should be particularly well appropriated in the case of real time applications.

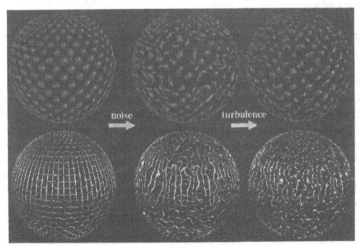

Fig. 5: Texture distortions are possible, which increases the graphical possibilities.

Since the BTF represents actually a 2D map as with usual texture mapping, it may also be distorted using stochastic functions such as noise or turbulence [3] for example. This considerably increases the graphical possibilities. Figure 5 illustrates two examples of distortion. The left pictures shows the original texture. The middle pictures show a distortion using a white noise and the right ones using a turbulence function [3]. In all cases, the illusion of geometry on the surface is preserved. Texture animation is even possible for example by making the mapping transformation M (see section 2) time dependent. In fact, all of the possibilities that usual 2D texture mapping offers can be directly transposed and applied to BTFs.

6. Discussion and conclusion

The fast and realistic rendering of complex macro-geometric surface structures is a difficult problem, already existing since many years in computer graphics. In this paper, we have introduced the new concept of bi-directional texture function (BTF), that represents a kind of view dependent extension of bump mapping, including also colour, transparency, material, etc. modulation. BTFs are, in fact, at coarse scale, what BRDFs are at fine scale. We believe that the BTF concept is an important new concept in the field of rendering. It can be thought of as a generalised formulation of the well-known 2D texture mapping technique. The basic principle of BTF is extremely simple. It can be implemented with ease. The computational requirements are low and many visual effects, beyond the possibilities of usual fast texturing techniques become realisable.

Current trends in Computer Graphics outline that it becomes more and more important to reproduce accurately complex geometric surface structures *at any precision scale*. Texture- and BRDF-synthesis represent usually disjoint domains and mainly "smooth" transitions between both were investigated [35]. However, both are related to the same underlying process (in fact, only at different scales). An accurate transition as developed in [35] is crucial. The multi-resolution approach of BTFs that we have presented, provides a kind of step-wise relation-ship between coarse (pixel-scale) details and fine (subpixel-scale) details of surfaces. This concept is actually very important to guarantee an "accurate" rendering at all levels of detail. In our case BRDFs and textures are very closely related.

Our multi-resolution scheme permits antialiasing and the merging of textures with BRDFs. Nevertheless, it remains unsatisfactory in physical terms, mainly with regard to light inter-reflection, because of the restrictive assumptions that we made in section 4. In future works, we will need to consider also these effects. Furthermore, we will need to integrate our BTFs into a complete global illumination rendering technique, for example based on radiosity. This, in turn, will first require the development of a method beyond the radiosity approach of [39]. In fact, the latter allows one an efficient processing of bump maps only. But BTFs are more complex and consequently also more difficult to be integrated into the "usual" radiosity simulation. We are currently working on such a radiosity-based global illumination rendering system.

References

[1] Blinn J.F., "Simulation of wrinkled surfaces", Computer Graphics 12, 1978, pp. 286-292.

[2] Peachey D., "Solid Texturing on complex Surfaces", Computer Graphics 19(3), 1985, pp. 279-286.

[3] Perlin K., "An Image Synthesizer", Computer Graphics, 19(3), July 1985, pp. 287-296.

[4] Gardner G., "Visual Simulation of clouds", Computer Graphics, 19(3). 1985, pp. 297-303.

[5] Kajiya J. and Kay T., "Rendering Fur with three dimensional textures", Computer Graphics, 23(3), 1989, pp. 271-298.

[6] Perlin K. ,"Hypertextures", Computer Graphics, 23(3), 1989, pp. 253-262.

[7] Debevec P.E., Taylor C.J. and Malik J., "Modeling and Rendering Architecture from Photographs: A hybrid geometry- and image-based approach", Computer Graphics, 30, 1996, pp. 11-20.

[8] Brodatz Ph., "Textures: a photographic album for artists and designers", Dover Publications, Inc., New York, 1966.

[9] Kajiya J., "The Rendering Equation", Computer Graphics 20(4), 1986, pp. 143-150.

[10] Cook R.L., "Shade Trees", Computer Graphics, 18(3), 1984, pp. 223-231.

[11] Max N., "Horizon mapping: shadows for bump-mapped surfaces", The Visual Computer, 4, 1988, pp. 109-117.

[12] Dischler J.M. and Ghazanfarpour D., "A Geometrical Based Method for Highly Complex Structured Textures Generation", Computer Graphics Forum, 14(4), 1995, pp 203-215.

[13] Neyret F., "A general and multiscale method for volumetric textures", Graphics Interface'95 Proceedings, May 1995, pp. 83-91.

180

[14] Neyret F., "Synthesizing verdant landscapes using volumetric textures", Proceedings of Eurographics Workshop on Rendering, June 1996, pp. 215-224.

[15] Groeller E., Rau R.T. and Strasser W., "Modeling textiles as three dimensional textures", Proceedings of Eurographics Workshop on Rendering, Porto, June 1996, pp. 205-214.

[16] Goldman D., "Fake Fur Rendering", Computer Graphics, 31, 1997, pp. 127-134

[17] Dischler J.-M. and Ghazanfarpour D., "A Procedural Description of Geometric Textures by Spectral and Spatial Analysis of Profiles", Computer Graphics Forum, 16(3), Eurographics 97, 1997, pp 129-139.

[18] Westin S. H., Arvo J. R. and Torrance K. E., "Predicting Reflectance Functions from Complex Surfaces", Computer Graphics, 26(2), 1992, pp. 255-263.

[19] Dana K.J., Nayar S.K, van Ginneken B. and Koenderink J.J., "Reflectance and Texture of Real-World Surfaces", IEEE Conf. on Comp. Vision and Pattern Recognition, 1997.

[20] Kajiya J., "Anisotropic reflectance models", Computer Graphics 19(3), 1985, pp. 15-21.

[21] Cabral B., Max N. and Springmeyer R., "Bidirectional reflection-functions from surface bump maps", Computer Graphics, 21(4), 1987, pp. 273-281.

[22] Ghazanfarpour D. and Dischler J.-M., "Spectral Analysis for Automatic 3D texture Generation", Computers & Graphics, 19(3), 1995, pp. 413-422.

[23] Ghazanfarpour D. and Dischler J.-M., "Generation of 3D Texture Using Multiple 2D Models Analysis", Computer Graphics Forum (Eurogr.'96) 15, pp. 311-323.

[24] Dischler J.-M. and Ghazanfarpour D., "Anisotropic Solid Texture Synthesis Using Orthogonal 2D Views", Computer Graphics Forum (to appear), 17(3), 1998.

[25] Bennis C., Vézien J.-M. and Iglésias G., "Piecewise Surface Flattening for Non-Distorted Texture Mapping", Computer Graphics, 25(4), 1991, pp 237-246.

[26] Maillot J., Yahia H. and Verroust A., "Interactive Texture Mapping", Computer Graphics, 27, 1993, pp 27-34.

[27] Arad N. and Elber G., "Isometric Texture Mapping for Free-form Surfaces", Computer Graphics Forum, 16(5), 1997, pp 247-256.

[28] Hanrahan P. and Krueger W., "Reflection from layered surfaces due to subsurface scattering", Computer Graphics, 27, 1993, pp. 165-174.

[29] Gondek J.S., Meyer G.W. and Newman J.G., "Wavelength dependent reflectance functions", Computer Graphics, 28, 1994, pp. 213-220.

[30] Norton A., Rockwood A. and Skolmoski P., "Clamping: a method of antialiasing textured surfaces", Computer Graphics, 16(3), 1982, pp. 1-8.

[31] Williams L., "Pyramidal parametrics", Computer Graphics, 17(3), 1983, pp. 1-11.

[32] Crow F., "Summed-area tables for texture mapping", Computer Graphics, 18(3), 1984, pp. 207-212.

[33] Glassner A., "Adaptative precision in texture mapping", Computer Graphics, 20(4), 1986, pp. 297-306.

[34] Ghazanfarpour D. and Peroche B., "A high quality filtering using forward texture mapping", Computers & Graphics, 15(4), 1991, pp. 569-577.

[35] Becker B. G. and Max N., "Smooth Transition between Bump Rendering Algorithms", Computer Graphics, 27, 1993, pp 183-190.

[36] Gortler S.J., Grzeszczuk R., Szeliski R. and Cohen M.F., "The Lumigraph", Computer Graphics, Annual Conference Series, 1996, pp. 43-54.

[37] Levoy M. and Hanrahan P., "Light Field Rendering", Computer Graphics, 1996, pp 31-42.

[38] Pharr M., Kolb C., Gershbein R. and Hanrahan P.,"Rendering Complex Scenes with Memory-Coherent Ray Tracing", Computer Graphics, 1997, pp. 101-108.

[39] Chen H. and Wu E.-H., "An Efficient Radiosity Solution for Bump Texture Generation", Computer Graphics, 24(4), 1990, pp 125-134.

Editors' Note: see Appendix, p. 332 for colored figures of this paper

Point Sample Rendering

J.P. Grossman[1] and William J. Dally[2]

Abstract

We present an algorithm suitable for real-time, high quality rendering of complex objects. Objects are represented as a dense set of surface point samples which contain colour, depth and normal information. These point samples are obtained by sampling orthographic views on an equilateral triangle lattice. They are rendered directly and independently without any knowledge of surface topology. We introduce a novel solution to the problem of surface reconstruction using a hierarchy of Z-buffers to detect tears. Our algorithm is fast and requires only modest resources.

1 Introduction

The research presented in this paper was motivated by a desire to render complex objects in real time without the use of expensive hardware. This ubiquitous goal has recently led to the use of images, rather than polygons, as primitives. Unlike traditional polygon systems in which rendering time is proportional to scene complexity, in image based systems rendering time is proportional to the number of pixels in the source images or the output image. The techniques of view interpolation [Chen93, Chen95], epipolar geometry [Laveau94], plenoptic modeling [McMillan95], light fields [Levoy96], lumigraphs [Gortler96], view based rendering [Pulli97], nailboards [Schaufler97] and post rendering 3D warping [Mark97] have all demonstrated the speed and viability of image based graphics. Despite the success of these approaches, however, they suffer from a large appetite for memory, noticeable artifacts from many viewing directions, and an inability to handle dynamic lighting. In [Wong97] the light field system was extended to support dynamic lighting, but only at a significant cost in memory and with an unspecified degradation of rendering speed.

Point Sample Rendering is an algorithm which features the speed of image based graphics with quality, flexibility and memory requirements approaching those of traditional polygon graphics. Objects are modeled as a dense set of surface point samples which are obtained from orthographic views and stored with colour, depth and normal information, enabling Z-buffer composition, Phong shading, and other effects such as shadows. Motivated by image based rendering, point sample rendering is similar in that it also takes as input existing views of an object and then uses these to synthesize novel views. It is different in that (1) the point samples contain additional geometric information, and (2) these samples are *view independent* - that is, the colour of a sample does not depend on the direction from which it was obtained.

Points have often been used to model 'soft' objects such as smoke, clouds, dust, fire, water and trees [Reeves83, Smith84, Reeves85]. The idea of using points to model solid objects was mentioned nearly two decades ago by Csuri et. al. [Csuri79] and was briefly investigated by Levoy and Whitted some years later for the special case of continuous, differentiable surfaces [Levoy85]. More recently Max and Ohsaki used point samples, obtained from orthographic views and stored with colour, depth

[1] Dept. of EECS, Massachusetts Institute of Technology. Email: jpg@ai.mit.edu
[2] Department of Computer Science, Stanford University. Email: billd@csl.stanford.edu

and normal information, to model and render trees [Max95]. However, almost all work to date dealing with a point sample representation has focused on using it to generate a more standard representation such as a polygon mesh or a volumetric model [Hoppe92, Turk94, Curless96].

The fundamental challenge of point sample rendering is the on-screen reconstruction of continuous surfaces. Surfaces rendered using point samples should not contain holes (figure 1). Furthermore, this issue should be addressed without seriously impacting rendering speed. The algorithm presented in [Levoy85] makes use of surface derivatives and is therefore applicable only to differentiable surfaces. In [Max95] no attempt is made to address this problem at render time, though it is alleviated by sampling the trees at a higher resolution than that at which they are displayed.

In this paper we will present a novel solution to the problem of surface reconstruction which is critical to the success of our rendering algorithm. To obtain a more uniform sampling, orthographic views are sampled on an equilateral triangle lattice rather than a traditional square lattice. A pyramid of Z-buffers is used to detect gaps in the output image; these gaps are filled in screen space as the final rendering step. The use of a hierarchical Z-buffer incurs very little computational overhead, and the screen space operation of filling gaps is quite fast.

We have implemented a software test system to demonstrate the capabilities of point sample rendering. The system takes as input object descriptions in Open Inventor format and produces point sample models as output. These models are then displayed in a window where they may be manipulated (rotated, translated and scaled) in real time by the user.

2 Surface Reconstruction

Given a set of surface point samples, it is easy enough to transform them to screen space and map them to pixels. When a point sample is mapped to a pixel, we say that the pixel is *hit* by the point sample. When a pixel is hit by points from multiple surfaces a Z-buffer can be used to resolve visibility. The difficulty is that, in general, not all pixels belonging to a surface will be hit by some point from that surface. This leaves holes through which background surfaces are visible (figure 1).

One approach to this problem is to treat the point samples as vertices of a triangular mesh which can be scan-converted. However, this method is extremely slow, requiring a large number of operations per point sample. Another approach is the use of 'splatting', whereby a single point is mapped to multiple pixels on the screen, and the colour of a pixel is the weighted average of the colours of the contributing points. This method was used by Levoy and Whitted [Levoy85]; it has also been used in the

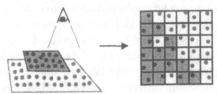

Figure 1: A dark surface is closer to the viewer than a light surface. When the surfaces are rendered using point samples, the points on the dark surface are more spread out. Some screen pixels are 'missed', and the light surface peeps through.

context of volume rendering [Westover90]. However, this is also too slow for our needs, again requiring a large number of operations per point.

In order to maximize speed, our solution essentially ignores the problem altogether at render time, as in [Max95]. To see how this can be achieved without introducing holes into the image, we start with the simplifying assumptions that the object will be viewed orthographically, and that the target resolution and magnification at which it will be viewed are known in advance. We can then, in principle, choose a set of surface point samples which are dense enough so that when the object is viewed there will be no holes, independent of the viewing angle. We say that an object or surface is *adequately sampled* (at the given resolution and magnification) if this condition is met. To fashion this idea into a working point sample rendering algorithm, we need to answer the following three questions:

- How can we generate an adequate sampling using as few points as possible?
- How do we display the object when the magnification or resolution is increased?
- How must we modify the algorithm to display perspective views of the object?

These questions will be addressed in the following sections.

2.1 Geometric Preliminaries

Suppose we overlay a finite triangular mesh on top of a regular array of square pixels. We say that a pixel is *contained* in the triangular mesh if its center lies within one of the triangles (figure 2a). As before, we say that a pixel is *hit* by a mesh point if the mesh point lies inside the pixel.

Theorem If the side length of each triangle in the mesh is less than the pixel side length, then every pixel which is contained in the mesh is hit by some mesh point.

Proof Our proof is by contradiction - suppose instead that some pixel P is contained in the mesh but is not hit by any mesh point. Then there is some mesh triangle ABC such that ABC contains the center of P, but A, B, C lie outside of P. Now we cannot have one vertex of ABC above (and possibly to the left/right of) P and another below P, as then the distance between these vertices would be greater than the side length of P (figure 2b). Similarly, we cannot have one vertex to the left of P and another vertex to the right of P. Without loss of generality, then, assume that A, B, C all lie to the left of and/or below P (figure 2c). The only way for a triangle with vertices in this region to contain the center of P is if one vertex is at the upper left corner of P and the other is at the lower right - but then the distance between these vertices is greater than the side length of P, contradiction.

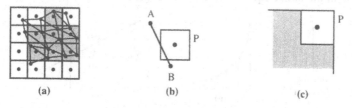

Figure 2: (a) Pixels contained in a mesh. (b) An impossible configuration. (c) A, B, C lie in the shaded region.

Corollary Suppose a surface is sampled at points which form a continuous triangular mesh on the surface. If the side length of every triangle in the mesh is less than the side length of a pixel at the target resolution (assuming unit magnification), then the surface is adequately sampled.

Proof This follows directly from the theorem and the fact that when the mesh is projected orthographically onto the screen, the projected side lengths of the triangles are less than or equal to their unprojected lengths.

2.2 Sampling

The previous section provides some direction as to how we can adequately sample an object. In particular, it suggests that to minimize the number of samples, the distance between adjacent samples on the surface of the object should be as large as possible but less than ds, the pixel side length at the target resolution (again assuming unit magnification). Now if we use a regular lattice to sample the orthographic views, which we consider a necessity for reasons of storage and rendering efficiency, then the distance between adjacent samples on the object could be arbitrarily large on surfaces which are nearly parallel to the viewing direction (figure 3a). However, suppose we restrict our attention to a surface (or partial surface) S whose normal differs by no more than θ from the projection direction at any point, where θ is a 'tolerance angle'. If we sample the orthographic view on an equilateral triangle lattice with side length $ds \cdot \cos\theta$, we obtain a continuous triangular mesh of samples on S in which each side length is at most $(ds \cdot \cos\theta) / \cos\theta = ds$ (figure 3b). Hence, S is adequately sampled.

If the object being modeled is smooth and convex, then all we need to do is choose a set of projections - which depend only on θ and not on the object itself - such that *every* part of the object's surface is adequately sampled by some projection. Unfortunately, most interesting objects exhibit a tendency to be neither smooth nor convex. We cannot, therefore, use the corollary of section 2.1 to 'prove' that an object is being adequately sampled. We simply use it to provide theoretical motivation for an algorithm which, ultimately, must be verified experimentally.

We conclude with two comments on sampling. First, although we have referred multiple times to 'surface meshes' we do *not* store any surface mesh information with the point samples, nor is it necessary to do so. It suffices for there to exist *some* continuous triangulation of the points on a surface which satisfies the conditions given in section 2. Second, the preceding discussion brings to light the advantage of using an equilateral triangle mesh. If we were to use a square lattice instead, we would need to decrease the spacing between lattice points in order to compensate for the longer diagonals within the lattice squares, which would increase the total number of samples.

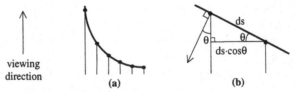

Figure 3: (a) Samples on the surface of the object can be arbitrarily far apart. (b) Using a tolerance angle to constrain the distance between adjacent samples.

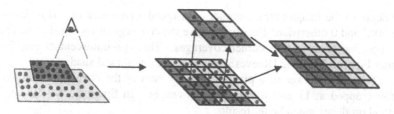

Figure 4: The dark surface is rendered using a higher level depth buffer in which there are no gaps. This depth buffer is used to detect and eliminate holes in the image.

2.3 Magnification and Perspective

An adequately sampled object can be rendered orthographically without holes at the target resolution/magnification. To deal with magnification and perspective (which magnifies different parts of the object by different amounts), we introduce a *hierarchy* of lower resolution depth buffers as in [Green93]. The lowest buffer in the hierarchy has the same resolution as the target image, and each buffer in the hierarchy has one half the resolution of the buffer below it. Then, in addition to mapping point samples into the destination image as usual, each part of the object is rendered in a depth buffer at a low enough resolution such that the points cannot spread out and leave holes. When all parts of the object have been rendered, we will have an image which, in general, will contain many gaps. However, it is now possible to detect and eliminate these holes by comparing the depths in the image to those in the hierarchy of depth buffers (figure 4). The next two sections provide details of this process.

2.4 Finding Gaps

Our strategy for detecting holes is to assign to each pixel a weight in [0, 1] indicating a 'confidence' in the pixel, where a 1 indicates that the pixel is definitely in the foreground and a 0 indicates that the pixel definitely represents a hole and its colour should be ignored. A simple way to assign these weights is to treat a pixel in the kth depth buffer as an opaque square which covers exactly 4^k image pixels. We can then make a binary decision for each image pixel, assigning it a weight of either 0 or 1, by determining whether or not it lies behind some depth buffer pixel. This, however, produces pronounced blocky artifacts, as can be seen in Figure 5a.

A better approach is to instead treat the kth depth buffer as a square mesh with vertices at the pixel centers. We then compute the amount by which the kth depth buffer 'covers' an image pixel as follows: The center of the image pixel lies inside one of the squares of the mesh. The depth at each corner of the square is compared to

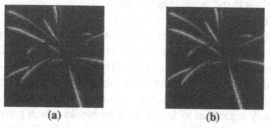

(a)   (b)

Figure 5: Naïve weight assignment (left) results in blocky artifacts.

the depth of the image pixel; a corner is assigned a *coverage* of 1 if it lies in front of the pixel and 0 otherwise. Finally, we take the coverage of the pixel to be the bilinear interpolation of these four corner coverages. This is reminiscent of the "Percentage Closer Filtering" used in [Reeves87] to generate antialiased shadows.

The total coverage of a pixel is then the sum of the coverages from each depth buffer (capped at 1), and *weight* = 1 - *coverage*. In figure 5b we can see that this method produces much better results.

2.5 Filling Gaps

To fill the gaps in the final image we must compute a colour for pixels with weight less than 1. This problem is not specific to point sample rendering; it is the general problem of image reconstruction given incomplete information. Our solution is a variation of the two phase "pull-push" algorithm described in [Gortler96].

The 'pull' phase of the algorithm consists of computing a succession of lower resolution approximations of the image, each one having half the resolution of the previous one. We do this by repeatedly averaging 2x2 blocks of pixels, taking the weight of the new pixel to be the sum of the weights of the four pixels, capped at 1. In the 'push' phase, we use lower resolution images to fill gaps in the higher resolution images. For a pixel with colour c and weight $w < 1$, we compute an interpolated colour c_i using as interpolants the three closest lower resolution pixels with weights ½, ¼ and ¼ (figure 6a). The pixel's colour is then taken to be $wc + (1-w)c_i$.

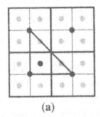

(a)

(b)

Figure 6: (a) An interpolated colour is computed using the three closest lower resolution pixels as shown. (b) Before gaps are filled (left) and after (right).

3   Object Modeling

Our software system is capable of automatically generating point sample representations for arbitrary synthetic objects; there is no user intervention. A target resolution is specified, and the object is scaled to fit inside a square window having this resolution. This implicitly sets the target magnification.

A fixed set (i.e. independent of the object being modeled) of 32 orthographic projections are sampled on an equilateral triangle lattice. The density of this lattice is determined by the tolerance angle described in section 2.2; we used 25° for all models in this paper. The samples in a projection are divided into 8x8 *blocks* which allows us to compress the database by retaining only those blocks which are needed, while maintaining the rendering speed and storage efficiency afforded by a regular lattice.

We use a greedy algorithm to find a set S of blocks which provides an adequate sampling of the object with as little redundancy as possible. We start with no blocks

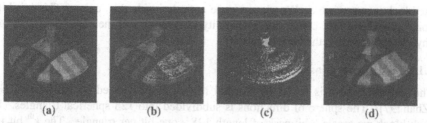

Figure 7: Block Selection. **(a)** Ray traced projection. **(b)** Projection reconstructed from current set of blocks. **(c)** Pixels where (a) and (b) have different depths. **(d)** Blocks added to set.

$(S = \phi)$. We then step through the list of orthographic projections; from each one we select blocks to add to S. A block is selected if it corresponds to a part of the object which is not yet adequately sampled by S. Rather than attempt to exactly determine which surfaces are adequately sampled, which is very difficult, we employ the heuristic of both ray tracing the orthographic projection and reconstructing it from S, then comparing the depths of the two images thus obtained (figure 7). This tests for undersampled surfaces by searching for pixels which are 'missed' by the reconstruction.

4 Rendering

There are five steps in the point sample rendering pipeline:

Visibility Testing We conservatively estimate a set of visible blocks using the view frustum and two additional data structures described in section 4.1.
Block Warping Each potentially visible block is 'warped' [Wolberg90] into the destination image. This is described in section 4.2.
Finding Gaps We assign weights to the image pixels as described in section 2.4.
Shade Each non-background pixel is shaded according to the current lighting model.
Filling Gaps Pixels with weight less than one are coloured as described in section 2.5.

4.1 Visibility Data Structures

In addition to the view frustum, we make use of two per-block data structures to speed up rendering by conservatively estimating a set of visible blocks.

4.1.1 Visibility Cones

The tangent plane at each point sample defines a half space from which the point is never visible. Taking the intersection of these half spaces over all the points in a block, we obtain a region in space, bounded by up to 64 planes, from which no part of

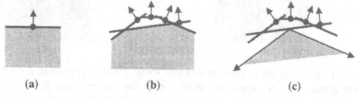

Figure 8: **(a)** Half space from which point is not visible **(b)** Region from which block is not visible **(c)** Cone inside region.

the block is ever visible. We place a cone entirely inside this region. This 'visibility cone' gives us a fast, conservative visibility test; no part of the block is visible from any viewpoint within the cone (figure 8).

4.1.2 Visibility Masks

The visibility mask is based on Normal Masks as introduced by Zhang and Hoff [Zhang97]. The sphere of directions is subdivided into 128 spherical triangles. For each block we create a bitmask of length 128 - one bit per triangle. The kth bit in a block's bitmask is 1 if and only if the block is visible (from a viewpoint outside the convex hull of the object) from some direction which lies inside the kth triangle. We call this bitmask the *visibility mask* for that block.

The simplest way to make use of these masks at render time is to construct a single visibility mask for the entire screen. In this mask, the kth bit is set if and only if some direction from the eye to a screen pixel lies inside the kth triangle on the triangulated sphere of directions. We then AND this mask with each block's mask; a block is definitely not visible if the result of this test is zero.

As explained in [Zhang97], however, this is overly conservative. We are, in effect, using the entire set of directions from the eye to the screen to approximate the set of directions subtended by a single block. In reality the latter set is almost certainly much smaller, so if we can improve the approximation then we will improve the rate of culling. To this end we subdivide the screen into $2^n \times 2^n$ square regions. We compute a mask for each region, and we use a BSP tree to quickly determine which region a block lies in, and therefore which mask to use to test the block's visibility.

There is naturally a tradeoff involved in choosing n - larger n will result in improved culling, but more overhead. We have found that $n = 2$ gives the best performance on average (i.e. a 4 x 4 subdivision of the screen).

4.2 Block Warping

The points in each potentially visible block must be transformed to screen space and mapped to image pixels. This procedure uses incremental calculations and is very fast. For each block, we also select a depth buffer with a low enough resolution such that the block's points cannot spread out and leave holes. Rather than storing the points' exact depths in this buffer, we first subtract a small threshold. The reason for this is that the depth buffer will be used to filter out points which lie behind the fore-

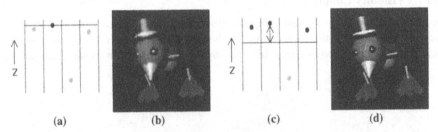

(a)   (b)   (c)   (d)

Figure 10: Four points are mapped to different image pixels contained in a single depth map pixel. **(a)** No depth threshold - only the nearest point is retained. **(b)** Corresponding image. **(c)** The use of a small threshold prevents points on the same surface as the nearest point from being discarded. **(d)** Corresponding image.

ground surface; the use of a threshold prevents points which lie *on* the foreground surface from being accidentally discarded. Figure 10 illustrates this in one dimension and shows the striking difference in image quality which results from the use of a threshold. We have found a good threshold to be three times the image pixel width.

4.3 Shading

In our test system we have implemented Phong shading with shadows. As shading is performed in screen space after all blocks have been warped, points' normals and shininess must be copied into the image buffer in addition to their colours.

There are two points worth noting here. First, since the specular component of illumination is computed at each pixel, the resulting highlights are of ray-trace quality (figure 11a). Second, since the creation of the shadow maps [Williams78] used to compute shadows involves only fast block warping without the overhead of copying normals, they can be constructed very quickly. For example, figure 11b was rendered at 256x256 with three light sources. It took 204ms to render without shadows and 356ms with shadows - an overhead of less than 25% per light source

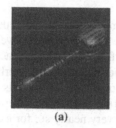 

(a)   (b)

Figure 11: (a) High quality specular highlights (b) fast shadows

5 Results

Our test system was run as a single process on a 333MHz dual Pentium II with 512MB RAM. Table 1 gives file size, number of blocks and construction time for the four models shown in the appendix. The models were rendered with no shadows and a single directional light at 256x256 pixels; table 2 gives timing statistics for these settings. The largest of the models, 'pine tree', was obtained from an Inventor model which contains 15,018 cones and ellipsoids and takes over 5 seconds to render using the Inventor Scene Viewer running on the same machine.

Table 3 shows the extent to which the visibility cone and visibility mask are able to eliminate blocks. Note that the visibility cone works quite well for objects with low curvature (e.g. Top) but poorly for objects with high curvature (e.g. Chair, Pine Tree). This is due to the fact that surfaces with high curvature are visible from a larger set of directions. In all cases the visibility cone is outperformed by the visibility mask, but it is still able to provide some additional culling.

Table 1: Model statistics

	Chair	Pine Tree	Sailboat	Top
file size (MB)	1.32	7.12	1.00	2.96
number of blocks	2666	11051	1786	4473
construction time (hrs : min)	0:15	1:35	0:13	0:43

Table 2: Timing for a software only implementation on a 333MHz Pentium II.

time (ms)	Chair	Pine Tree	Sailboat	Top
initialize buffers & screen masks	9	9	9	9
visibility (cone and mask)	3	9	2	6
warp blocks	44	226	33	61
find gaps	3	4	6	3
shade	14	25	18	21
fill gaps	9	9	9	8
total	**82**	**282**	**77**	**108**

Table 3: % blocks eliminated as a function of which visibility tests are used.

% culled	Chair	Pine Tree	Sailboat	Top
neither	0	0	0	0
cone only	13.9	1.1	24.9	42.9
mask only	23.1	9.5	33.4	43.3
both	26.1	9.9	35.8	51.5

6 Discussion and Future Work

The key to the speed of our algorithm is the simplicity of the procedure used to warp points into the destination image. All of the relatively expensive computations - shading, finding gaps, and filling gaps - are performed in screen space after all points have been warped. This places an absolute upper bound on the overhead of these operations, independent of the number and/or complexity of models being rendered. Ideally, one would also like the overall rendering time to be independent of model complexity. Although this is not the case, it is very nearly so; for example, the 'pine tree' inventor model is two orders of magnitude more complex than 'sailboat', yet the point sample rendering time is less than four times greater.

One limitation of the algorithm as described is that it is difficult to properly sample thin spoke-like structures which are on the order of one pixel wide. As a result, such structures often appear ragged in rendered images (figure 12a). One possible solution to this problem, which is itself an interesting area of investigation, is to have objects generate their own point samples rather than relying on some external sampling process. Each geometric primitive would produce a set of points which form an adequate sampling of the primitive, much in the same way that primitives 'dice' themselves into micropolygons in the Reyes rendering architecture [Cook87]. The difficulty lies in organizing these generated points into blocks or some other structure so that they may be stored and rendered efficiently.

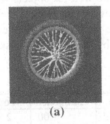  

(a) (b) (c)

Figure 12: (a) Ragged spokes (b) Rough edges (c) Rendered at 256x256 and displayed at 128x128 (2x2 subpixel anti-aliasing).

Another weakness of point sample rendering is that without the precise geometric information available in polygon rendering, it is difficult to perform any sort of sub-pixel anti-aliasing, particularly at silhouette edges. A brute force solution is to increase the sampling density and render objects at double the image resolution. This works well (figure 12b,c), but has the obvious disadvantage of being significantly slower. A more sophisticated variant of this solution would be to store both low and high resolution copies of each block, using the slower high resolution copy only for those blocks that lie on a silhouette edge.

A topic which invariably arises in any discussion of modeling paradigms is the acquisition of models from real objects. In order to construct a point sample model for an object, it is necessary to somehow obtain both shape and reflectance information for that object. This is a challenging problem, but it has been addressed quite successfully by Sato, Wheeler and Ikeuchi [Sato97a]. In fact, one of the problems they encountered was that the models they generated - which included dense grids containing colour, normal and specular information - could not be rendered using standard software or hardware. It was therefore necessary for them to write software to display these models and image generation was slow [Sato97b]. This problem would be solved by generating point sample representations of these models which would make full use of the reflectance information and could be rendered in real time.

Acknowledgments

We would like to thank the following people for granting us permission to use their Inventor models: Barbara Cutler ('sailboat'), Brendon Glazer ('toybird'), Michael Golding ('gavel'), Stephen Ho ('tree' and 'pine tree'), and Tara Schenkel ('chair'). These models were created as part of a computer graphics course taught by Seth Teller at M.I.T. We would also like to thank Seth Teller, Pat Hanrahan, John Owens, Scott Rixner and Shana Nichols for their valuable comments. This work was sponsored by the Defense Advanced Research Projects Agency (DARPA) under DARPA Order E254 and monitored by the Army under Contract DABT63-96-C-0037.

References

[Chen93]   Shenchang Eric Chen, Lance Williams, "View Interpolation for Image Synthesis", SIGGRAPH '93 Proceedings, pp. 279-285

[Chen95]   Shenchang Eric Chen, "QuickTime® - An Image-Based Approach to Virtual Environment Navigation", SIGGRAPH '95 Proceedings, pp. 29-37

[Cook87]   Robert L. Cook, Loren Carpenter, Edwin Catmull, "The Reyes Image Rendering Architecture", SIGGRAPH '87 Proceedings, pp. 95-102

[Csuri79]   C. Csuri, R. Hackathorn, R. Parent, W. Carlson, M. Howard, "Towards an Interactive High Visual Complexity Animation System", SIGGRAPH '79 Proceedings, pp. 289-299

[Curless96] Brian Curless, Marc Levoy, "A Volumetric Method for Building Complex Models from Range Images", SIGGRAPH '96 Proceedings, pp. 303-312

[Dally96]   William J. Dally, Leonard McMillan, Gary Bishop, Henry Fuchs, "The Delta Tree: An Object-Centered Approach to Image-Based Rendering", Artificial Intelligence memo 1604, Massachusetts Institute of Technology, May 1996.

[Gortler96] Stephen J. Gortler, Radek Grzeszczuk, Richard Szeliski, Michael F. Cohen, "The Lumigraph", SIGGRAPH '96 Proceedings, pp. 43-54

[Greene93]   Ned Greene, Michael Kass, Gavin Miller, "Hierarchical Z-buffer Visibility", Proc. SIGGRAPH '93 Proceedings, pp. 231-238

[Grossman98] J.P. Grossman, "Point Sample Rendering", Massachusetts Institute of Technology Dept. of EECS Master's Thesis, 1998

[Heckbert87] Paul S. Heckbert, "Ray Tracing Jell-O® Brand Gelatin", *Computer Graphics* (SIGGRAPH '87 Proceedings), Vol. 21, No. 4, July 1987, pp. 73-74

[Hoppe92]   H. Hoppe, T. DeRose, T. Duchamp, J. McDonald, W. Stuetzle, "Surface Reconstruction from Unorganized Points", SIGGRAPH '92 Proceedings, pp. 71-78

[Laveau94]   S. Laveau, O.D. Faugeras, "3-D Scene Representation as a Collection of Images and Fundamental Matrices", INRIA Technical Report No. 2205, February 1994

[Levoy85]   Mark Levoy, Turner Whitted, "The Use of Points as a Display Primitive", Technical Report TR 85-022, The University of North Carolina at Chapel Hill, Department of Computer Science, 1985

[Levoy96]   Mark Levoy, Pat Hanrahan, "Light Field Rendering", SIGGRAPH '96 Proceedings, ACM SIGGRAPH, pp. 31-42

[Mark97]   William R. Mark, Leonard McMillan, Gary Bishop, "Post-Rendering 3D Warping", Proc. 1997 Symposium on Interactive 3D Graphics, pp. 7-16

[Max95]   Nelson Max, Keiichi Ohsaki, "Rendering Trees from Precomputed Z-Buffer Views", 6th Eurographics Workshop on Rendering, June 1995, pp. 45-54

[McMillan95] Leonard McMillan, Gary Bishop, "Plenoptic Modeling: An Image-Based Rendering System", SIGGRAPH '95 Proceedings, pp. 39-46

[Pulli97]   K. Pulli, M. Cohen, T. Duchamp, H. Hoppe, L. Shapiro, W. Stuetzle, "View-based Rendering: Visualizing Real Objects from Scanned Range and Color Data", *Rendering Techniques '97*, Proc. Eurographics Workshop, pp. 23-34

[Reeves83]   William T. Reeves, "Particle Systems - A Technique for Modeling a Class of Fuzzy Objects", SIGGRAPH '83 Proceedings, pp. 359-376

[Reeves85]   W.T. Reeves, "Approximate and Probabilistic Algorithms for Shading and Rendering Structured Particle Systems", SIGGRAPH '85 Proceedings, pp. 313-322

[Reeves87]   William T. Reeves David H. Salesin, Robert L. Cook, "Rendering Antialiased Shadows with Depth Maps", SIGGRAPH '87 Proceedings, pp. 283-291

[Smith84]   Alvy Ray Smith, "Plants, Fractals and Formal Languages", *Computer Graphics* (SIGGRAPH '84 Proceedings), Vol. 18, No. 3, July 1984, pp. 1-10

[Sato97a]   Yoichi Sato, Mark D. Wheeler, Katsushi Ikeuchi, "Object Shape and Reflectance Modeling from Observation", SIGGRAPH '97 Proceedings, pp. 379-387

[Sato97b]   Yoichi Sato, Personal Communication (Question period following presentation of [Sato97a], SIGGRAPH 1997)

[Schaufler97] Gernot Schaufler, "Nailboards: A Rendering Primitive for Image Caching in Dynamic Scenes", 8th Eurographics Workshop on Rendering, pp. 151-162

[Turk94]   Greg Turk and Marc Levoy, "Zippered Polygon Meshes from Range Images", Proc. SIGGRAPH '94 Proceedings, pp. 311-318

[Westover90] Lee Westover, "Footprint Evaluation for Volume Rendering", *Computer Graphics* (SIGGRAPH '90 Proceedings), Vol. 24, No. 4, 1990, pp. 367-376.

[Williams78] Lance Williams, "Casting Curved Shadows on Curved Surfaces", SIGGRAPH '78 Proceedings, pp. 270-274

[Wolberg90] G. Wolberg, **Digital Image Warping,** IEEE Computer Society Press, Los Alamitos, California, 1990.

[Wong97]   Tien-Tsin Wong, Pheng-Ann Heng, Siu-Hang Or, Wai-Yin Ng, "Image-based Rendering with Controllable Illumination", *Rendering Techniques '97*, Proc. Eurographics Workshop, pp. 13-22

[Zhang97]   Hansong Zhang, Kenneth E. Hoff III, "Fast Backface Culling Using Normal Masks", Proc. 1997 Symposium on Interactive 3D Graphics, pp. 103-106

Editors' Note: see Appendix, p. 331 for colored figure of this paper

Rendering hyper-sprites in real time

Dr. Gavin Miller, Interval Research Corporation
Marc Mondesir, Stanford University
miller@interval.com

Abstract. This paper describes efficient algorithms for rendering sprites with the appearance of ray-traced objects. These sprites refract what is behind them and reflect the user as seen by a video camera. Applications include interactive multimedia as well as previewing glossy and metallic ink effects.

1 Introduction

Most graphical user interfaces, and many multi-media titles, are based on a composited sprite model. Images are layered on top of each other using a color value and an alpha value at each pixel. As such, sprites are only capable of supporting a limited range of shading effects, usually diffuse surfaces with fixed lighting, or reflective objects with fixed reflections. The continuing utility of sprites raises the question of whether they may be generalized to a more extended class of effects. The Talisman architecture [4] used warped sprites to amortize the cost of 3-D rendering over a number of frames in an animation sequence. This paper, on the other hand, explores the combination of sprite-based techniques with the visual realism of ray-tracing algorithms. The new interfaces allowed by such an extension include tools which have a more realistic appearance. Examples of these include a paint dropper which looks as if it is made of glass, and a magnifying glass which depicts convincing optical refraction and reflection effects.

It was proposed in [2], that a truly realistic computer display would create the illusion of objects which reflect the optical surroundings of the display. Looking at the image of a shiny object, the user should see his or her face reflected in it. This effect was dubbed hyper-reality, the theoretically optimal device for which would employ a 2-dimensional array of cameras. A simplified model is chosen here in which a single camera is pointed at the user. Sprites of reflective objects are used to re-map a video stream to create the illusion of real-time reflections. For this reason the sprites will be referred to as "hyper-sprites".

It is possible to imagine future printing technologies capable of creating a variety of surface effects such as metallic inks and localized glossy finishes. Previewing such effects in existing graphics systems would not give a very realistic depiction of the final printed product. If, instead, we combined video-based real-time reflections with convincing refraction effects, then we could visualize such composite structures in a realistic manner, but to do so would require real-time ray-tracing.

Ray-tracing has been used to generate images of great realism supporting refraction and reflection effects as well as shadows [5]. Unfortunately, ray-tracing a scene is computationally intensive. Great efforts have been put into reducing the number of ray-object intersections which need to be computed. However, despite the best efforts of algorithm development, frame times for general ray-traced images have remained outside the scope of real-time rendering on personal computers.

An alternative approach to reducing ray-object intersections is to cache image-space data at every pixel. This allows rapid changes in the shading while the camera and geometry are fixed. In [3] a scene was ray-traced once as intermediate results were cached at every pixel. The same scene could be rendered with a new set of lights in a small fraction of the original rendering time. While impressive as an acceleration method, the range of fast interactions was limited. This paper attempts to broaden Séquin's work to create a series of sprite-like objects. The sprites can be translated over a background image and each other, while maintaining the illusion of correct ray-tracing.

2 Pre-ray-tracing a model

To optimize the speed of the final graphical interaction, we pre-compute as much as possible in a step called "pre-ray-tracing". For the sake of speed and simplicity, it will be assumed that the lights are fixed relative to the models in the scene. It is also assumed that the objects are rendered in orthographic projection. To pre-ray-trace the model we use a modified form of recursive ray-tracing. Each ray is split at a dielectric boundary into a reflected and a refracted ray. As the rays propagate they are attenuated by the reflection and refraction coefficients. The attenuation coefficients are concatenated with those of parent rays to give a "contribution coefficient" to the original pixel. Rays will eventually intersect a diffuse surface, in which case their brightness is computed, or they will exit the bounding volume and intersect with the environment.

For a given parent ray entering the model, this results in a total diffuse brightness along with a list of environment ray vectors and their contribution coefficients. A hyper-sprite is a data-structure which stores these rays and intensities for a two-dimensional array of pixels.

2.1 Rendering layered hyper-sprites

To create an image of a hyper-sprite we need to have a way of intersecting the environment rays with other objects in the scene. In the general case, of course, this requires a ray-tracer. However, we may create approximations to a useful special case.

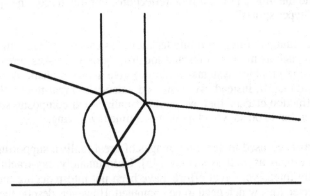

Fig. 1 Ray-traced sphere above a background plane.

Figure 1 shows a cross-section of a sphere sitting above a plane. Rays incident on the sphere will split and end up being reflected and refracted such that they either intersect the background plane or intersect the foreground environment. If we translate the hyper-sprite parallel to the background then the intersection points will be translated by the same amount. Thus we may recast the hyper-sprite as a list of samples of intersection points plus the diffuse intensities. A more complex example is shown in Figure 2, where two sprites are being rendered, starting with the one closest to the background called Hyper-sprite A.

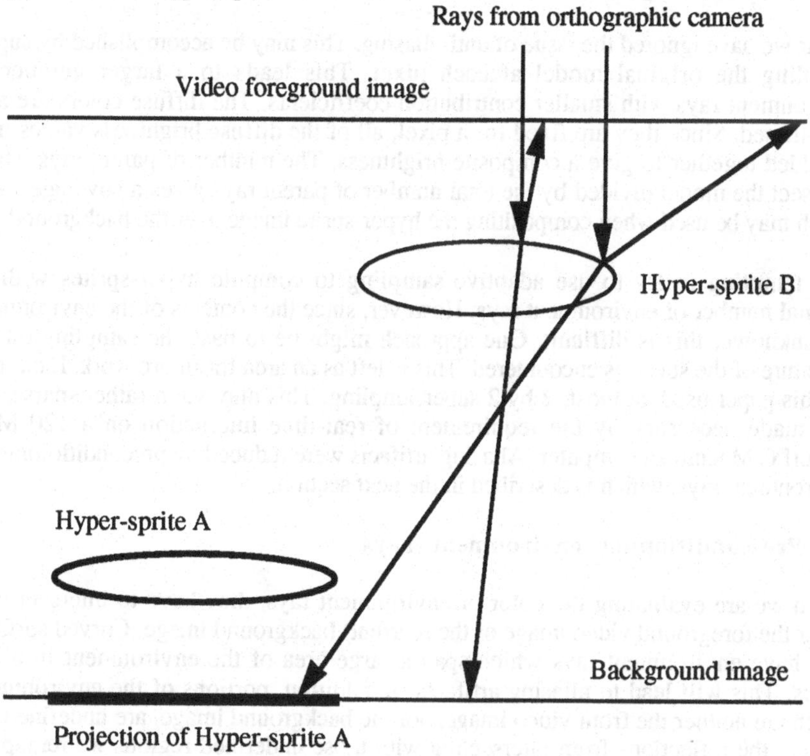

Rays from orthographic camera

Video foreground image

Hyper-sprite B

Hyper-sprite A

Background image

Projection of Hyper-sprite A

Fig. 2. Hyper-sprite B being rendered after A was projected to the background.

The algorithm for rendering a collection of sprites is as follows:

1. Initialize the display image to a background picture.

2. Initialize the foreground image to a frame of video from the camera pointing at the user.

3. For the farthest away sprite, compute the translated intersection points, evaluating the image brightness at these points, and summing the intensities for that pixel in the hyper-sprite.

4. When all of the hyper-sprite's pixels have been computed, composite the hyper-sprite image over the display image.

5. Repeat for the next hyper-sprite.

This algorithm makes the approximation that the hyper-sprite, once drawn, exists at the plane of the background image. Nearer sprites do not show the refraction of the more distant sprites directly, instead they refract the projection of the sprites onto the background. This is not exactly correct, but it does give acceptable results and prevents the need for general ray-object intersections during the interactive portion of the algorithm.

2.2 Anti-aliasing

So far we have ignored the issue of anti-aliasing. This may be accomplished by super-sampling the original model at each pixel. This leads to a larger number of environment rays with smaller contribution coefficients. The diffuse colors are also anti-aliased. Since they are fixed for a pixel, all of the diffuse brightness values may be added together to give a composite brightness. The number of parent rays which intersect the model divided by the total number of parent rays gives a coverage value which may be used when compositing the hyper-sprite image over the background.

It is tempting to try to use adaptive sampling to compute hyper-sprites with an optimal number of environment rays. However, since the contents of the environment are unknown, this is difficult. One approach might be to base the sampling on the curvature of the surfaces encountered. This is left as an area for future work. Examples for this paper used, at most, 2 by 2 supersampling. This may seem rather sparse, but was made necessary by the requirement of real-time interaction on a 120 MHz PowerPC Macintosh computer. Aliasing artifacts were reduced by preconditioning the environment rays, which is described in the next section.

2.3 Pre-conditioning environment rays

When we are evaluating the color of environment rays, they have to intersect with either the foreground video image or the (current) background image. Curved surfaces may have environment rays which span a large area of the environment in a few pixels. This will lead to aliasing artifacts. In addition, portions of the environment, which are neither the front video image, nor the background image, are undefined. To prevent the reflections from intersecting with those undefined regions we remap the environment rays to be more closely aligned to the Z-axis. This also has the effect of deflecting the environment rays so that their intersection points are clustered more closely around the location of the hyper-sprite which helps to reduce aliasing artifacts.

If the display image is on an x, y grid, then the ray z component of an environment ray is modified as follows:

$$r_z' = \begin{cases} r_z + s & \text{if } r_z > 0 \\ r_z - s & \text{if } r_z < 0 \end{cases}$$

where r is the environment ray direction and s is the deflection coefficient. The larger the value of s the more localized the intersection points of the deflected rays.

To compute the intersection with the background we use:

$$u = x - z\frac{r_x}{r'_z}$$

$$v = y - z\frac{r_y}{r'_z}$$

where u, v are the intersection parameters, and x, y, z is the environment ray origin and r is the environment ray direction. Note that the coordinate origin is zero at the background plane. When the values of u, v are outside the pixel range of the foreground or background image, they are made to wrap with mirror symmetry at the boundaries.

Figure 3 (see Appendix) shows a metal and a diffuse sphere. Figure 3a shows the scene with 1 by 1 supersampling. Figure 3b shows the same scene with 2 by 2 supersampling. Note that the edges of the diffuse sphere as well as the reflections in the metallic sphere are antialiased. The user is visible reflected in the central portion of the metallic sphere, whereas the background is seen reflected in the outer portion of the reflective sphere. The highlight on the reflective sphere was cached as part of the shading computation at the surface.

Figure 4 (see Appendix) shows metallic and glass text. The metallic text is derived from a font and resembles embossed metal. The same geometry used with both reflection and refraction gives the impression of glass. This example illustrates the potential for using hyper-sprites to preview the metallic and glossy effects that might be possible with future generations of computer printers. As the user moves his or her head the impression of the actual presence of metallic and glossy materials is convincing. The shallow nature of the geometry prevents parallax cues from destroying the visual illusion of embossed specular objects.

2.4 Shadows

Since the lighting is fixed relative to the geometry, we may compute a shadow projected onto the background image. This is stored as an intensity map. If the sprite is translated parallel to the surface, the shadow will be translated by the same amount. If the sprite is moved in depth, the shadow will be translated in the plane of the background.

Once again, if we have several sprites, the task of casting shadows onto one sprite from another requires general ray-tracing. We chose to eliminate all shadows except those on the background image. A brightness map is initialized to maximum brightness. Then the shadow for each sprite is scan-converted into the brightness map. The algorithm takes the minimum of the intensity in the brightness map and in the corresponding shadow image. This combines multiple overlapping shadows in a reasonable way. The brightness map is then used to attenuate the background image before using it to render the hyper-sprites. (Since we are trying to optimize performance, the shadow image is cleared using the location of the shadows from the previous frame. In this way we do not need to initialize the whole image each frame.)

Figure 5a (see Appendix) shows a dropper sprite in front of a map. The dropper's shadow is visible through the glass. In Figure 5b, a magnifying glass is being used in combination with the dropper. The shadows of both are visible, although the magnifying glass does not cast a shadow onto the dropper handle, and the magnifying glass cannot be seen reflected in the dropper.

3 Areas for future work

In the future, hyper-sprites could be combined with head-tracking in the following way: The position of the user's head could be used to select the version of the hyper-sprite that was rendered from the direction from which it is being viewed. As the user moved, he or she would observe correct motion parallax effects (such as those discussed in [1]) as well as changes in the reflected and refracted ray directions. Unfortunately, this would require enormous amounts of memory since a 2-D array of hyper-sprites would be required for each object. Another problem that might arise is discontinuous changes in the order in which the hyper-sprites would be rendered. This is a problem with all composited sprite-based systems.

4 Conclusions

Objects may be pre-ray-traced and then composited like sprites while still giving the illusion of correct shadows and refraction effects. Feeding such sprites with a video image of the user gives rise to visually convincing reflection effects. A software implementation of these ideas was capable of rendering quite large sprites at a few frames per second on a 120 MHz PowerPC personal computer.

5 Acknowledgments

This work was completed at the Graphics Research Group of Apple Research Laboratories at Apple Computer, Inc. in Cupertino. Thanks to Douglass Turner for use of his ray-tracing code, and to Steven Rubin for programming support.

References

1. Deering, Michael, "High Resolution Virtual Reality", Computer Graphics, 26, 2, July 1992, pp195-201.

2. Miller, Gavin S. P., "Volumetric Hyper Reality: A Computer Graphics Holy Grail for the 21st Century", Proceedings of Graphics Interface '95.

3. Séquin, Carlo H., and Eliot K. Smyrl, "Parameterized Ray Tracing", Computer Graphics, Vol. 23, No. 3, July 1989, pp307-314.

4. Torborg, Jay, James T. Kajiya, "Talisman: Commodity Realtime 3D Graphics for the PC", Computer Graphics Proceedings, Annual Conference Series, 1996. pp. 353-364.

5. Whitted, Turner, "An Improved Illumination Model for Shaded Display", CACM, vol. 23, no. 6, June 1980, pp343-349.

Editors' Note: see Appendix, p. 333 for colored figures of this paper

Automatic Calculation of Soft Shadow Textures for Fast, High Quality Radiosity

Cyril Soler and François X. Sillion

iMAGIS [1] – GRAVIR/IMAG

Abstract.
We propose a new method for greatly accelerating the computation of complex, detailed shadows in a radiosity solution. Radiosity is computed using a "standard" hierarchical radiosity algorithm with clustering, but the rapid illumination variations over some large regions receiving complex shadows are computed on the fly using an efficient convolution operation, and displayed as textures. This allows the representation of complex shadowed radiosity functions on a single large polygon. We address the main issues of efficiently and consistently integrating the soft shadow calculation in the hierarchical radiosity framework. These include the identification of the most appropriate mode of calculation for each particular configuration of energy exchange, the development of adequate refinement criteria for error-driven simulation, and appropriate data structures and algorithms for radiosity representation and display. We demonstrate the efficiency of the algorithm with examples involving complex scenes, and a comparison to a clustering algorithm.

1 Introduction

Despite the impressive progress in realistic rendering techniques over the last two decades, the production of very accurate images involving the simulation of light exchanges remains an expensive process, especially for complex scenes (more than a few thousands of objects).

Hierarchical radiosity techniques[4, 9, 7] were very successful in introducing the ability to trade accuracy for speed in a controlled manner. Unneeded form factor calculations can be avoided, and the solution to the global illumination problem can be estimated with the desired accuracy. In practice however, these techniques are mostly used to compute fast approximations of the solution. While possible in theory, the calculation of very high quality solutions using hierarchical radiosity is still expensive for complex scenes.

Of course this depends on the definition of "quality". For many applications, including lighting design, it may suffice to correctly represent the overall balance and flow of light. For applications in realistic rendering, however, shadows should be of very high quality[6]. This goal pushes hierarchical radiosity to its limits, because it requires a very fine subdivision of shadow variation areas. This is a direct consequence of the simple shape assumption of the radiosity function –constant or linear–across each mesh element.

In this paper, we introduce a new approach where complex shadowed radiosity functions can be represented across large elements using textures. Textures have previously

---

[1]iMAGIS is a joint research project of CNRS, INRIA, INPG and UJF within the GRAVIR/IMAG laboratory. Contact: csoler@imag.fr, http://W3imagis.imag.fr/Membres/Cyril.Soler/csoler.gb.html

been used for the representation of precomputed radiosity[6], or the storage of sampled illumination values[5, 1], but in our approach, textures are computed as a whole, and on the fly as an alternative method for simulating the energy transfer between two hierarchical elements. Basically, the textures are obtained by modulating the direct irradiance calculation by the "soft shadow map" that accounts for the portion of the source visible from any point on the receiver. This idea was used in[11, 8] but at the expense of computing time due to the large number of visibility tests necessary to obtain the shadow map.

We have recently introduced a new tool for the calculation of soft shadow maps, based on the convolution between images of a source and a blocker[10]. This tool has definite strengths: it intrinsically captures fine soft shadow details, integrates an error control tool and produces artifact-free soft shadow maps at interactive rates. It also has an associated cost, and should only be used in appropriate configurations. In a general radiosity simulation, we encounter a variety of situations, some of which may be suited for the use of this tool. In this paper we show

- that the convolution tool can dramatically accelerate the calculation of slightly approximate soft shadows across surfaces, consistently with hierarchical radiosity.
- how to select configurations where the convolution tool can/should be used
- how to choose appropriate convolution parameters automatically
- how to combine soft shadow textures with traditional hierarchical radiosity representations
- how to control the hierarchical simulation process, using and balancing the variety of possible refinement techniques

The resulting hierarchical radiosity system generates global illumination solutions, including very accurate soft shadows, in a matter of seconds for scenes containing several tens of thousands of polygons.

The paper is organized as follows: we begin with a rapid summary of the convolution technique for the generation of soft shadows (Section 2). Section 3 discusses the issues raised by the combination of hierarchical radiosity and the convolution method. In Section 4 we propose a number of solutions to these issues, in particular for the combination of different radiosity representations. Refinement criteria for the hierarchical simulation are discussed in Section 5. We then present some results (Section 6) and conclude.

2 Creating soft shadow textures using convolution

This section is intended to provide the reader with the essential principles for the computation of soft shadow maps using a convolution operation. A detailed presentation of this technique can be found in [10].

2.1 Basic principle

The *convolution method* provides a means to compute the illumination on a selected receiving object, subject to the light emitted by an extended light source, in the presence of a set of occluding objects. We refer to these occluders collectively as "the blocker" since it is considered as a cluster.

We first obtain an image of the source and the blocker by rendering the associated geometry in an offscreen buffer. For this projection, the source is viewed using an orthogonal transformation and the blocker is viewed using a perspective projection from the center of the source (See Fig.1)

Secondly, we compute the convolution of the two images by multiplication in the spectral domain, using *Fast Fourier Transforms*.

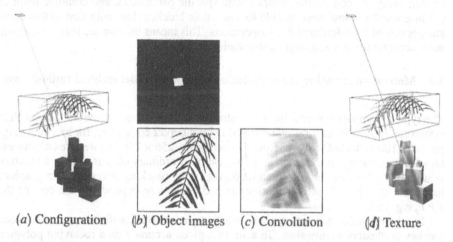

(a) Configuration   (b) Object images   (c) Convolution   (d) Texture

Fig. 1. Three basic steps of the convolution method: To compute the shadow cast by a extended light source on a receiver (here a cluster of cubes) through a blocker (a) we first compute offscreen images of the source and blocker (b), then compute the convolution of these images (c). The result is used as a texture to modulate illumination on the receiver (d)

The resulting image can be shown to be a valuable approximation of the portion of the source that is visible from any given point on the receiver. We call such an image a *soft shadow map*. We simulate the illumination on the receiver by using this image as a texture that we modulate by the reflectance of the receiver and the unoccluded form factor to the source.

During offscreen rendering of the source and blocker, both projection viewing frustum are adjusted so as to achieve a particular aspect ratio between the two images. This ratio accounts for the relative distance of the three components: the further the source, the smaller its size in the image, and the sharper the shadow.

This principle can be extended to non planar or non uniform sources, and more general receivers such as object clusters [10].

2.2 Hierarchical combination of soft shadow textures from different blockers

The above convolution method only provides a mathematically exact soft shadow map when the source, blocker and receiver are planar and parallel. In this particular case, using the convolution to compute the illumination on the receiver only amounts to neglecting the correlation between the radiosity kernel and the visibility terms.

To better understand the approximation involved when the blocker is non-planar, note that two blockers that look identical as viewed from the source when generating the blocker image will produce the same shadow on the receiver, if their relative distances from the source and the receiver are identical. Thus the actual position and size of the blocker between the source and receiver is not totally accounted for. The most noticeable consequence is that in a large blocker, all polygons produce shadows of equal sharpness, though their relative distance from the source and the receiver may vary.

In a general configuration, the thicker the blocker, the larger is the error. We give in [10] an estimator of this error, and a way of controlling it: to obtain a more accurate result, we separately compute soft shadow maps for N child clusters in the blocking cluster, using the convolution method with specific parameters, and combine them to produce a soft shadow map suitable for the whole blocker. The main cost of this operation is that of $2N$ additional FFT operations. This means that we are able to compute more accurate images at a larger computation cost.

2.3 Motivation for using the convolution method in a hierarchical radiosity system

The typical computation cost for a soft shadow texture using our method is less than $400ms$ for 256×256 textures on both SGI O^2 and $Onyx2$ computers, for blockers ranging from 100 to 40,000 polygons. As a general rule, 256×256 textures are only necessary for very large surfaces such as walls. On most ordinary objects, a 64×64 texture is sufficient. Because the offscreen rendering of the blocking polygons uses graphics hardware, the cost of the convolution method essentially depends on the cost of the FFT, e.g. on texture size.

In a hierarchical radiosity system, energy transfers are usually computed between patches or clusters of any size. To achieve a given accuracy on a receiving polygon, complex visibility events may impose a fine subdivision to this polygon, and therefore the computation of many form factors including visibility tests.

The preceding arguments suggest that computing the energy transfers on sufficiently large receivers using the convolution method can be far cheaper than performing a classical subdivision, in the case where a large number of complex blocking objects participate to the shadow on the receiver.

We discuss in the next section the main avenues of incorporating the convolution method into a hierarchical radiosity system.

3 Principles for using soft shadow maps in a hierarchical radiosity algorithm

Considering the potential benefits of using the convolution method for the calculation of shadow maps, we would like to leverage the power of the technique in the context of hierarchical radiosity. This should allow a hybrid system to automatically select the convolution method whenever appropriate, while maintaining the benefits of radiosity (global energy balance, indirect illumination effects, ability to provide solutions with various cost/accuracy trade-offs, generation of view-independent solutions for diffuse scenes, etc.).

Therefore we wish to be able to compute the illumination on some potentially large receivers using the convolution method, during a hierarchical radiosity iteration. This can be accomplished by introducing two distinct types of *links* in a hierarchical radiosity system. In addition to the usual type, carrying a form factor estimate between two objects, *convolution links* are created for appropriate configurations of the source and receiver. The radiosity contribution of a convolution link on the receiver is represented by the association of a soft shadow texture and the kernel values used to modulate it (representing unoccluded irradiance estimates at selected points on the receiver).

Specific algorithms will be presented in the next section: In particular, care must be taken to derive a consistent algorithm for the representation and combination of energy contributions, coming either from "normal" radiosity calculations or from the convolution tool. Therefore, the gathering operation and the Push/Pull step must be modified to ensure consistency and efficiency.

Proper rendering of the computed solution also requires some care, as we must merge uniform or interpolated radiosity values with several textures modulated by arbitrary factors.

Since we are introducing an alternative method for the representation of light exchanges, the hierarchical refinement phase must be extended to properly incorporate this additional choice. We have designed an algorithm for link refinement that estimates the error associated with a given link and then chooses the best way to reduce this error among the different options for each kind of link. This algorithm will be presented in detail in Section 5

4   A complete hierarchical radiosity system with accurate shadows

We now describe a possible implementation of the above ideas, realizing a consistent combination of hierarchical radiosity with clustering and soft shadow maps. We explain how to store and gather energy across convolution links, how to perform a consistent push/pull using the information computed during the gathering operation, and how to display the resulting radiosity functions.

4.1   Gather

For standard links, gathering means multiplying a single radiosity value of the source element by the form factor associated to the link, which gives its irradiance contribution. As shown in the push/pull section, the source radiosity value is the sum of standard radiosity and average values of all textures assigned to that source.

For a convolution link, the source polygon is rendered with all its radiosity textures to form the source image, which accounts for the correlation between the source radiosity fonction and the visibility through the blocker. For that kind of link, a complete gathering operation would ideally consist in computing the soft shadow texture and modulate it by the unoccluded illumination values. For practical reasons (especially concerning the display) we have chosen to store soft shadow textures equipped with illumination values at the leaves of the hierarchy only. One reason for this is that the illumination values strongly depends on the geometry of the leaf itself, rather than on the parent on which the convolution link is attached, which in certain cases may be a cluster.

Therefore, the soft shadow textures computed during the gather step are just stored in an *soft shadow texture list* at the current hierarchical level, waiting to be pushed to the leaves and equipped with adequate unoccluded illumination values during the push/pull operation.

Note that just as in a standard clustering algorithm, the intra-cluster visibility of the receiver is not accounted for, whereas is can be achieved when rendering, using a two-pass shadowing technique.

4.2 Push/Pull

During push/pull, uniform irradiance values coming from standard links are added together down to the leaves where they are multiplied by the local reflectance value. At the same time, soft shadow textures stored during the gathering are collected on a *texture stack* and pushed down to the leaves of the hierarchy.

Once on a leaf, for each of the collected soft shadow textures, a proper set of modulation values is computed from reflectance and unoccluded form factor values. Practically, these values are computed at the vertices of a raw *display mesh*, which size depends on the geometry of the leaf. We also store of each of these points adequate texture coordinates in the plane of the soft shadow texture, that are used when rendering. We call the resulting object, combining the soft shadow map, the irradiance values and texture coordinates, a *complete* texture. Thus the Push/Pull operation essentially transforms the soft shadow maps into complete textures at the leaf nodes of the hierarchy.

Using the information computed at the same mesh nodes, we also compute an approximative mean value of the radiosity contribution of the texture on the leaf. This value is added to the uniform radiosity value due to standard links and pulled up in the hierarchy in a classical way.

Since we only store the complete soft shadow textures on leaves of the hierarchy, we must also push down the complete soft shadow textures possibly existing as a result of a preceding push/pull pass on elements that are no longer leaves of the hierarchy. In that case, the shadow texture is separated from its display mesh, and the texture itself is added to the stack of pushed down textures.

Using this algorithm, the array containing the convolution image is shared between all leaves of an element that receives a convolution link. Only display meshes are specific to each leaf.

The pseudo-code presented in Fig.2 shows the entire push/pull process.

4.3 Display

To render the computed solution, we need to display the total radiosity value of each leaf of the hierarchy. This means that we must add to the standard radiosity of each leaf, the list of soft shadow textures that can be stored on it.

We first draw the polygon with its uniform radiosity value and then the different radiosity textures using *OpenGL* blending capabilities (with an add operation).

When rendering the texture modulated by the direct illumination values at each node of the texture mesh, *OpenGL* clamps these color values to the range $[0, 1]$ before using them to modulate the texture. This means that in a situation where the direct illumination exceed 1.0 but the modulated texture still lies into $[0, 1]$, the complete texture won't be rendered correctly. To achieve the adequate texture modulation by an illumination map L whose maximum value M is larger than 1.0, we render the texture n times and modulate it by $\frac{1}{n}L$, where n is the smallest integer larger than M. This technique partially solves the problem for it also scales by a factor n the steps in the shadow texture (These steps come from the fact that textures are stored in 8 bits arrays). Except for very high illumination gradients due to extreme light values, this effect is not noticeable.

Note that because each texture is drawn on the same polygon that the one used to display the uniform radiosity values (i.e leaves of the hierarchy), setting the $z-$buffer comparison function to *less or equal* is sufficient to avoid $z-$buffer comparison artifacts in the resulting image.

We summarize in Fig.3 the different steps of rendering algorithm.

```
PushPull(node H, Spectrum IrradDown, Stack TextureStack, Spectrum RadUp)
    RadUp= 0
    TextureStack→Push(H.SoftShadowTexList())

    if(H.Is_A_Leaf())
        RadUp = (H.Irradiance() + IrradDown) * H.Reflectance()
        ForAll T∈ TextureStack
            RadUp += ComputeTexAverage(H,T)
    else
        TextureStack→Push(H.CompleteTextures())
        Spectrum R
        ForAllChildren c of H
            PushPull(c,IrradDown+H.Irrad(),TextureStack,R)
            RadUp += R * c.Area()
        TextureStack→Pop(H.CompleteTextures())
        H.DeleteCompleteTextures()
        RadUp /= H.Area()

    H.SetRadiosity(RadUp)
    TextureStack→Pop(H.SoftShadowTexList())
    H.ResetSoftShadowTexList()
```

Fig. 2. Pseudo code for a Push/Pull with radiosity textures

5 Controlling refinement

We present in this section the refinement algorithm we use in our method.

5.1 Principle

The general purpose of a refiner is to decide if computing an energy transfer between a given source S and receiver R can be done while keeping the approximation errors under a user-defined bound ε. If so, it establishes a link, that contains the necessary information to compute the transfer (for instance form factors). If not, it assumes that a more precise result can be obtained when computing transfers between sub-regions of the source and/or the receiver, and subdivides the element that most reduces this error.

Since we deal with two kinds of links, the refiner must be able to decide where it can use a convolution link or not. This requires a criterion for the identification of source/receiver pairs for which using a convolution is possible and useful. We propose that the configuration must fulfill the following conditions:

1. The energy transfer between the source and the receiver must be "significant".
2. The receiver must be large enough to justify the use of a texture with non-trivial size.
3. Blockers must be present between the source and receiver, and cast noticeable shadows: too smooth a shadow can be easily represented by smooth shading a small collection of large sub-areas of the receiver.

```
DrawPoly(Polygon P)
    SetColor(P.Radiosity() - P.Textures().MeanValue() )
    DrawSimplePoly(P)
    Enable(BLENDING)
    SetZBufferFunction(≤)
    ForAll T∈ P.Textures()
        n = SmallestIntegerLargerThan(T.MaximumLuminance)
        for i = 1to n
            DrawPolyWithTexture(P,T,n)
    Disable(BLENDING)
```

Fig. 3. Pseudo code for rendering a polygon with radiosity textures

4. The computation of the shadow map using convolution must reach a given accuracy at a reasonable cost.

We have seen in Section 2.2 that when computing a transfer using the convolution method, an alternative way of reducing the error is to subdivide the blocker. Therefore, in the general case, the refiner must answer the following questions:

- do I establish a link at this level, and what kind of link do I use ?
- if not, I choose between the three following refinement alternatives: source refinement, blocker refinement, or receiver refinement.

We have combined these requirements to form the refinement algorithm described in the next section.

5.2 Refinement algorithm

Our strategy is to use a convolution whenever possible, provided that the approximation error and the computation cost lie some user-defined intervals.

Thus, our refinement algorithm consists in successively checking for all the convolution site selection conditions in order of their computation cost. If using a convolution is not technically possible, or can't satisfy the refinement error condition, we transfer the refinement control to a standard refiner that tries to establish a standard link or split the source or receiver (See Fig.4 for a general overview of this). As a standard refinement algorithm, we use a method similar to the algorithm described in [2].

We first compute an upper bound B on the energy transfer from the source to the receiver, If B is smaller than a user-defined threshold, we do not refine the current link (or we create a standard link, if no link is already present).

The next test concerns the size of the receiver: complex scenes are usually filled with fine polygonal models, and we want to avoid computing too large a number of convolutions for very small polygons. For more generality, we want this test to apply to clusters as well as surfaces. Thus, we compare the dimensions of the object bounding box to a fixed value.

At that point, a little more work must be done to examine the blockers that occlude visibility between the source and receiver. The blocking clusters are first selected using an algorithm described below. If no blockers are found, we do not apply a convolution.

Otherwise, we compute the number of sub-blockers to use in the convolution method necessary to be able to compute an illumination map whose error (Given by the error criterion from [10]) is below an error threshold E_{max}. We set the variable E_{rr} to the best error estimation reached. In both cases N is the number of sub-blockers to use.

If $E_{rr} > E_{max}$, no convolution can be used. We transfer control to the standard refiner. Otherwise, the required precision can be satisfied using N blockers, i.e $2 + 2N$ FFT transformation. We have experimentally determined a constant C for which the computation time for a FFT on a $n \times n$ image is approximatively

$$t(n) = C\, n^2 \log(n)^2$$

Thus, the computation time for the illumination map can be estimated as

$$t = (2 + 2N)C\, n^2 \log(n)^2$$

If t is larger than a given computation time threshold, we give up the convolution. In the case where t is small enough, we create a convolution link between S and R, using N sub-blockers.

When one of the convolution sites identification criteria fails, some information can nevertheless be extracted from the various computations: The blocker selection process helps classifying the visibility configuration (full or partial visibility). While computing E_{rr} and N, a bound α_{min} on the sharpness of the penumbra regions on the receiver can be computed. We use this bound for determining a dimension δ under which it is not interesting to split the receiver.

In summary, we make use of four user defined thresholds to control refinement:

a_{min} : minimum receiver area for a convolution link.
E_{max} : maximum blocker error tolerance for a convolution link.
t_{max} : maximum computation time allowed for a single convolution.
ε : minimum energy transfer value for a link.

Blocker selection. The selection of blocking clusters between the source and the receiver is done using a shaft culling method [3]. We first build the shaft between the source and the receiver. Then, a recursive procedure selects the largest clusters that intersect the shaft without containing the source or the receiver. Generally, a single blocker is selected. When two or more clusters are selected, we pack them in a single virtual blocker.

The non-containing test ensures that neither the source nor the receiver would participate to the resulting umbra, as a part of the blocker hierarchy. A more careful examination of the problem leads to turning this test into a non-overlapping test, to also discarding possible blocking objects that hide behind the receiver with respect to the source or hide behind the source with respect to the receiver.

As a consequence, this test must be sufficiently numerical tolerant to allow a cluster lying on a polygon (Typically an object lying on the floor) to remain selected as a occluder/receiver configuration.

6 Results

We show on the left hand side of Figure 6 (See also color plates) an image of a classical interior scene computed with 16 convolution links (from each source to each of the

208

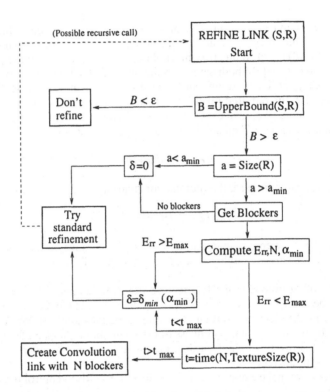

Fig. 4. Overview of the refinement algorithm for a link between a source S and receiver R

two walls and the floor, and from some of the sources to clusters containing the desks). Three iterations were performed for a total computation time of 20.2 seconds on an *Onyx2 workstation - Reality engine*. On the right hand side is a solution obtained with only classical hierarchical radiosity with clustering in the same computation time. We show the HR mesh and the smooth shaded version of the result as a comparison to the previous image. Very fine shadows have been captured by our method with only a few convolution links in a short time, resulting in a high quality radiosity solution.

Figure 5 (See color plates) shows the effect of changing the refinement parameters on the proportion of convolution links (e.g textures) used by the refiner. The blocker time parameter t_{max} is fixed at 1sec. In (a) the use of convolution links is favored by a large blocker error E_{max}. The floor receives a convolution link (512×512 texture, 6 sub-blockers) at the highest level. In (b) we see that reducing the blocker error tolerance forces the refiner to switch to standard links in the center of the floor, where textures can no longer satisfy the error criteria E_{max} in less time than t_{max}. The progressive meshing around this region is a constraint in the quadtree. Convolution links arrive at the highest possible level in this region. Finally, in (c) too small a blocker error threshold totally prevents convolution links. The minimum receiving area for standard links has been set so as to obtain the same resolution with standard refinement as the texture in (a).

Note that some significant part of the computing time is used to compute transfers between the cubes or clusters that contain them. Some of the cubes also receive textures from convolution links from the source to the cluster containing them.

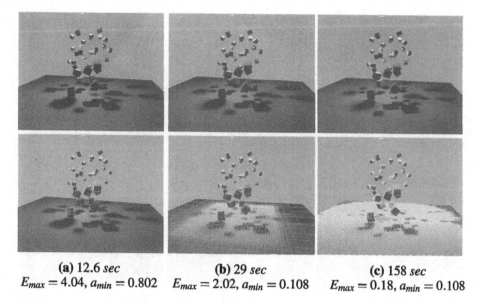

(a) 12.6 *sec*   **(b)** 29 *sec*   **(c)** 158 *sec*

$E_{max} = 4.04, a_{min} = 0.802$ $E_{max} = 2.02, a_{min} = 0.108$ $E_{max} = 0.18, a_{min} = 0.108$

Fig. 5. The effects of changing refinement parameters on the kind of link used and the computation time

7 Conclusions and future work

We have described a new radiosity algorithm based on the use of a convolution technique to compute shadow maps where complex blockers may cause a classical hierarchical radiosity algorithm to get lost in mesh subdivision. The algorithm basically merges two radiosity storage methods: spectrum and textures. It could be used with another texture computation method than the convolution with hardly any modification provided that the replacing algorithm gives a bound on the error incurred and a way to decrease it.

Storing radiosity values as textures is not compatible with *already-textured* environments, unless we manually compute the pixel-by-pixel product of radiosity textures and scene textures, which is a costly operation. Nevertheless, we can reasonably expect that this capability may be provided by future graphics hardware.

The intra-cluster visibility problem is not specifically treated by our method but it can be addressed by refining the receiver for textured links arriving at a cluster. This causes some parts of the cluster to become treated as occluders, thus producing the desired internal shadows.

Finally, a single user defined error threshold for controlling the refinement would be appreciated. The difficulty is here to find comparable error estimators for both kinds of links. We think that this gap can be filled using perceptually based error criteria.

References

1. James R. Arvo. Backward Ray Tracing. In *ACM SIGGRAPH '86 Course Notes - Developments in Ray Tracing*, volume 12, August 1986.

210

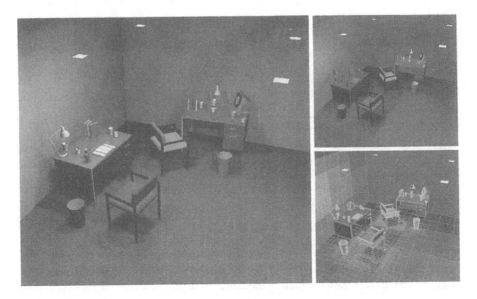

Fig. 6. Comparison between our method (*left*) and a standard HR algorithm (*right*) for equivalent computation times (20 *sec*, 5380 input polygons).

2. S. Gibson and R. J. Hubbold. Efficient hierarchical refinement and clustering for radiosity in complex environments. *Computer Graphics Forum*, 15(5):297–310, December 1996.
3. Eric Haines and John Wallace. Shaft culling for efficient ray-traced radiosity. In *Eurographics Workshop on Rendering*, 1991.
4. Pat Hanrahan, David Salzman, and Larry Aupperle. A rapid hierarchical radiosity algorithm. In Thomas W. Sederberg, editor, *Computer Graphics (SIGGRAPH '91 Proceedings)*, volume 25, pages 197–206, July 1991.
5. Paul Heckbert. Adaptive Radiosity Textures for Bidirectional Ray Tracing. In *Computer Graphics (ACM SIGGRAPH '90 Proceedings)*, volume 24, pages 145–154, August 1990.
6. Karol Myszkowski and Tosiyasu L. Kunii. Texture mapping as an alternative for meshing during walkthrough animation. In *Fifth Eurographics Workshop on Rendering*, pages 375–388, Darmstadt, Germany, June 1994.
7. François Sillion. A unified hierarchical algorithm for global illumination with scattering volumes and object clusters. *IEEE Transactions on Visualization and Computer Graphics*, 1(3), September 1995. (a preliminary version appeared in the fifth Eurographics workshop on rendering, Darmstadt, Germany, June 1994).
8. Philipp Slusallek, Michael Schroder, Marc Stamminger, and Hans-Peter Seidel. Smart Links and Efficient Reconstruction for Wavelet Radiosity. In P. M. Hanrahan and W. Purgathofer, editors, *Rendering Techniques '95 (Proceedings of the Sixth Eurographics Workshop on Rendering)*, pages 240–251, New York, NY, 1995. Springer-Verlag.
9. Brian Smits, James Arvo, and Donald Greenberg. A clustering algorithm for radiosity in complex environments. In Andrew Glassner, editor, *Proceedings of SIGGRAPH '94 (Orlando, Florida, July 24–29, 1994)*, Computer Graphics Proceedings, Annual Conference Series, pages 435–442. ACM SIGGRAPH, ACM Press, July 1994. ISBN 0-89791-667-0.
10. Cyril Soler and François Sillion. Fast calculation of soft shadows textures using convolution, 1998. to appear in SIGGRAPH 98 conference proceedings.
11. Harold R. Zatz. Galerkin radiosity: A higher-order solution method for global illumination. In *Computer Graphics Proceedings, Annual Conference Series: SIGGRAPH '93 (Anaheim, CA, USA)*, pages 213–220. ACM SIGGRAPH, New York, August 1993.

Editors' Note: see Appendix, p. 333 for colored figure of this paper

Three Point Clustering for Radiance Computations

Marc Stamminger, Philipp Slusallek, and Hans-Peter Seidel

Computer Graphics Group, University of Erlangen
{stamminger,slusallek,seidel}@informatik.uni-erlangen.de

Abstract. There has been great success in speeding up global illumination computation in diffuse environments. The concept of clustering allows radiosity computations even for scenes of high complexity. However, for lighting simulations in complex non-diffuse scenes, Monte-Carlo sampling methods are currently the first choice, because non-diffuse finite-element approaches still exhibit enormous computation times and are thus only applicable to scenes of very modest complexity.

In this paper we present a novel clustering approach for radiance computations, by which we overcome some of the problems of previous methods. The algorithm computes a radiance solution within a line space hierarchy, that allows us to efficiently represent light propagation and reflection between arbitrary non-diffuse surfaces and clusters.

1 Introduction

1.1 Rendering Equation

All global illumination algorithms compute an approximate solution of the rendering equation [Kaj86]. In this equation, a scene is assumed to be composed of surface patches in a non-participating medium. The reflection properties of the surfaces are described by a bidirectional reflectance distribution function, or BRDF for short. Furthermore, parts of the surfaces can emit light and thus act as light sources. In this paper, we will use the so called three-point parameterization of the rendering equation

$$L(y, z) = L^e(y, z) + \int_S f(x, y, z)\, G(x, y)\, L(x, y) dA_x,$$

where S is the set of all surfaces in the scene, $L^e(y, z)$ is the emission of surface point y in direction to z, $f(x, y, z)$ is the BRDF at a point y for light coming from x being reflected towards z. $G(x, y)$ is a purely geometrical term describing the geometric relation between the surfaces at x and y

$$G(x, y) = \frac{\cos(n(x), y - x)\, \cos(n(y), x - y)}{\pi \|x - y\|^2} V(x, y),$$

where $V(x, y)$ denotes the visibility between x and y.

The rendering equation describes the light exchange between scene objects. Its solution is the equilibrium radiance distribution $L(x, y)$, which we approximate by global illumination computations.

The light transport integral of the rendering equation can be separated into two operators $\mathcal{G}$ and $\mathcal{K}$ [ATS94], which allows for a better understanding of the process of light transport. $\mathcal{G}$ is the so called *field radiance operator* that computes the incident radiance field at some point from the exitant radiance at other visible surfaces, i.e. it propagates light through the scene. The *local reflection operator* $\mathcal{K}$ reflects this incident light and computes exitant radiance again. The rendering equation can then simply be written as

$$L = L^e + \mathcal{K}\mathcal{G}L.$$

$\mathcal{K}$ and $\mathcal{G}$ have very different properties and should therefore be computed separately.

A very common restriction for global illumination is to consider diffuse surfaces only, which simplifies the computation and allows for handling even complex scenes. For diffuse surfaces, the BRDF is constant, so the reflection computation is trivial. Instead of the four dimensional radiance distribution, only a two dimensional radiosity function on the surfaces has to be computed. Of course, most environments are hardly diffuse, which mandates a solution to the general rendering equation.

1.2 Previous Work

For global illumination computations in arbitrary environments, Monte-Carlo methods have often been preferred, despite of their often very slow convergence and noisy solutions. But there have also been several approaches for solving the general rendering equation in a finite-element context, which we briefly summarize below.

Immel et al. proposed a straight forward generalization of the radiosity approach [ICG86]. Directions are globally discretized similar to the hemicube algorithm. For each patch and each direction the outgoing energy is computed by gathering energy from all other directions, each weighted by the BRDF of the surface. Since the method is not hierarchical or adaptive, an enormous number of directional values needs to be computed and stored, posing severe time and memory constraints.

In [AH93] and [SH94] the three point parameterization of the rendering equation is solved using extended form factors, that describe the light exchange from a sender via a reflector to a receiver. The algorithms are hierarchical and exhibit a time complexity of $\mathcal{O}(k^3)$, where k is the number of initial patches in the scene. The resulting solutions are of high quality, but the large computation times make the method applicable to small scenes only.

Clustering for radiance computations is presented in [SDS95]. Clusters are approximated as point sources. Incident radiance is either directly added to the children of the cluster or an incident light field is computed explicitly in a finite basis. Coupling coefficients between these basis functions, similar to radiosity form factors, are used to transport the light through the scene.

A similar approach is followed in [CLSS97]. Again, a cluster is approximated by a point light source with a directionally varying exitant radiance distribution, which is represented in a Haar-wavelet basis. For every cluster, the incident light is approximated in a similar representation and then reflected in a single step. Importance is used to adapt the accuracy of the computations with respect to their contribution to the final result. A costly final gathering step is used to create the final image.

1.3 Radiance Versus Radiosity

All the algorithms described above exhibit a run time behavior that is significantly worse than that of a comparable radiosity computation. Having a closer look at the rendering equation, it can be seen that this behavior is not inherent to the problem. Because the only difference between diffuse and non-diffuse scenes is the local reflection operator, one should expect that the main difference in computation time also arises from the increased complexity of the reflectance computations. However, in contrast to light propagation, the reflection operator is purely local. Therefore, there is hope to find a finite-element algorithm for the non-diffuse case that exhibits a similar time behavior as a comparable radiosity computation.

One reason that radiance algorithms often take significantly longer than radiosity methods is due to representing directional information. For radiosity computations the incoming light can be sufficiently described by a single color value, namely the irradiance of the patch. The direction of incidence is of no interest. For non-diffuse surfaces, the direction of incident light is required to compute the reflection. Therefore, the propagation computation must provide this directional information and forward it to the reflection computation.

Two main approaches have been suggested for handling directional information. First of all, we can compute the incident light field explicitly. This approach exhibits the best time complexity, because the incident light at a cluster can be computed without considering the cluster's content. On the other hand, representing the incident light explicitly requires additional memory and may reduce accuracy due to projection into a finite basis. In addition, accurately computing reflected light from such a light field is non-trivial [LF97].

For the second solution, we do not represent the incident light directly, but instead immediately reflect each small incoming light portion that is computed by the propagation algorithm. This approach saves memory, but problems arise in the context of hierarchical representations of the reflected light.

The problem is demonstrated in Figure 1. A glossy patch is illuminated by four senders. The incident light from each sender is reflected, resulting in an exitant radiance response proportional to the BRDF. Whereas in the top row the senders cover a large solid angle with respect to the reflector, in the bottom row the senders are smaller. So the resulting reflection in the top case is rather constant, whereas in the bottom example the reflection has a strong variation.

Note that for the hierarchical representation of the top case a coarser resolution suffices than for the bottom row. However, in the case of immediate reflection only a single incident light impulse is known, and the algorithm cannot differentiate between the two cases. If the complete illumination is known the entire reflected light distribution can be computed and projected in a single step. This allows to differentiate between the two cases from the example.

To put it in a more formal way: instead of summing the projected responses it is obviously better to project the sum. Note that this difference is only important for hierarchical projections. In the case of uniform basis functions only the numerical error would differ slightly.

1.4   Three Point Clustering

The *three point clustering* algorithm presented in this paper was developed to avoid the problems described above. In this section we describe only the basic ideas of the algorithm, details are described in the following sections.

In three point clustering a hierarchical representation of L is computed in a *line space hierarchy* [DS97] of all unoccluded connections between scene objects. We denote the set of line segments connecting a sender s and a receiver t as $[s, t]$, which is depicted in Figure 2. Subdividing one side of the interaction, for instance t, the set $[s, t]$ is split into two subsets $[s, t_1]$ and $[s, t_2]$ (left image). This way a hierarchy on the set of line segments between s and t is created (center image), which is essentially the same as in [DS97]. Of course this is also possible if the scene contains clusters (right image). Note, that the figure suggests a close relationship between this data structure and light fields [LH96] or Lumigraphs [GGSC96]. In essence, our data structure is very similar and serves the same purposes. However, it does not necessarily have the same nice parameterization and regularity as the other data structures.

Since radiance does not change along a ray, the exitant radiance from an object s to an object t is equal to the incident radiance at t from s. Due to this symmetry the description of the exitant light can also be interpreted as an incident light description. Using the notation of operators $\mathcal{G}$ and $\mathcal{K}$, this means that $\mathcal{G}$ is trivial and that the transport computation is essentially shifted to the local reflection operator $\mathcal{K}$.

The algorithm starts by computing a hierarchical representation of the self emission of the objects within a line space hierarchy. An oracle, similar to HR, is used to steer the depth of the refinement. Starting with this radiance distribution $\tilde{L}^0$, Jacobi or Gauss-Seidel iteration are used to approximate the global solution: $\tilde{L}^{i+1} \approx \mathcal{K}\mathcal{G}\tilde{L}^i + \tilde{L}^0$.

Assuming we want to compute a single value $\tilde{L}^{i+1}(x,y) = (\mathcal{K}\mathcal{G}\tilde{L}^i)(x,y)$, first all incident light at surface point x has to be collected from the line space hierarchy. All line space nodes contributing to the illumination of x are a subtree of the line space hierarchy that can be traversed quickly. The incident light portions then have to be reflected towards y and added to the sum $(\mathcal{K}\mathcal{G}\tilde{L}^i)(x,y)$.

This can be seen as getting an adaptive representation of the incident light at x, as depicted in Figure 3. The figure shows the patch s containing x, which is hierarchically illuminated by three walls of a box. The dotted arrows show the line space nodes illuminating the grandfather, father and s itself. The box on the right shows how the set of all interactions adaptively partitions the set of all directions around s.

Based on this pointwise evaluation, a complete representation of $\tilde{L}^{i+1}$ can be computed from $\tilde{L}^i$ in a top-down manner. Starting at the line space root $[s^0, s^0]$, the value of $\tilde{L}^{i+1}$ is approximated over that line space range. If a representation within that range by a constant value is not possible, a representation of $\tilde{L}^{i+1}$ is determined recursively on the level of the children. Otherwise, recursion ends.

Note, that using this scheme there is an important difference to other three-point algorithms. Whereas refinement is usually steered by looking at a single three-point form factor, in our algorithm refinement is made on basis of the resulting exitant radiance only. From an abstract point of view, one can say we do not project the kernel but the radiance. A discussion of this property follows at the end of the paper.

In order to obtain accurate results, we do not only compute exitant radiance values, but bounds on them. This way, sampling problems (e.g. for computing the BRDF) can mostly be avoided. This concept proved to be very helpful in the context of radiosity computations [SSS97]. For the non-diffuse case it is mandatory, because highly specular BRDFs exhibit very small details, which are likely to be missed by any sampling approach.

In the following we describe the important aspects of the algorithm and its implementation in detail and discuss the benefits of the new approach.

2 Line Space Hierarchy

As mentioned in the introduction, we want to parameterize the radiance in a scene by the set of all mutually visible pairs of surface points. Two such points x and y define a *line segment*. The set of all oriented line segments **R** is the set of all mutually visible

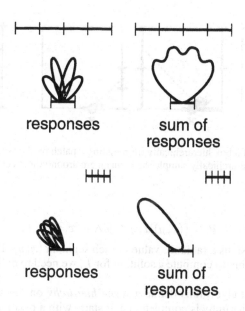

Fig. 1. Incident light portions at a reflector from four senders for two sender configurations (top and bottom row). Although the single responses are similar for each configuration, the sum differs significantly. A refiner computing a hierarchical representation for each single response independently cannot exploit the smoothness of the response sum in the top row.

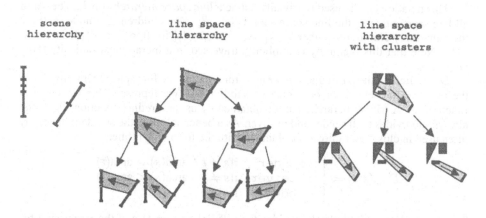

Fig. 2. Creating a line space hierarchy. Left: Simple scene of two patches and a hierarchical subdivision on it. Center: Hierarchy on the set of connections. The root is the set of all connections, a subdivision can be achieved by subdividing the receiver or sender. Of course this idea can be transformed to clusters (Right).

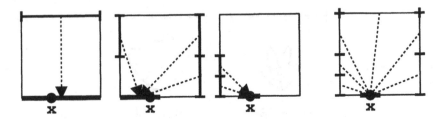

Fig. 3. Three walls of a box hierarchically illuminating a patch on the bottom line. These illuminating interactions hierarchically sample the hemisphere around the receiving patch.

surface points:

$$\mathbf{R} = \{(x, y) | x, y \in S \wedge V(x, y) = 1\}$$

$L(x, y)$ then assigns a radiance value to each segment (x, y). If we want to define a hierarchical algorithm to compute a solution for L, we need to define a hierarchy on its domain $\mathbf{R}$.

All hierarchical algorithms define a *scene hierarchy* on the surfaces. In the case of clustering this hierarchy is complete, i.e. it starts with a root node s^0 containing the whole scene. If s is a node in that hierarchy, it has a number $n(s)$ of children. We will denote the children of s using an index: $s_1, \ldots, s_{n(s)}$. Each node in the hierarchy can thus be described by the path from the root node.

Any two nodes s and t of the scene hierarchy, which are at least partially visible to each other, define a subset of the scene's line segments: $[s, t]$ is the set of all segments from s to t, which are not occluded. In the following, we will refer to s as the sender and to t as the receiver of $[s, t]$.

This notation can be used to trivially define a *line space hierarchy* on R. The set of all line segments, i.e. the line space root, is $[s^0, s^0]$. The children of a node $[s, t]$ are the non-empty children of either $[s, t_j]$, $[s_i, t]$, or $[s_i, t_j]$ for $0 < i < n(s), 0 < j < n(t)$. Note that this hierarchy is implicitly traversed in a hierarchical radiosity (HR) algorithm.

Depending on the particular *refiner* used for creating the line space hierarchy, either the receiver, the sender, or both may be subdivided. We represent this refiner by a function D. A trivial hierarchy can be obtained by using a refiner function D_{both} that always subdivides both sender and receiver. As a better solution, the standard hierarchy often used in clustered radiosity is obtained with the following refiner:

$$D_{\text{area}}(s, t) = \begin{cases} \text{sender} & \text{if} s \neq t \wedge \text{area}(s) \geq \text{area}(t) \\ \text{receiver} & \text{if} s \neq t \wedge \text{area}(s) < \text{area}(t) \\ \text{both} & \text{if} s = t \end{cases}$$

The refiner D_{area} subdivides both sides for self links ($s = t$) and the larger side for non-self links.

For the sake of radiance representations, both subdivision schemes are certainly not optimal. If for a line space node $[s, t]$ the sender s has a constant exitant radiance and is visible from the entire t, then $L_{s \rightarrow t}$ does not vary over the receiver t. Therefore, subdividing t does not improve the quality of the radiance representation, because all children would contain the same value.

Better refiners should take this and similar situations into account and finding them is a very interesting area for future research. For our implementation, we use the simple

D_{area} refiner. However, selection of a good refiner is rather orthogonal to the remaining algorithm described below.

In order to avoid confusion of the two hierarchies that are created during computations, we must differentiate between the scene hierarchy and the line space hierarchy. The nodes in these hierarchies will be called *scene nodes* (or *objects*) and *line space nodes*, respectively.

2.1 Measuring Line Space

In order to compute the amount of light transported by a line space node, we define a measure on the line segments $\mathbf{R}$. This measure assigns the weight $d\mu(r) = dA\,d\omega$ to each bundle of line segments $r \in \mathbf{R}$ emanating from a perpendicular infinitesimal area dA into a solid angle $d\omega$.

Thus, the line segments between two scene object nodes s and t have the measure

$$||[s,t]|| = \int_{[s,t]} d\mu(r) = \int_s \int_t \frac{\cos(n_{s'}, t' - s') \cos(n_{t'}, s' - t')}{||t' - s'||^2} V(s', t') ds'\, dt'.$$

Note, that this measure is equivalent to the form factor from s to t times the area of the receiver t. Thus, if the radiance L between s and t is assumed to be constant, the flux transported from s to t is

$$\Phi(s,t) = L \cdot ||[s,t]||$$

and the irradiance at t due to s is

$$E^s(t) = L \cdot \frac{||[s,t]||}{\text{area}(t)}$$

3 Implementation

3.1 Data Structures

At the beginning of the algorithm, a scene hierarchy is built by clustering the initial scene objects into a complete hierarchy. This hierarchy will be refined during the algorithm by subdividing leaf objects in order to allow for a better representation of L.

Parallel to the scene hierarchy, a line space hierarchy is created. Each of these line space nodes corresponds to a link in hierarchical radiosity. However, we store illumination in the line space nodes and not with the object nodes as in hierarchical radiosity. Furthermore, each of the ray nodes contains the measure $||[s,t]||$ of the represented line segments as well as the visibility information.

3.2 Main Loop

Light transport in the scene is computed by a top-down traversal of the line space hierarchy. Each full traversal computes one iteration of a Jacobi- or Gauss-Seidel iteration. During one traversal, incident light is reflected in the scene and an adaptive representation of the reflected light is computed. This new representation is then the new input for the next iteration. If the new results are directly used as input to the current iteration, this corresponds to Gauss-Seidel instead of Jacobi iteration. The main loop can be described in pseudo code as follows:

```
0 main loop:
1    until converged
2       traverse line space hierarchy
3          recompute radiance L of current node [s,t]
4          if refinement criterion fulfilled for [s,t]
5             refine
6             set L to average radiance of children nodes
7          else
8             store L in [s,t]
```

Line 8 corresponds to the pull step of hierarchical radiosity. Because radiance is stored in the line space hierarchy and not on the objects a push operation is not necessary. The essential work is done in line 3, where the radiance for the current node $[s, t]$ is recomputed. This radiance is the sum of the emission of the sender towards the receiver plus the radiance reflected off the sender towards the receiver:

```
0 recompute radiance of current node [s,t]
1    L = emission of s towards t
2    traverse line space hierarchy for all
                    nodes [r,s'] illuminating s
3       L += reflection of [r,s'] towards t
4    return L
```

In order to compute the radiance of the current line space node $[s, t]$, we reflect *all* incident light at the sender by a new partial traversal of the line space hierarchy accessing all nodes that contribute to the illumination of s, i.e. all nodes $[r, s']$ for which s' is an ancestor of s or s itself. These nodes form a subtree of the line space hierarchy and can be traversed efficiently. During this traversal, we sum the reflected light for each incident illumination at s.

In order not to miss important detail, we bound the set of all directions from s to t with a cone [SSS97]. Using this cone, we can then obtain bounds on the emission of the sender towards the receiver using a special emission query interface of light sources. With this approach, light sources that emit light in a very small directional range can be used without sampling problems. A similar approach is used for obtaining the value of the BRDF. Cones for the directions from s to t as well as from the illuminating objects r to s are computed and used to bound the BRDF within that range of interaction. This approach is mandatory for capturing all details of a strongly varying BRDF.

3.3 Computing the Line Space Measure

The measure of a line space node is required during reflection computations at the receiver. This is computed in two phases: First, the measure is approximated without considering the visibility term $V(x, y)$. This is equivalent to the normal form factor computations in HR. Applying the ideas of Bounded Radiosity [SSS97], not only an approximated measure, but also an upper and a lower bound are computed. Thus, we obtain reliable bounds on the unoccluded measure and can be sure not to miss important illumination detail due to sampling problems. The method works for arbitrary objects, e.g. for curved surfaces and clusters.

In the second phase, the occlusion between the two objects is approximated separately by shooting a number of sample rays between the objects. The product of the sampled visibility factor and the unoccluded line space measure is then used as measure for the line space node. In case of partial visibility the refiner may choose to further subdivide the node.

3.4 Diffuse Objects

One significant optimization can be achieved by differentiating between diffuse and non-diffuse objects. We added an irradiance field to any diffuse object. If the radiance of a line space node containing a diffuse receiver is recomputed, the irradiance at the receiver can directly be updated, similar to HR. This allows us to compute the reflected radiance at diffuse objects by simply weighting the irradiance with the diffuse reflection coefficient. Thus, the partial traversal to gather incident light can be omitted for diffuse surfaces. For a completely diffuse scene, the algorithm then behaves very much like HR.

3.5 Discussion

Comparing this approach to a HR algorithm we see that the algorithm traverses the line space hierarchy in the same way as HR does. For each traversed node, its measure $\|[s, t]\|$ is computed, which is equivalent to computing the HR form factor and visibility. The essential difference is in the reflection computation. In our algorithm, a partial traversal of the line space hierarchy is necessary to get the incident light. However, this traversal is cheap: beginning at the root we can quickly traverse the subtree containing nodes illuminating s. Thus we reuse the directional information implicitly stored in the line space hierarchy for computing reflections.

There is an important difference of our approach to most other hierarchical methods. In our method the refinement decision is made considering the computed reflected radiance of a line space node $[s, t]$, i.e. the radiance is directly projected to the hierarchical basis. In contrast, previous algorithms decide about refinement looking at a single three point transport between a triple of patches. This is equivalent to projecting the kernel of the rendering equation to a hierarchical basis and computing the transport with the projected kernel. This kernel projection leads to the 'sum of projection' problem described in Section 1.3. Projecting the radiance directly as in three point clustering results in a lower granularity of the refinement, but projecting the target function directly allows for much better control of the representation error.

4 Results

4.1 Caustic

For a simple test we computed a caustic from a specular half cylinder illuminated by a small light source (Figure 4). The scene consists of only three surfaces, i.e. the cylinder is modeled as a single surface. The example mainly shows the benefit from using bounded computations. Most other algorithms would exhibit sampling problems being applied to the shown configuration. The shown solution took 206 seconds on a 196 MHz R10k and built a line space hierarchy of 82.000 nodes.

4.2 Mirror Ball

As a more complex test scene we used a room with a "disco" ball decorated with small, highly glossy patches (see Appendix). This sphere is illuminated by an area light source

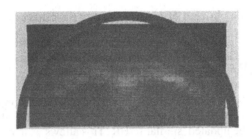

Fig. 4. A caustic from a cylinder illuminated by a small light source. Computation time: 206s, 82.000 links.

(one side of the cube on the left). The reflections appear as bright spots on the diffuse walls of the room.

The computation of this particular solution took 800 seconds on an SGI Onyx with a R10k CPU at 195 MHz. It is interesting to notice that about half of this time has been spent on computing the line space measure, i.e. on propagation computations. The other half was used for reflection computations, of which the largest part (about 80 percent) is spent on the bounded BRDF evaluation, in particular for computations on clusters. In our implementation BRDF queries to clusters are simply passed to the patches within the cluster and the results are combined. Finding a better solution for this computation would speed up the solution significantly.

4.3 Room Scene

As a third test scene we used the well known dining room provided by Peter Shirley. To obtain glossy effects, the material of the table and chairs was changed to be a metal-like with a Phong exponent of 5. In particular the glossiness of the table results in a strong highlight at the ceiling of the room (Figure 5). This scene contains a total of 389 specular surfaces.

The solution shown in Figure 5 took 500 seconds. About two thirds of this time have been spent on reflection computations, and again about 80 percent of this time are used within bounded BRDF queries. As for the last example, our measurements show that the high cost for BRDF queries to clusters slow down the simulation in particular.

4.4 Diffuse Room Scene

Finally, we compared the computation of a solution for a diffuse scene with three point clustering and with a standard clustering algorithm. In order to get useful results, we turned off the diffuse object optimizations of three point clustering, i.e. diffuse objects were handled as specular ones. Choosing parameters that produced comparable solutions, it turned out that the three point clustering solution took less than twice as long as the radiosity solution. Using the diffuse object optimization, the algorithms behaves as HR.

Fig. 5. Shirley's room test scene. The table and the chairs are specular with a Phong exponent of 5, producing a significant highlight on the ceiling.

5 Conclusion

We proposed a novel approach to non-diffuse global illumination simulations, which was developed to overcome problems identified in previous methods. Due to transporting and representing illumination in a line space hierarchy, there is no need to compute and maintain separate representations of incident and exitant radiance. For each object, the entire incident light is reflected in a single step before being projected into the line space hierarchy. Clustering comes naturally in this approach and is used to make the algorithm less dependent on the input complexity of the scene. Bounded form factor computations and BRDF evaluations allow us to use arbitrary curved objects and shader functions without the danger of running into sampling problems.

Tests showed that in our implementation the evaluation of non-diffuse shaders takes up to two times longer than computing the light propagation. In particular, the high number of BRDF evaluations is very costly. However, we are currently working on an optimized hierarchical version of the reflection computations for clusters, which should result in large improvements in efficiency.

References

AH93.   L. Aupperle and Pat Hanrahan. A hierarchical illumination algorithm for surfaces with glossy reflection. In *Computer Graphics (SIGGRAPH '93 Proceedings)*, pages 155–162, August 1993.

ATS94.   James Arvo, Kenneth Torrance, and Brian Smith. A framework for the analysis of error in global illumination algorithms. *Computer Graphics (SIGGRAPH '94 Proceedings)*, pages 75–84, 1994.

CLSS97. Per H. Christensen, Dani Lischinski, Eric Stollnitz, and David H. Salesin. Clustering for glossy global illumination. *ACM Transactions on Graphics*, 16(1):3–33, January 1997.

222

DS97.   George Drettakis and François Sillion. Interactive update of global illumination using a line-space hierarchy. *"Computer Graphics (SIGGRAPH '97 Proceedings)*, pages 57–64, aug 1997.

GGSC96. Steven J. Gortler, Radek Grzeszczuk, Richard Szelinski, and Michael F. Cohen. The Lumigraph. *Computer Graphics (SIGGRAPH '96 Proceedings)*, pages 43–54, August 1996.

ICG86. David S. Immel, Michael F. Cohen, and Donald P. Greenberg. A radiosity method for non-diffuse environments. *Computer Graphics (SIGGRAPH '86 Proceedings)*, pages 133–142, August 1986.

Kaj86. James T. Kajiya. The rendering equation. *Computer Graphics (SIGGRAPH '86 Proceedings)*, 20(4):143–150, August 1986.

LF97.   Paul Lalonde and Alain Fournier. Filtered local shading in the wavelet domain. In *Eurographics Rendering Workshop 1997*, pages 163–174. Springer, June 1997.

LH96.   Marc Levoy and Pat Hanrahan. Light field rendering. *Computer Graphics (SIGGRAPH '96 Proceedings)*, pages 31–45, August 1996.

SDS95. François Sillion, George Drettakis, and Cyril Soler. A clustering algorithm for radiance calculation in general environments. In *Rendering Techniques '95 (Proceedings of Sixth Eurographics Workshop on Rendering)*, pages 196–205. Springer, August 1995.

SH94.   Peter Schröder and Pat Hanrahan. Wavelet methods for radiance computations. In *Photorealistic Rendering Techniques (Proceedings Fifth Eurographics Workshop on Rendering)*, pages 303–311, Darmstadt, June 1994. Springer.

SSS97. Marc Stamminger, Philipp Slusallek, and Hans-Peter Seidel. Bounded radiosity – illumination on general surfaces and clusters. *Computer Graphics Forum (EUROGRAPHICS '97 Proceedings)*, 16(3), September 1997.

Editors' Note: see Appendix, p. 334 for colored figure of this paper

The Visible Differences Predictor: applications to global illumination problems

Karol Myszkowski

The University of Aizu; Aizu-Wakamatsu 965-8580; Japan

Abstract

In this study of global illumination computations, we investigate the applications of the perceptually-based Visual Difference Predictor (VDP) developed by Daly [5]. First, we validate the performance of this predictor in shadow masking by texture and luminance contrast experiments. We also experiment with Contrast Sensitivity Functions (CSFs) derived from the results of various psychophysical experiments, various spatial frequency and orientation channel decomposition schemes, and contrast definitions, in order to check predictor integrity and sensitivity to differing models of visual mechanisms. We show applications of the VDP to monitor the perceived quality of the progressive radiosity and Monte Carlo solutions, and decide upon their stopping conditions. Also, based on the local error metric provided by the predictor we show some initial attempts to drive adaptive mesh subdivision in radiosity computations.

1 Introduction

Since the beginning of computer graphic studies, the quest for development of perception-driven techniques has been and continues to be an important issue. This is mostly because of the expected dramatic performance gains by focusing computations on scene features which can be readily perceived by human observers under given viewing conditions. This seems especially attractive for global illumination solutions [10], which deal with very costly computations of light interactions between the surfaces modeling a scene. Current algorithms usually rely on energy-based metrics of solution errors, which do not necessarily correspond to the visible improvements of the image quality [13]. Ideally, one may advocate the development of perceptually-based error metrics which can control the accuracy of every light interaction between surfaces. This can be done by predicting the visual impact those errors may have on the perceived fidelity of the rendered images. In practice, apart from the conceptual difficulties in elaborating such a low level error metric, it could be costly in evaluation, and could depend strongly upon a specific global illumination algorithm.

Another approach is to develop a perceptual metric which operates directly on the rendered images. Of course, such a metric provides a summary of the algorithm performance as a whole rather than giving a detailed insight into the work of its particular elements. However, the metric and *a priori* knowledge of the current stage of computations can be used to obtain more specific measures of adaptive meshing performance, accuracy of shadow reconstruction, convergence of the solution for indirect lighting, and so on. If the goal of rendering is just a still frame, then the image-based error metric is adequate. In the case of view-independent solutions, the application of the metric becomes more complex because a number of "representative" views should be chosen. In practice, instead of measuring the image quality in the absolute terms, it is much easier to derive a relative metric which predicts the perceived differences between a pair of images [21]. (It is well-known that a common mean-squared error metric usually fails in such a task [5, 24, 21, 8].) A single numeric value might be adequate for some applications such as monitoring of the perceptually measured simulation progress; however, for more specific guiding of the lighting computations, a local metric operating at the pixel level is required. In this work, we investigate the application of such local metrics in the framework of the **H**ierarchical (link-less), cluster-based **P**rogressive **R**adiosity (HPR) [18] and **M**onte **C**arlo **P**ath **T**racing (MCPT) [22, 12] algorithms.

2 Modeling Human Visual System

So far, a comprehensive and complete model of Human Visual System (HVS) has not been developed. While the higher order visual processing remains mostly unexplored, the early stages in the visual pathway (beginning with the retina, ending with the visual cortex [V1], and including various intervening nuclei) are relatively well understood [6]. Since other cortical visual areas of the brain appear to receive visual information mostly via V1 neurons, an important issue becomes the representation of an image by cells in the V1. In this section, we outline such an internal representation based on spatial frequency and orientation channels, which provide the core of the most recent HVS models that attempt to describe spatial vision. Also, we discuss the existing HVS models specialized in predicting the perceived differences between images.

2.1 Cortical image representations

Though we ultimately require only a model of how visual information is processed, it is instructive to examine results of studies on organization of the retina and visual cortex, and these studies have exerted a great influence upon attempt to model human spatial vision. The anatomical, physiological and psychophysical findings suggest that receptive fields of the V1 simple cells (about 50% of all V1 cells) can be well modeled by a set of differently oriented and narrowly tuned spatial filters (channels) [16, 26, 28]. A model of these channels can be used to predict the perceptual response to visual images, and most of the HVS computational models assume that channels form the basis of internal image representation. Indeed, the channel model explains well many visual phenomena such as: the overall behavioral Contrast Sensitivity Function (CSF) (visual system sensitivity is a function of the spatial frequency and orientation content of the stimulus pattern), spatial masking (detectability of a particular pattern is reduced by the presence of a second pattern of similar frequency content), and sub-threshold summation (adding two patterns of sub-threshold contrast together can improve detectability within a common channel). See [7] for an excellent discussion of these phenomena and further references.

The characteristics of these channels subserving spatial vision are difficult to establish precisely because different psychophysical measurement methods often lead to significantly different results, and the measurement error within such experiments can be quite large. Indeed, significant day-to-day variation of measurements for a single subject have been observed. (See [6], page 205 for channel bandwidth estimates derived using various psychophysical and electrophysiological measurement techniques.) In practice, some HVS models place more emphasis upon the resemblance of the channels to the action of the visual cortex, while others place more attention on computational efficiency. In the first group, Gabor functions [16] are widely used (see [23] for further references), because of their close resemblance to receptive fields of the primate V1 cells. Also, a good match with physiological measurements are filters shaped according to DOG (differences of Gaussians) [28, 14, 7] and DOOG (differences of offset differences of Gaussians) [15] functions. Watson [26, 5] proposed to build a model, termed the Cortex transform, from filters whose shape roughly approximate 2D Gabor functions, and which can be organized into a computationally-efficient multiscale pyramid. In addition, wavelet transforms, though not targeted to approximate receptive fields of the V1 cells, have been used because of their computation efficiency and compact storage [17, 31, 8].

In practice, it seems that the choice of a particular filter function is not critical as long as it roughly approximates the point-spread function of the V1 cells [15], and the frequency and orientation bandwidths are not too broad. For all the above channel models, usually 4–6 bands are used, each with full frequency bandwidth at half height falling within the 1.0–1.6 octave range, and each exhibiting one of 3–6 different orientations. However, Teo and Heeger [24] reported a problem when trying to fit their HVS model to psychophysical data. A simple model based on only three orientation channels was not adequate, but six orientations gave a close fit. Of course, there are other factors that might be considered. For example, the HVS channels are not truly independent, but rather have been found to be mutually inhibitory [6], and the visual response to a complex stimulus (e.g., a realistic synthetic image) often involves significant inhibition between channels. While most of the existing HVS models ignore this problem, some heuristics extending the HVS models are available [15, 24]. What features are included in a given model must depend upon the application.

2.2 Visible differences predictors

Our application of the HVS model concerns how to predict whether a visible difference will be observed between two images. Therefore, we were most interested in the HVS models developed for similar tasks [30, 14, 17, 5, 24, 4, 27, 7, 8, 23], which arise from studying lossy image compression, evaluating dithering algorithms, designing CRT and flat-panel displays, and generating computer graphics. Let us now describe briefly the Visual Differences Predictor (VDP) developed by Daly [5] as a representative example, which was selected by us for our experiments on global illumination algorithms. Figure 1 contains a block diagram for the VDP. The two images to be compared (target and mask) undergo an identical initial processing. At first, the original pixel intensities are compressed by the amplitude non-linearity based on the local luminance adaptation and simulating the Weber's law-like behavior. Then processing of CSF is performed that takes into account the global state of luminance adaptation, orientation, image size and eccentricity from the fovea region. The resulting image is decomposed into the spatial frequency and orientation channels using the Cortex transform. For every channel and for every pixel, global contrast is computed (the image mean luminance is used), and elevation of the detection threshold based on masking is calculated. This detection threshold is then used to normalize the contrast dif-

226

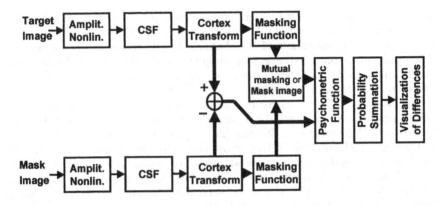

Fig. 1. Block diagram of the Visible Differences Predictor (heavy arrows indicate parallel processing of the spatial frequency and orientation channels)

ferences between target and mask images. The normalized differences are input to the psychometric function which estimates probability of detecting the differences for a given channel. This estimated probability value is summed across all channels for every pixel, and visualization of visible differences between the target and mask images is performed.

The main advantage of the VDP (and the main reason why we choose it) is prediction of local differences between images (on the pixel level), while most of methods, including that recently developed in [8, 7], did not show such functionality, and provided a single scalar value as a measure of the differences. This locality of predictions is instrumental in developing perception-driven refinements in global illumination calculations. The Daly model also takes into account the visual characteristics that we think are extremely important in our application: a Weber's law-like amplitude compression, advanced CSF model, and masking (the latter one is ignored by many HVS models [14, 17, 8, 23]). The Cortex transform is a pyramid-style, invertible, and computationally efficient image representation, and the shape of basic filters resembles the shape of Gabor functions (which appear to be the most accurate model of receptive fields of the V1 simple cells).

Disadvantages of the Daly model have been identified, particularly from the standpoint of integrity. As pointed out by Taylor *et al.* [23], a variety of psychophysical data that is used in the Daly model was derived from the results of various unrelated experiments, usually conducted using completely different tasks. Taylor *et al.* executed psychophysical experiments that directly determined the parameters of their model. They measured contrast sensitivity as a function of frequency, luminance, and orientation of Gabor patches, which are the same functions as those used by them in their image pyramid decomposition. Also, they pointed out that it is more appropriate to use contrast discrimination thresholds instead of the more commonly used detection thresholds for predicting visible differences between images. However, they had not yet included masking into their model, which was left for future work.

The HVS model proposed by Wilson and Gelb [29] also has many attractive properties. It has good integrity since it was derived from the averaged masking data for three subjects, and shows a good fit to some adaptation and subthreshold summation experiments. Wilson and Gelb adjust the absolute sensitivity of each channel to fit threshold-sensitivity data for each of the different types of patterns and viewing conditions. However, in our application such procedure does not seem to be practical because we deal with diversified and complex stimuli in rendered images, and the results obtained in experiments employing simple and isolated stimuli are typically hard to generalize [6]. The quest for a more automatic procedure motivated Ferwerda *et al.* to extend Wilson's HVS model relying on the the results from disparate studies (perhaps at expense of the model integrity). Ferwerda *et al.* incorporated a general CSF [17] in the selection of specific channel sensitivities in order to take into account luminance adaptation under the given viewing conditions for a particular image.

The original Daly model did not support color, although its extension seems straightforward, based upon solutions proposed in [14, 17, 2, 7]. In our research, we use the VDP on an achromatic channel only. We did not want to add more heuristics to this model, which would be necessary for color processing because of a rather small volume of experimental data on the chromatic mechanisms (e.g., the luminance dependence of the chromatic CSF [17]). Also, in global illumination calculations many important effects such as shadow quality, can be relatively well captured by the achromatic mechanism, the most sensitive mechanism subserving human spatial vision.

The Daly model clearly is not perfect for reasons discussed above. Also, it ignores local luminance adaptation of the eye (only global contrast is computed), mutual channel inhibition, and suphrathreshold vision (for the most part). Still, it seems to be one of the best existing choices for our applications. One of the goals of this work was to check suitability of Daly's model to detect differences between images which arise in typical global illumination and rendering computations. What follows is a description of our experiments designed to test the validity of Daly's model in these applications.

3 VDP validation experiments

Before attempting to use Daly's VDP model in the actual global illumination computations, we chose to execute pilot studies to validate its performance in comparing realistic synthetic images. In all the experiments reported in this section, we analyzed the predicted visible differences in shadow quality for two subsequent stages of the adaptive mesh subdivision algorithm. The shadow reconstruction was a part of the view-independent illumination map computations performed using our progressive radiosity method, HPR, which begins with direct lighting computations and then proceeds with indirect lighting computations [18].

Perceived shadow quality vs. masking by texture. The goal of the first experiment was to investigate shadow masking by texture. Figure 2a and b show the reconstructed shadows cast on the floor at two different stages of mesh computations (Figure 4d in Appendix/Color Section shows the complete scene from which this region of interest was extracted). We experimented with various texture patterns and texture scales. We used images of resolution 512×512, and we assumed that the CRT was observed from the distance of 0.5 meter. Figure 2c shows the output image of the VDP when images in Figure 2a and b were used as the target and mask. Via alpha blending, color is added to each pixel in the original grey-scale image in order to indicate its difference-detection probability value (refer to [1] for the full resolution color images); red pixels indicate probability values greater than 0.75 (standard threshold value for discrimination tasks [28]), while shades of yellow correspond to probabilities below this threshold. Figure 3a summarizes the numeric VDP responses (percentage of pixels for which the probability of the difference detection is over 0.75) as a function of texture scale (the generic texture pattern is the same as that in the top row of Figure 2). The greatest shadow masking was predicted for similar scales of major texture and shadow elements (scale may be quantified in terms of spatial frequency content). Figure 3b shows VDP response as a function of texture orientation. Here, the greatest masking was predicted for texture elements aligned with major shadow elements. On the Web page [1], we provide more extended documentation of these tests for three textures. It is also possible to download the representative original images for which the experiment was performed. As it can be seen in [1], the VDP predicted less masking for textures containing more randomly oriented patterns (without dominant elements that might be aligned with shadows).

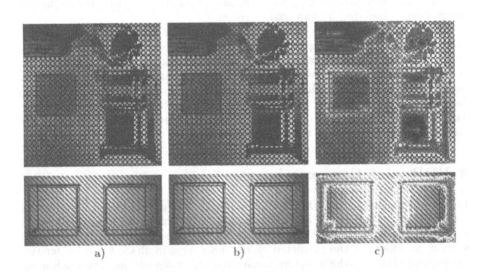

a)   b)   c)

Fig. 2. Perceived shadows artifacts: a) mask, and b) target images, c) the VDP output. Refer to [1] for the full resolution color images.

Perceived shadows quality vs. observation distance. When a typical desktop CRT of resolution 1280×1024 is observed from the distance of 0.5 meter, the maximum number of cycles per visual degree (cpd) that can be properly reconstructed is about 15, according to the Nyquist law. This means that the distance of about 1.5 meters is needed to observe the effects of the high frequency cut-off of the CSF, which falls in the range 40–50 cpd under typical CRT viewing conditions. Figure 3c shows a graph with the VDP responses as a function of distance (see [1] for visualization of the VDP output).

Perceived shadows quality vs. solution convergence. We tested performance of the VDP in predicting visible differences between shadows for a range of illuminations resulting from progressive stages of indirect lighting computation. Figure 3d shows a graph of the VDP responses for stages ranging from the direct lighting solution (at time zero), through subsequent stages of the progressive radiosity solution. The differences in shadow visibility decrease as the solution converges, with a dramatic jump in contrast masking as the first iteration of indirect lighting is completed (see [1] for more details). Figure 3d shows for comparison the corresponding RMS error (dashed line), which changes only slightly, and poorly predicts the perceived differences. This experiment quantitatively demonstrates that important savings in terms of the number of mesh elements can be achieved if the meshing algorithm is based upon an approximated global illumination solution, even if the approximation is quite rough. This is in contrast with usual practice, in which meshing is often done based only upon the direct lighting distribution. One notable exception is a meshing technique proposed by Gibson and Hubbold [9], that uses an "ambient correction term" to compensate for the lack of knowledge of global illumination during the initial stages of the progressive radiosity solution.

4 VDP integrity

We chose to examine the Daly model integrity in terms of how critical given components were to maintaining a useful output. By replacing various model components with functionally similar components derived from results on different tasks, we were able to show how robust this model is. We experimented with three types of CSF used in the following HVS models: [5], [17, 7], and [8]. The response of the VDP was very similar in the former two cases, while for the latter one discrepancies were more significant. A possible reason for such discrepancies is that the CSF used in [8] does not take into account luminance adaptation for our test, which could differ from the conditions under which the CSF was originally measured.

We were also curious how the channel decomposition method influenced the VDP responses. We compared the Cortex transform [5], having 6 spatial frequency and 6 orientation channels, with the DOG-like pyramid construction of Burt [3], having 6 spatial frequency channels, and extended to include 4 orientation channels. While the quantitative results are different, the distribution of probabilities of detection differences between images corresponds quite well. The quantitative differences can be reduced by an appropriate scaling of the VDP responses.

Daly's original VDP model used an average image mean to compute the global contrast for every channel of the Cortex transform. We experimented with the local contrast using a low-pass filter on the input image to provide an estimate of luminance adaptation for every pixel. This made the VDP more sensitive to differences in dark image regions, and we found that in many cases the VDP responses matched better our subjective impressions (i.e., predicted

230

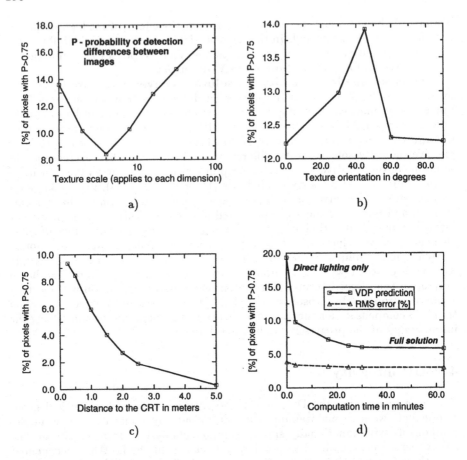

Fig. 3. Perceived shadow artifacts: masking by textures of various a) scales, and b) orientations; as a function of c) distance to the CRT (optical blur is ignored), and d) convergence of the indirect lighting solution (Note that the symbols plotted on the y axis, at time zero, indicate the probability for the direct lighting solution only, providing a reference for the subsequent stages of indirect lighting).

image differences corresponded well with actual perceived differences). In [1] we provide a variety of selected experimental results that support the conclusions reported in this section.

5 VDP applications

We applied the VDP to more practical tasks arising in global illumination calculations. In this section, we present results of comparison of the "perceptual" convergence of two progressive indirect lighting simulation techniques. Then we discuss perceptually-driven stopping conditions for such progressive computations. Finally, we experiment with simplification of the adaptive mesh subdivision by direct embedding of the VDP into decision process, which controls the mesh refinement.

Comparing progression of indirect lighting solution. A practical rendering system must be fast, and provide progressively updated images to show the current status of computations. A meaningful metric of the rendering progression can be expressed as perceived differences between the intermediate and final images as a function of time. In developing new or experimental illumination algorithms, we can use this metric to compare their performance to that of other solutions. Traditionally, such comparisons required the use of human observers examining the differences between the images. Using the VDP, quantitative measures of such differences are automatically generated, and also the image regions are identified in which such differences will be seen by the human observer at some high probability.

We applied the VDP to compare progressivity of indirect lighting solution for two different algorithms: the hierarchical (link-less) cluster-based progressive radiosity (HPR) [18] and Monte Carlo Path Tracing (MCPT) [12]. Both methods are based on the same ray-tracing kernel which is used to compute form factors (visibility computations) and tracing photons. We compared the ray traced images that take into account indirect lighting computed using the two methods, because we wanted to capture the influence of indirect lighting on the final appearance of specular and view-dependent effects. Figure 4a (see Appendix/Color Section) shows the intermediate images with indirect lighting computed at 1 and 30 minutes, and at 1 and 4 minutes, using the HPR and MCPT techniques (CPU R10000; the rendered scene is built of about 90,000 meshed triangles). Images in the second column (Figure 4b) show the absolute differences (normalized by the image mean) between the intermediate images (Figure 4a) and the final images (Figure 4d and e). The last column of images (Figure 4c) shows the corresponding responses of the VDP, which provide a more reliable picture of the perceived differences. This becomes especially evident for the MCPT images, which locally exhibit many fluctuations of indirect lighting that are usually well masked by textures. All such fluctuations are immediately reported by the absolute differences metrics, while the VDP responds more selectively by taking into account their contrast, spatial frequencies and masking textures. The graph in Figure 5a summarizes results of our comparisons, and shows that in our implementation the HPR technique converges much more slowly than the MCPT. (We did not use the ambient light approximation or overshooting techniques, because we are interested in physically sound intermediate results.)

Stopping conditions for lighting computations. When global illumination calculations are performed, the problem of their stopping conditions naturally presents itself. In an ideal case, computations should be stopped immediately when the image quality becomes indistinguishable from that of the fully converged solution for the human observer. (This assumption is valid if the only goal of computations is a computer image, not numerical values of illumination which can be important in some engineering applications.) Typically, the fully converged solution is not known. In practice, to stop computations, some estimate of the simulation error in terms of energy is provided by a lighting simulation algorithm, and compared against a threshold value imposed by the user. However, energy-based error measures are not reliable in predicting possible differences in appearance of the intermediate and final images. For example, the MCPT has a stochastically derived estimate of the RMS energy error [12], which

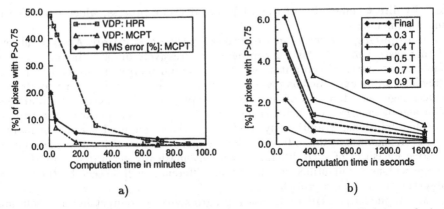

Fig. 5. a) The VDP predicted "perceptual" convergence of indirect lighting solution for HPR and MCPT techniques. b) The VDP predicted differences between I_C and I_T, and I_T and $I_{\alpha T}$ images.

is shown in Figure 5a (solid line) for a scene in Figure 4. The actual RMS error measured between the intermediate and final images is usually significantly different because of the compressive power of the Tone Mapping Operator (TMO) [25] used to convert luminance to displayable RGB. Also, the RMS measure is global for the whole scene, and the actual error may be locally much higher. This observation led us to use images directly, and compute the differences between intermediate images using the VDP, which captures such local errors much better. If the VDP does not report significant differences between intermediate images, we assume that computations can be stopped. A question arises: *which images should be compared against the current image to get robust stopping conditions?*

Let us assume that the current image I_T is obtained after the computation time T, and let us denote by $VDP(I_T, I_{\alpha T})$ the VDP response for a pair of images I_T and $I_{\alpha T}$ where $0 < \alpha < 1$. We should find an α to get a reasonable match between $VDP(I_T, I_{\alpha T})$ and $VDP(I_C, I_T)$, where I_C is an image for the fully converged solution. We tried to address this algorithm-dependent question experimentally for the MCPT technique. Figure 5b shows the numerical values of $VDP(I_C, I_T)$ and $VDP(I_T, I_{\alpha T})$ for $T = \{100, 400, 1600\}$ seconds and various α, (the same scene as in Figure 4, but built of 17,500 meshed triangles). While the numerical values of $VDP(I_T, I_{0.5T})$ provide the upper bound for $VDP(I_C, I_T)$ over all investigated T, it is even more important that the image regions with the perceivable differences are similar in both cases (refer to [1] for color images with $VDP(I_C, I_T)$ and $VDP(I_T, I_{0.5T})$). This means that for certain regions of $I_{0.5T}$ and I_T the variance of the luminance estimate is very small (below the perceived level), and it is likely that it will be so for I_C. For other regions such variance is high, and it is likely that luminance estimates for $I_{0.5T}$ and I_T which fluctuate around the converged values for I_C will be different, and can be captured by the VDP. Thus, the choice of α is a trade-off. The α should be small enough to capture such perceivable fluctuations. However, it cannot be too small because $I_{\alpha T}$ may exhibit high variance in the regions in which the solution for I_T converged to that of I_C, with luminance differences below the noticeable level. In our experiments with stopping conditions for the MCPT technique for various scenes we found that $\alpha = 0.5$ (50% of photons are the same for I_T and $I_{0.5T}$) is such a reasonable trade-off.

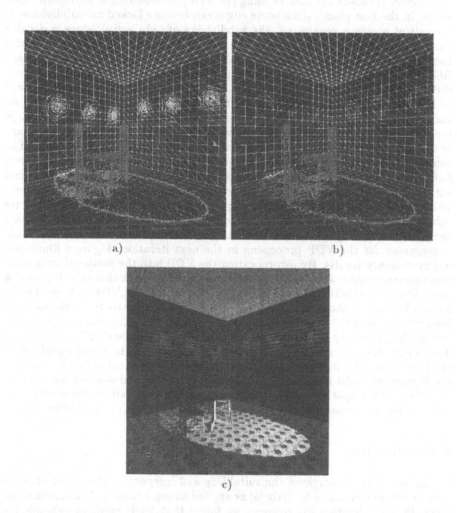

Fig. 6. Mesh simplification: a) original technique, b) the VDP-enhanced solution, c) the resulting shading quality.

Adaptive mesh subdivision. The goal of the adaptive meshing is to reconstruct lighting function without visible artifacts using the minimal number of mesh elements. The reported perception-based meshing solutions focus on application of Tone Mapping Operator [25] and perceptually uniform color spaces to compute perceivable differences guiding the mesh subdivision [19, 9, 11]. However, masking lighting patterns by textures, which can lead to important savings in the number of mesh elements, seems to be mostly unexplored in the existing solutions. We applied the VDP to support decisions of mesh splitting by expanding our earlier iterative mesh subdivision technique [19]. Our current implementation consists of three phases, which are repeated for every iteration: (1) choosing candidate edges for subdivision using the traditional approach [19], (2) verification of chosen candidates using the VDP, (3) restoring unnecessarily split edges. In the first phase, since some edges can become locked for subdivision as a result of previous processing, the subdivision of only active edges is decided based on differences in RGB along the edge. (We use the TMO [25] to derive the displayable RGB from luminance values, and we choose different, predefined RGB thresholds for dark and bright image regions to decide upon an edge subdivision [19].) If the differences are too small, the edge becomes locked and will be skipped in lighting calculations. At this stage, textures are ignored, and we use "representative" (averaged) color of a texture to compute RGB values. In the second phase, the mesh based on subdivided edges is built, and an image of the resulting luminance map with textures is generated. The VDP is used to compare this image with the final image from the previous iteration. The item buffer is applied to identify all triangles with visible differences (at least one pixel with $P > 0.75\%$), and whose unlocked edges are marked as candidates for subdivision. In the third phase, all active edges are re-scanned: the marked edges are subdivided, and the remaining edges are locked for future subdivisions. Finally, the perceptually adjusted mesh is built, and the final image for this iteration is generated for the VDP processing in the next iteration. Figure 6 illustrates our preliminary results. By incorporating the VDP into the mesh-splitting decision process, usually we are able to reduce the number mesh elements. Figures 6a and b show the subdivided mesh using our original and the VDP-enhanced techniques. Figures 6c shows the resulting shading which appears to be the same in both cases. The VDP-based meshing works well for smoothly changing lighting, but in the proximity of lighting discontinuities mesh reduction is rather poor due to excessive spread of predicted differences to pixels in the discontinuities neighborhood. (At high contrast, phase shifts in the channel responses can occur because of slight dislocation of the mesh-reconstructed shadow boundaries, and much wider spatial extent than the boundary dislocations themselves are required to compensate for the resulting differences in the channel responses.)

6 Conclusions

In this work, we investigated the suitability and integrity of the Visible Difference Predictor developed by Daly [5] as applied to some basic global illumination tasks. In our validation experiments we found that VDP predicts well the detectability of various target patterns (lighting solution elements such as shadows) imposed on background patterns (textures) of various scales and orientations. The integrity of the model was shown via comparisons of its predictions when the model incorporated a variety of contrast definitions, spatial and orientation channel decomposition methods, and CSFs derived from different psychophysical experiments. Though it is difficult to generalize from our results to a wide range of input images and potential VDP applications, prediction was quite robust across the tasks we examined and variations in the configuration of VDP modules. The application of the VDP to monitor "perceptual" convergence of global illumination solutions was shown to provide a meaningful metric to compare their performance and decide upon computation stopping conditions. Also, we exploited shadow masking by texture and contrast to simplify the final mesh resulting from the mesh adaptive subdivision computations.

In future work, we plan to perform more systematic validation and calibration of the VDP in human psychophysical experiments designed specifically for our task. We also wish to incorporate the modeling of suphrathreshold vision [20] and mutual channel inhibition [15, 24], which should further improve the fit of VDP responses to the results of psychophysical experiments.

7 Acknowledgments

The author would like to thank Takehiro Tawara for help in organizing the experimental material into Web pages [1]. Special thanks to Bill Martens for stimulating discussions, and for reviewing the manuscript.

References

1. http://www.u-aizu.ac.jp/labs/csel/vdp/ - the Web page accompanying this paper.
2. Mark R. Bolin and Gary W. Meyer. A frequency based ray tracer. In *SIGGRAPH 95 Conference Proceedings*, pages 409–418, 1995.
3. P.J. Burt. Fast filter transforms for image processing. *Computer Vision, Graphics and Image Processing*, 21:368–382, 1983.
4. S. Comes, O. Bruyndonckx, and B. Macq. Image quality criterion based on the cancellation of the masked noise. In *Proc. of IEEE Int'l Conference on Acoustics, Speech and Signal Processing*, pages 2635–2638, 1995.
5. S. Daly. The Visible Differences Predictor: An algorithm for the assessment of image fidelity. In A.B. Watson, editor, *Digital Image and Human Vision*, pages 179–206. Cambridge, MA: MIT Press, 1993.
6. R.L. De Valois and DeValois K.K. *Spatial vision.* Oxford University Press, Oxford, 1990.
7. J.A. Ferwerda, S. Pattanaik, P. Shirley, and D.P. Greenberg. A model of visual masking for computer graphics. In *SIGGRAPH 97 Conference Proceedings*, pages 143–152, 1997.
8. A. Gaddipatti, R. Machiraju, and R. Yagel. Steering image generation with wavelet based perceptual metric. *Computer Graphics Forum (Eurographics '97)*, 16(3):241–251, September 1997.
9. S. Gibson and R. J. Hubbold. Perceptually-driven radiosity. *Computer Graphics Forum*, 16(2):129–141, 1997.
10. D.P. Greenberg, K.E. Torrance, P. Shirley, J. Arvo, J.A. Ferwerda, S.N. Pattanaik, E.P.F. Lafortune, B. Walter, S.C. Foo, and B. Trumbore. A framework for realistic image synthesis. In *SIGGRAPH 97 Conference Proceedings*, pages 477–494, 1997.
11. David Hedley, Adam Worrall, and Derek Paddon. Selective culling of discontinuity lines. In *8th Eurographics Workshop on Rendering*, pages 69–80, 1997.
12. A. B. Khodulev and E. A. Kopylov. Physically accurate lighting simulation in computer graphics software. In *Graphicon'96*, pages 111–119, 1996.

236

13. D. Lischinski, B. Smits, and D.P. Greenberg. Bounds and error estimates for radiosity. In *SIGGRAPH 94 Conference Proceedings*, pages 67–74, 1994.
14. C. Lloyd and R.J. Beaton. Design of spatial-chromatic human vision model for evaluating full-color display systems. In *Human Vision and Electronic Imaging: Models, Methods, and Appl.*, pages 23–37. SPIE Vol. 1249, 1990.
15. J. Malik and P. Perona. Preattentive texture discrimination with early vision mechanisms. *J. Opt. Soc. Am. A*, 7(5):923–932, 1990.
16. S. Marcelja. Mathematical description of the responses of simple cortical cells. *J. Opt. Soc. Am.*, 70:1297–1300, 1980.
17. R.A. Martin, A.J. Ahumada, and J.O. Larimer. Color matrix display simulation based upon luminance and chrominance contrast sensitivity of early vision. In *Human Vision, Visual Processing, and Digital Display III*, pages 336–342. SPIE Vol. 1666, 1992.
18. K. Myszkowski and T.L. Kunii. An efficient cluster-based hierarchical progressive radiosity algorithm. In *ICSC '95*, volume 1024 of *Lecture Notes in Computer Science*, pages 292–303. Springer-Verlag, 1995.
19. K. Myszkowski, A. Wojdala, and K. Wicynski. Non-uniform adaptive meshing for global illumination. *Machine Graphics and Vision*, 3(4):601–610, 1994.
20. E. Peli. Contrast in complex images. *J. Opt. Soc. Am. A*, 7(10):2033–2040, 1990.
21. H. Rushmeier, G. Ward, C. Piatko, P. Sanders, and B. Rust. Comparing real and synthetic images: some ideas about metrics. In *Sixth Eurographics Workshop on Rendering*, pages 82–91. Eurographics, June 1995.
22. P. Shirley, B. Wade, P. M. Hubbard, D. Zareski, B. Walter, and D. P. Greenberg. Global Illumination via Density Estimation. In *Sixth Eurographics Workshop on Rendering*, pages 219–230, 1995.
23. C.C. Taylor, Z. Pizlo, J. P. Allebach, and C.A. Bouman. Image quality assessment with a Gabor pyramid model of the human visual system. In *Human Vision and Electronic Imaging III*. SPIE Vol. 3299, 1998.
24. P.C. Teo and D.J. Heeger. Perceptual image distortion. pages 127–141. SPIE Vol. 2179, 1994.
25. J. Tumblin and H.E. Rushmeier. Tone reproduction for realistic images. *IEEE Computer Graphics and Applications*, 13(6):42–48, 1993.
26. A.B. Watson. The Cortex transform: rapid computation of simulated neural images. *Comp. Vision Graphics and Image Processing*, 39:311–327, 1987.
27. S.J.P. Westen, R.L. Lagendijk, and J. Biemond. Perceptual image quality based on a multiple channel HVS model. In *Proc. of IEEE Int'l Conference on Acoustics, Speech and Signal Processing*, pages 2351–2354, 1995.
28. H.R. Wilson. Psychophysical models of spatial vision and hyperacuity. In D. Regan, editor, *Spatial vision, Vol. 10, Vision and Visual Disfunction*, pages 179–206. Cambridge, MA: MIT Press, 1991.
29. H.R. Wilson and D.J. Gelb. Modified line-element theory for spatial-frequency and width discrimination. *J. Opt. Soc. Am. A*, 1(1):124–131, 1984.
30. C. Zetzsche and Hauske G. Multiple channel model for the prediction of subjective image quality. In *Human Vision, Visual Processing, and Digital Display*, pages 209–216. SPIE Vol. 1077, 1989.
31. B. Zhu, A.H. Tewfik, and O.N. Gerek. Image coding with mixed representations and visual masking. In *Proc. of IEEE Int'l Conference on Acoustics, Speech and Signal Processing*, pages 2327–2330, 1995.

Editors' Note: see Appendix, p: 335 for colored figure of this paper

Fidelity of Graphics Reconstructions : A Psychophysical Investigation

Ann McNamara[1]   Alan Chalmers[2]   Tom Troscianko[3]   Erik Reinhard[4]

[1,2,4]Department of Computer Science,
[3]Department of Experimental Psychology,

University of Bristol, United Kingdom

Abstract. In this paper we develop a technique for measuring the perceptual equivalence of a graphical scene to a real scene. Ability to compare images is valuable in computer graphics for a number of reasons but the main motivation is to enable us to compare different rendering algorithms and to bring us closer to a system for validating lighting simulation algorithms against measurements.
In this study we conduct a series of psychophysical experiments to assess the fidelity of graphical reconstruction of real scenes. Methods developed for the study of human visual perception are used to provide evidence for a perceptual, rather than a mere physical, match between the original scene and its computer representation. Results show that the rendered scene has high perceptual fidelity compared to the original scene, which implies that a rendered image can convey albedo. This investigation is a step toward providing a quantitative answer to the question of just how "real" photo-realism actually is.

1   Introduction

Realistic image synthesis focuses on generating images that emulate the impression of real scenes. It is now possible to accurately simulate the distribution of light energy in a scene, however, physical accuracy in rendering does not ensure that the displayed images will have authentic visual appearance. Even if we assume the lighting simulation is correct to within a given tolerance, problems exist with the manner in which human observers perceive the resulting images. Reasons for this include the limited range of intensities that can be displayed on a display device, and the fact that in general, most state of the art renderers don't compensate for these limitations. Some encouraging research, which models the parameters of perceptual response [4, 11], has begun, but this poses some new questions; for example how can we compare these images which claim to be realistic, to real scenes. Furthermore, many of these techniques are based on algorithms which make assumptions about the functioning of the human visual system; such assumptions often apply to restricted viewing conditions rather than complex scenes. In general, the question of how to compare images with other images to determine how alike they are, remains largely unanswered.

Several computational methods for assessing the quality of such computer generated images have been proposed. The most effective of these methods compare images based on perceptual appearance rather than photometric accuracy. Rushmeier *et al.* explored a number of such perceptually based metrics and concluded that perceptual metrics may be used to numerically compare renderings and captured images in a man-

---

[1]E-mail: mcnamara@cs.bris.ac.uk

ner that approximately corresponds to human contrast perception[9]. An overview of these metrics is given in the next section.

There are many ways in which perceptual fidelity could be measured; in fact, it is likely that no single measure can fully deal with what is a complex issue. For this reason we use actual human response rather than perceptual data when attempting to compare images. We chose a particular task - that of matching materials in the scene against a display of originals - because the task has a number of attractive features. First, Gilchrist [6, 7] has shown that the perception of lightness (the perceptual correlate of reflectance) is strongly dependent on the human visual system's rendition of both illumination and 3-D geometry. These are key features of perception of any scene and are in themselves complex attributes. However, the simple matching procedure used here depends critically on the correct representation of the above parameters. Therefore, the task should be sensitive to any mismatch between the original and the rendered scene. Secondly, the matching procedure is a standard psychophysical task and allows excellent control over the stimulus and the subject's response. The task chosen here corresponds closely to the methodology of Gilchrist [2, 6, 7] which permits simple measures (of lightness) to be made at locations in complex scenes. Ultimately, the task was chosen to be simple while also being sensitive to perceptual distortions in the scene. In addition, this measure of perceived contrast may be used to provide a simple way to predict object visibility within a scene [2, 7].

2   Previous Work

The following image comparison metrics were derived from [3, 5, 8] in a study which compared real and synthetic images by Rushmeier *et al* [9]. Each is based on ideas taken from image compression techniques. The goal of this work was to obtain results from comparing two images using these models that were large if large differences between the images exist, and small when they are almost the same. These suggested metrics include some basic characteristics of human vision described in image compression literature. First, within a broad band of luminances, the eye senses relative rather than absolute luminances. For this reason a metric should account for luminance variations, not absolute values. Second, the response of the eye is non-linear. The perceived "brightness" or "lightness" is a non-linear function of luminance. The particular non-linear relationship is not well established and is likely to depend on complex issues such as perceived lighting and 3-D geometry. Third, the sensitivity of the eye depends on the spatial frequency of luminance variations. The following methods attempt to model these three effects. Each model uses a different Contrast Sensitivity Function (CSF) to model the sensitivity to spatial frequencies.

Model 1 After Mannos and Sakrison [8]: First, all the luminance values are normalised by the mean luminance. The non linearity in perception is accounted for by taking the cubed root of each normalised luminance. A Fast Fourier Transform (FFT) is computed of the resulting values, and the magnitude of the resulting values are filtered with a CSF to an array of values. Finally the distance between the two images is computed by finding the Mean Square Error (MSE) of the values for each of the two images. This technique therefore measures similarity in Fourier amplitude between images.

Model 2 After Gervais et al [5]: This model includes the effect of phase as well as magnitude in the frequency space representation of the image. Once again the luminances are normalised by dividing by the mean luminance. An FFT is computed producing an array of phases and magnitudes. These magnitudes are then filtered with an anisotropic CSF filter function constructed by fitting splines to psychophysical data. The distance between two images is computed using methods described in [5].

Model 3 After Daly:adapted from [3]: In this model the effects of adaptation and non-linearity are combined in one transformation, which acts on each pixel individually. In the first two models each pixel has significant global effect in the normalisation by contributing to the image mean. Each luminance is transformed by an amplitude non-linearity value. An FFT is applied to each transformed luminance and then they are filtered by a CSF (computed for a level of 50 cd/m^2). The distance between the two images is then computed using MSE as in model 1.

One of the goals of our psychophysical investigation is to validate or refute existing metrics. In general, metrics of perceptual fidelity, such as the afore mentioned models [9], depend on factors such as the Fourier composition of the scene. We do not expect a moderate change in Fourier composition to affect the perceived illumination and 3-D properties of a scene. Thus, we seek to show a dissociation between our method and the other, algorithmic, approaches. In the case of moderate Fourier filtering (eg low-pass) we would expect "perfect" results from our method, but low apparent fidelity from the other metrics. A reverse situation is also possible, in which the 3-D geometry is corrupted without a change in Fourier content; here, our method would signal poor fidelity whereas existing metrics would indicate good fidelity. Such a double dissociation would clearly show that perceptual fidelity cannot be described by a single number; rather, psychophysical measures must be made to establish its degree.

Our technique looks for a similar set of matching data in the original and the rendered scene. If such an equality is obtained, we can conclude that the illumination and 3-D qualities of both scenes are perceptually equivalent. The psychophysical experiments we are performing aim to validate or refute existing metrics for image comparison. Our tests make no assumption about the functioning of the human visual system other than its ability to judge equality. This lack of prior assumptions makes it a potentially very robust technique. The disadvantage is the lack of a single algorithm to account for the psychophysical data.

3 Method

This study required an experimental set-up comprising of a real environment and a computer representation of that environment. Here we describe the equipment used to construct the real world test environment, along with the physical measurements performed to attain the necessary input for the synthetic representation.

3.1 The Real Scene

The test environment was a small box of 557 mm high, 408 mm wide and 507 mm deep, with an opening on one side, figure 1. All interior surfaces of the box were painted with white matt house paint.

Fig. 1 Experimental Set up

To the right of this enclosure a chart showing thirty gray level patches, labelled as in figure 2, were positioned on the wall to act as reference. The thirty patches were chosen to provide perceptually spaced levels of reflectance from black to white, according to the Munsell Renotation System [13].

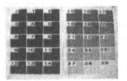

Fig. 2 Reference patches

A series of fifteen of these gray level patches were chosen at random, reshaped, and placed in no particular order within the physical environment. A small front-silvered, high quality mirror was incorporated into the set up to allow the viewing conditions to be changed to two settings, viewing of the original scene or viewing of the modelled scene on the computer monitor. When the optical mirror was in position, subjects viewed the original scene. In the absence of the optical mirror the computer representation of the original scene was viewed. The angular subtenses of the two displays were equalised, and the fact that the display monitor had to be closer to the subject for this to occur, was allowed for by the inclusion of a +2 diopter lens in its optical path; the lens equated the optical distances of the two displays.

The light source consisted of a 24 volt quartz halogen bulb mounted on optical bench fittings at the top of the test environment. This was supplied by a stabilised 10 amp DC power supply, stable to 30 parts per million in current. The light shone through a 70 mm by 115 mm opening at the top of the enclosure. Black masks, constructed of matt cardboard sheets, were placed framing the screen and the open wall of the enclosure, a separate black cardboard sheet was used to define the eye position. An aperture in this mask was used to enforce monocular vision, since the VDU display did not permit stereoscopic viewing.

The entire experimental set-up resides in an enclosed dark laboratory in which the only light sources are the DC bulb (shielded from direct view) or illumination from the monitor. Gilchrist [2, 6, 7] has shown that such an experimental environment is sufficient for our purposes.

3.2 The Graphical Representation

The geometric model of the real environment was created using Alias Wavefront [1]. The photometric instrument used throughout the course of the experiments was the Minolta Spot Chroma meter CS-100. The Minolta chroma meter is a compact, tristimulus colorimeter for non contact measurements of light sources or reflective surfaces. The one degree acceptance angle and through the lens viewing system enables accurate targeting of the subject. The chroma meter was used to measure the chromaticity and luminance values of the materials in the original scene and from the screen simulation. The luminance meter was also used to take similar readings of the thirty reference patches. For input into the graphical modelling process the following measurements were taken.

Geometry: A tape measure was used to measure the geometry of the test environment. Length measurements were made with an accuracy of the order of one millimetre.

Materials: The chroma meter was used for material chromaticity measurements. To ensure accuracy of the measurements five measurements were recorded for each material, the highest and lowest luminance magnitude recorded for each material discarded and an average was taken of the remaining three values. The CIE (1931) x, y chromaticity co-ordinates of each primary were obtained and the relative luminances for each phosphor were recorded using the chroma meter. Following Travis [10], these values were then transformed to screen RGB tristimulus values as input to the renderer by applying a matrix based on the chromaticity coordinates of the monitor phosphors and a monitor white point.

Illumination: The illuminant was measured by illuminating an Eastman Kodak Standard White powder, pressed into a circular cavity, which reflects 99% of incident light in a diffuse manner. The chroma meter was then used to determine the illuminant tristimulus values.

The rendered image was created using the Radiance Lighting simulation package [12] to generate our graphical representation of the real scene. Radiance is a physically based renderer, which means that physically meaningful results may be expected, provided the input to the renderer is meaningful.

4 Image Comparison Experiment

In this work we inspect perceptual, as opposed to physical, correspondence between a real and graphical scene by performing tests of visual function in both situations and establishing that both sets of results are identical (allowing for measurement noise). Such an identity will provide strong evidence that no significantly different set of perceptual parameters exist in each situation.

The subjects' task was to match gray level patches within a physical environment to a set of control patches. Then subjects were asked to repeat the same task with the original environment replaced by its computer representation, including some slight variations of the computer representation, such as changes in Fourier composition (blurring), see figure 3.

Fig. 3 Rendered Image (left) with blurring (right)

In the *Original Scene*, physical stimuli were presented in the test environment, described in the previous section. Subjects viewed the screen monocularly through a fixed viewing position. The experiment was undertaken under constant and controlled illumination conditions.

While viewing the *Computer Simulated Scene*, representation of the stimuli, rendered using Radiance, were presented on the monitor of a Silicon Graphics 02 machine. Again, subjects viewed the screen monocularly through a fixed viewing position.

Our use of the psychophysical lightness-matching procedure is chosen because it is sensitive to errors in perceived depth. Lightness constancy depends on a correct representation of the three dimensional structure of the scene [6, 7]. Any errors in depth perception, when viewing the computer model, will result in errors of constancy, and thus a poor psychophysical matching performance.

5 Results

Data were obtained for 15 subjects; 14 of these were naive as to the purpose of the experiment, and one was an author. Subjects had either normal or corrected-to-normal vision. Each subject performed a number of conditions, in random order, and within each condition the subject's task was to match the fifteen gray test patches to a reference chart on the wall. Each patch was matched once only and the order in which each subject performed the matches was varied between subjects and conditions.

Figure 4 shows the results obtained for comparison of a rendered and a blurred scene to the real environment. The x-axis gives the *actual* Munsell value of each patch, the y-axis gives the *matched* Munsell value, averaged across the subjects.

A perfect set of data would lie along a 45^0 diagonal line. The experimental data for the real environment lie close to this line, with some small but systematic deviations for specific test patches. These deviations show that lightness constance is not perfect for the original scene. What this means is as follows: when observing a given scene, small (but significant) errors of lightness perception are likely to occur. A perceptually-perfect reconstruction of the scene should produce a very similar pattern of errors if it

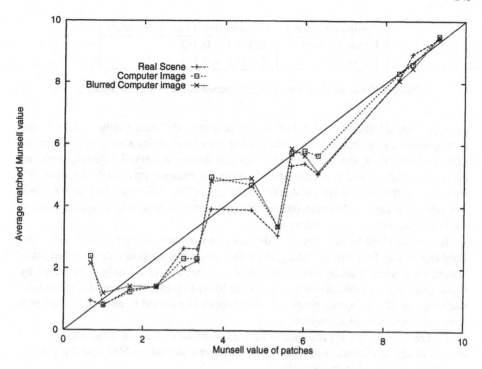

Fig. 4 Comparison of average matchings of Munsell values

is perceptually similar to the original.

The two other graphs relating to the rendered and the blurred rendered images are plotted on the same axes. In general, it can be seen that the matched values are very similar to those of the original scene, in other words, the same (small) failures of constancy apply both to the rendered and the blurred rendered images.

This, in turn suggests that there is no significant perceptual difference between the original scene and both the rendered version and the blurred rendered version. This is in spite of the fact that the mean luminance of the rendered versions was lower by a factor of about 30 compared to the original; also, under our conditions the blurred version looked very different subjectively, but again similar data were obtained.

It is possible to reduce the pattern of results to a single value as follows :

- taking the matches to the original scene as reference, calculate the mean signed deviation for the rendered and blurred rendered functions.

- Compute the mean and standard deviation of these

Table 1 shows the results obtained. A value of zero in this table would indicate *perceptually perfect* match; the actual values given come close to this and are statistically not significantly different from zero. This, therefore, again indicates high perceptual fidelity in both versions of the rendered scene

How do these values compare to other methods? Using the algorithm of Daly [3], we found a 5.04% difference between the rendered and blurred rendered images. As

Compared to Real	Mean Munsell Value Deviation
Rendered Scene	-0.37 ($\sigma = 0.44$)
Blurred Scene	-0.23 ($\sigma = 0.57$)

Table 1 Comparison of Rendered and Blurred Scene to Real Environment

a comparison, a left-right reversal of the image gives a difference value of 3.71%; and a comparison of the image with a white noise grey level image results in a difference value of 72%. Thus, the algorithm suggests that there is a marked difference between the rendered image and blurred rendered image; for example this is a 36% greater difference than that with a left-right reversed image. (This difference increases for less symmetrical images). However, our method suggests that these two scenes are perceptually equivalent in terms of our task.

It *may* therefore be that there is a dissociation between our method and that of the Daly technique. In addition, the algorithmic method cannot give a direct comparison between the original scene and the rendered version; this could only be achieved by frame grabbing the original which is a process likely to introduce errors due to the non-linear nature of the capture process. Further work is planned to attempt to capture a scene without errors of reproduction.

Figure 5 shows the results obtained for comparison of the real scene and rendered images of the environment after the depth has been altered by 50% and the patches specularity increased from 0% to 50%.

As can be seen from the graph, for simple scenes lightness constancy is extremely robust against changes in depth, specularity and blurring.

In summary: the results show that the rendered scenes used in this study have high perceptual fidelity compared to the original scene, and that other methods of assessing image fidelity yield results which are markedly different from ours. The results also imply that a rendered image can convey albedo.

6 Conclusions and Future Work

We have introduced a method for measuring the perceptual equivalence between a real scene and a computer simulation of the same scene. Because this model is based on psychophysical experiments, results are produced through study of vision from a human rather than a machine vision point of view. We have presented a method for modelling a real scene, then validating that model using the response of the human visual system.

By conducting a series of experiments, based on psychophysics, we can estimate how much alike a rendered image is to the original scene. Preliminary results show that given a real scene and a faithful representation of that scene, the visual response function in both cases is similar.

There is still much work to be done in this area, and numerous avenues for further research exist. While we have compared a real scene to the computer model we have yet to attempt the experiments using a captured image of our real scene. This has proved prohibitive in attempting to compare our method to those currently available. In particular the perceptual metrics should be used to compare our rendered image to

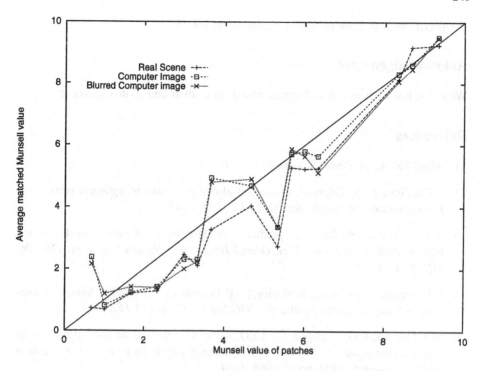

Fig. 5 Comparison of average matchings of Munsell values

a captured image of our real scene which does not suffer from distortions resulting from the capture process. This prevented us from being able to compare our method to those currently available. In an attempt to include comparison to existing metrics such as those outlined in section 2, we have compared the simulated scene to a blurred simulation of the measured scene. Results show that our method gives different results from the metric in the case of the blurred image. This is due to the likely lack of importance of high spatial frequency contents of the image to the task of matching lightness values.

Results from these experiments demonstrate how psychophysics may be used to compare real and synthetic images and in particular, how visual response in the original scene corresponds to visual response to the rendered scene. Our long range goal is to define a technique capable of identifying "perceptual" parameter space so that distances between images in a parameter space are close when humans perceive the images to be similar, and are not close otherwise. The utility being the capacity to evaluate the quality of photo-realistic rendering software, and develop techniques to improve such renderer's ability to produce high fidelity images.

Because the complexity of human perception and the compute expensive rendering algorithms that exist today, future work should focus on developing efficient methods from which resultant graphical representations of scenes yield the same perceptual effects as the original scene. To achieve this the full gamut of colour perception, as opposed to simply lightness, must be considered by introducing scenes of increasing complexity such as those including indirect illumination.

Acknowledgements

We would like to thank our colleagues who donated their time to do experiments.

References

1. Alias/Wavefront. *OpenAlias Manual*, 1996

2. J. Cataliotti & A. Gilchrist. Local and global processes in lightness perception. *Perception and Psychophysics, 57(2), 125-135, 1995.*

3. S. Daly. The visible difference predictor: an algorithm for the assessment of image fidelity. *In A. B. Watson Editor, Digital Images and Human Vision, pp 179-206, MIT Press, 1993*

4. J. Ferwerda, S. Pattanaik, P. Shirley, D. P. Greenberg. A Model of Visual Adaptation for Realistic Image Synthesis. *SIGGRAPH '97, pp 143-152, 1997.*

5. M.J. Gervais, L.O. Harvey,Jr., and J.O. Roberts. Identification confusions among letters of the alphabet. *Journal of Experimental Psychology: Human Perception and Performance, 10(5), pp 655-666, 1984.*

6. A. Gilchrist. Lightness, Brightness and Transparency. *Hillsdale: Lawerence Erlibaum Associates.*

7. A. Gilchrist. Lightness contrast and failures of lightness constancy: a common explanation. *Perception and Psychophysics, 43(5), 125-135, 1988.*

8. J.l. Mannos and D.J Sakrison. The effects of a visual fidelity criterion on the encoding of images. *IEEE Transactions on Information Theory, IT-20(4):525-536, 1974.*

9. H. Rushmeier et al. Comparing Real and Synthetic Images: Some Ideas About Metrics. *6th Eurographics Workshop on Rendering, pp 213-222, 1995*

10. D. Travis. Effective Colour Displays, Theory and Practice. *Academic Press, 1991*

11. G. Meyer, H. Rushmeier, M. Cohen, D. Greenberg, K. Torrence. An Experimental Evaluation of Computer Graphics Imagery *ACM Transactions on Graphics, Vol.5, No. 1, pp 30-50, 1986*

12. G. L. Ward. The RADIANCE lighting simulation and rendering system. *Computer Graphics, 1994.*

13. Wyszecki & Stiles. Colour Science: concepts and methods, quantitative data and formulae (2nd edition) *New York: Wiley*

Global Ray-bundle Tracing
with Hardware Acceleration

László Szirmay-Kalos [1], Werner Purgathofer

Department of Control Engineering and Information Technology, TU of Budapest
Budapest, Műegyetem rkp. 11, H-1111, HUNGARY
szirmay@fsz.bme.hu

Abstract. The paper presents a single-pass, view-dependent method to solve the general rendering equation, using a combined finite element and random walk approach. Applying finite element techniques, the surfaces are decomposed into planar patches that are assumed to have position independent, but not direction independent radiance. The direction dependent radiance function is then computed by random walk using bundles of parallel rays. In a single step of the walk, the radiance transfer is evaluated exploiting the hardware z-buffer of workstations, making the calculation fast. The proposed method is particularly efficient for scenes including not very specular materials illuminated by large area lightsources or sky-light. In order to increase the speed for difficult lighting situations, walks can be selected according to their importance. The importance can be explored adaptively by the Metropolis sampling method.

1 Introduction

The fundamental task of computer graphics is to solve a Fredholm type integral equation describing the light transport. This equation is called the *rendering equation* and has the following form:

$$L(\vec{x}, \omega) = L^e(\vec{x}, \omega) + \int_\Omega L(h(\vec{x}, -\omega'), \omega') \cdot \cos\theta' \cdot f_r(\vec{x}, \omega', \omega) \, d\omega' \quad (1)$$

where $L(\vec{x}, \omega)$ and $L^e(\vec{x}, \omega)$ are the radiance and emission of the surface in point $\vec{x}$ at direction ω, Ω is the directional sphere, $h(\vec{x}, \omega')$ is the visibility function defining the point that is visible from point $\vec{x}$ at direction ω', θ' is the angle between the surface normal and direction $-\omega'$, and $f_r(\vec{x}, \omega', \omega)$ is the bi-directional reflection/refraction function. Since the rendering equation contains the unknown radiance function both inside and outside the integral, in order to express the solution, this coupling should be resolved. Generally, two methods can be applied for this: finite element methods or random walk methods.

Finite element methods project the problem into a finite function base and approximate the solution here. The projection transforms the integral equation to a system of linear equations for which straightforward solution techniques are available. Finite element techniques that aim at the solution of the non-diffuse case can be traced back to the finite element approximation of the directional functions using *partitioned sphere*

---

[1]This work has been supported by the National Scientific Research Fund (OTKA), ref.No.: F 015884, the Austrian-Hungarian Action Fund, ref.No.: 29p4, 32öu9 and 34öu28, and the Spanish-Hungarian Fund, ref.No.: E9.

[8] or *spherical harmonics* [23], and to the application of *extended form factors* [22]. Since the radiance function is not smooth and is of 4-variate if non-diffuse reflection should also be considered, finite element methods require a great number of basis functions, and thus the system of linear equations will be very large. Although, hierarchical or multiresolution methods [4] and clustering [3] can help, the memory requirements are still prohibitive for complex scenes.

Random walk methods, on the other hand, resolve the coupling by expanding the integral equation into a Neumann series, and calculate the resulting high-dimensional integrals by numerical quadrature from discrete samples. A single discrete sample corresponds to a complete photon-path (called the *walk*) from a lightsource to the eye, which is usually built by d ray-shooting steps if the photon is reflected d times. Since classical quadrature rules are useless for the calculation of very high dimensional integrals, Monte-Carlo or quasi-Monte Carlo techniques must be applied.

In computer graphics the first Monte-Carlo random walk algorithm — called *distributed ray-tracing* — was proposed by Cook et al. [5], which spawned to a set of variations, including *path tracing* [9], *light-tracing* [7], *Monte-Carlo radiosity* [20][14][17], and two-pass methods which combine radiosity and ray-tracing [29].

The problem of naive generation of walks is that the majority of the paths do not contribute to the image at all, and their computation is simply waste of time.

Thus, on the one hand, random walk must be combined with a deterministic step that forces the walk to go to the eye and to find a lightsource. *Light tracing* connects each bounce position to the eye deterministically [7]. *Bi-directional path-tracing* methods start a walk from the eye and a walk from a lightsource and connect the bounce positions of the two walks [11],[27].

On the other hand, importance sampling [24] should be incorporated to prefer useful paths along which significant radiance is transferred. Note that although the contribution on the image is a function of the complete path, computer graphics applications usually assign estimated importance to individual steps of this path, which might be quite inaccurate. In a single step the importance is usually selected according to the BRDF [7] [11], or according to the direction of the direct lightsources [21]. Combined methods that find the important directions using both the BRDF and the incident illumination have been proposed in [26], [12]. Just recently, Veach and Guibas [28] proposed the Metropolis method to be used in the solution of the rendering equation. Unlike other approaches, Metropolis sampling [13] can assign importance to a complete walk not just to the steps of this walk, and it explores important regions of the domain adaptively while running the algorithm.

In order to reduce the noise of these methods, very many samples are required, especially when importance sampling cannot help significantly — that is when the lightsources are large and the surfaces are not very specular. One way of reducing the ray-object intersection calculation cost is storing this information in the form of *illumination networks* [1], but it has large memory requirements, and representing the light-transport of small number of predefined rays might introduce artifacts.

The proposed new method also combines the advantages of finite-element and random-walk approaches and can solve the general non-diffuse case. The method needs no preprocessing, the memory requirements are modest, and it is particularly efficient for scenes containing larger area lightsources and moderately specular surfaces — that is where other importance-sampling walk methods become inefficient.

2 Informal discussion of the new algorithm

Walk methods proposed so far use individual ray-paths as samples of the integrand of the rendering equation. However, ray-shooting may waste a lot of computation by ignoring all the intersections but the one closest to the start of the ray. Thus it seems worth using a set of *global directions* [18][14] for the complete scene instead of solving the visibility problem independently for different points $\vec{x}$. Moreover, ray-shooting is a simple but by no means the most effective visibility algorithm since it is unable to take advantage of image coherence. Other methods based on the exploitation of image coherence, such as the z-buffer, painter's, Warnock's, etc. algorithms can be considered as handling a bundle of parallel rays and solving the visibility problem for all of them simultaneously. Continuous (also called object-precision) methods can even determine the visibility problem independently of the resolution, which corresponds to tracing infinitely many parallel rays simultaneously. This paper aims at using ray-bundles instead of individual rays to solve the rendering equation.

These visibility algorithms assume that the surfaces are decomposed into planar patches, thus the proposed method also uses this assumption. On the other hand, the patch decomposition is also used for the finite-element structure, where each surface element is assumed to have uniform radiance (note that this does not mean that the radiance function itself is uniform or diffuse, just the same radiance function is valid anywhere inside a patch). In order to simulate multiple interreflections, ray-bundles should be traced several times in different directions. The series of directions is called a *global random walk*.

2.1 Computation of global ray-bundle walks

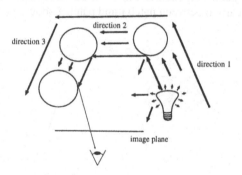

Fig. 1. A path of ray-bundles

The algorithm takes samples of these global walks and uses them in the quadrature. A single walk starts by selecting a direction either randomly or quasi-randomly, and the emission transfer of all patches is calculated into this direction (figure 1). Then a new direction is found, and the emission is transferred and the irradiance generated by the previous transfer is reflected from all patches into this new direction. The algorithm keeps doing this for a few times depending on how many bounces should be considered, then the emission is sent and the irradiance caused by the last transfer is reflected towards the eye. Averaging these contributions results in the final image. When the radiance reflection is calculated, the radiance is attenuated by the BRDF of the corresponding surface element.

In order to make this method work, efficient algorithms are needed that can compute the radiance transfer of all patches in a single direction. A new global visibility method capable of taking advantage of the built-in z-buffer is proposed for this task.

In the following sections, first the formal aspects of the combination of finite element and random walk techniques are discussed, then the global visibility method is presented.

3 Reformulation of the rendering equation using the finite-elements

According to the concept of finite-elements, the radiance, emission and the BRDF of patch i are assumed to be constant and are denoted by $L_i(\omega)$, $L_i^e(\omega)$ and $\tilde{f}_i(\omega, \omega')$, respectively. Using these assumptions, integrating the rendering equation on patch i we obtain:

$$L_i(\omega) \cdot A_i = L_i^e(\omega) \cdot A_i + \int\limits_{\Omega} \int\limits_{A_i} L(h(\vec{x}, -\omega'), \omega') \cdot \cos\theta' \cdot \tilde{f}_i(\omega', \omega) \, d\vec{x} \, d\omega'. \quad (2)$$

Taking into account that the integrand of the inner surface integral is piece-wise constant, it can also be presented in closed form:

$$\int\limits_{A_i} L(h(\vec{x}, -\omega'), \omega') \cdot \cos\theta' \cdot \tilde{f}_i(\omega', \omega) \, d\vec{x} = \sum_{j=1}^{n} \tilde{f}_i(\omega', \omega) \cdot A(i, j, \omega') \cdot L_j(\omega'), \quad (3)$$

where $A(i, j, \omega')$ expresses the projected area of patch j that is visible from patch i in direction ω'. In the unoccluded case this is the intersection of the projections of patch i and patch j onto a plane perpendicular to ω'. If occlusion occurs, the projected areas of other patches that are in between patch i and patch j should be subtracted as shown in figure 2.

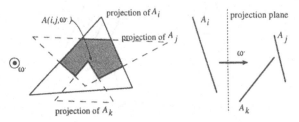

Fig. 2. Interpretation of $A(i, j, \omega')$

Using equation (3) and dividing both sides by A_i, the matrix form of the rendering equation can be obtained as:

$$\mathbf{L}(\omega) = \mathbf{L}^e(\omega) + \int\limits_{\Omega} \mathbf{F}(\omega', \omega) \cdot \mathbf{A}(\omega') \cdot \mathbf{L}(\omega') \, d\omega', \quad (4)$$

where $\mathbf{L}(\omega)$ is the vector of radiance values, $\mathbf{F}(\omega', \omega)$ is a diagonal matrix of BRDFs, and *geometry matrix* $\mathbf{A}$ contains the relative visible areas: $\mathbf{A}(\omega')|_{ij} = A(i, j, \omega')/A_i$.

Note that equation (4) is highly intuitive as well. The radiance of a patch is the sum of the emission and the reflection of all irradiances into this direction. The role of the patch-direction-patch "form-factors" is played by $A(i, j, \omega')/A_i$.

3.1 Random walk

The solution of integral equation (4) can be obtained in the form of a Neumann series:

$$L(\omega) = \sum_{i=0}^{\infty} \mathcal{T}^i L^e(\omega), \quad \text{where} \quad \mathcal{T}L(\omega) = \int_{\Omega} F(\omega',\omega) \cdot A(\omega') \cdot L(\omega') \, d\omega'. \quad (5)$$

The terms of this infinite Neumann series have intuitive meaning as well: $\mathcal{T}^0 L^e(\omega) = L^e(\omega)$ comes from the emission, $\mathcal{T}^1 L^e(\omega)$ comes from a single reflection (called 1-bounce), $\mathcal{T}^2 L^e(\omega)$ from two reflections (called 2-bounces), etc.

In practice the infinite sum of the Neumann series is always approximated by a finite sum. The number of required terms is determined by the contraction of the operator $\mathcal{T}$ — that is the overall reflectivity of the scene. Let us denote the maximum number of calculated bounces by D. The truncation of the Neumann series introduces a bias in the estimation, which can be tolerated if D is high enough.

In order to simplify the notations, we introduce the *max d-bounce irradiance* I_d for $d = 1, 2, \ldots$ as follows:

$$
\begin{aligned}
I_0 &= A(\omega_D') \cdot L^e(\omega_D') \\
I_d &= A(\omega_{D-d}') \cdot \left(L^e(\omega_{D-d}') + 4\pi \cdot F(\omega_{D-d+1}', \omega_{D-d}') \cdot I_{d-1} \right),
\end{aligned}
$$

where I_d is a $d + 1$ dimensional function of directions $(\omega_{D-d}', \omega_{D-d+1}' \ldots, \omega_D')$. The max d-bounce irradiance represents the irradiance arriving at each patch after completing a path of length d following the given directions, and gathering and then reflecting the emission of the patches.

Limiting the solution to consider at most $D + 1$ bounces, the solution of the rendering equation can be obtained as a $2D$-dimensional integral:

$$L(\omega) = (\frac{1}{4\pi})^D \int_{\Omega} \ldots \int_{\Omega} [L^e(\omega) + 4\pi \cdot F(\omega_1', \omega) \cdot I_D(\omega_1', \ldots, \omega_D')] \, d\omega_D' \ldots d\omega_1'.$$

$$(6)$$

This high-dimensional integral can be estimated by Monte-Carlo quadrature.

3.2 Simple Monte-Carlo, or quasi-Monte Carlo integration

In order to evaluate formula (6), M random or quasi-random [10] walks should be generated (the difference is that in Monte-Carlo walks the directions are sampled randomly while in quasi-random walks they are sampled from a $2D$-dimensional low-discrepancy sequence, such as the $2D$-dimensional *Halton* or *Hammersley* sequence [16]). When the D-bounce irradiance is available, it is multiplied by the BRDF defined by the last direction ω_1 and the viewing direction ω to find a Monte-Carlo estimate of the radiance that is visible from the eye position. Note that this step makes the algorithm view-dependent. There are basically two different methods to calculate the estimate of the image. On the one hand, evaluating the BRDF once for each patch, a radiance value is assigned to them, then in order to avoid "blocky" appearance, bi-linear smoothing can be applied. Using Phong interpolation, on the other hand, the radiance is evaluated at each point visible through a given pixel, based on the irradiance field, the surface normal and on the BRDF of the found point. In order to speed up this procedure, the surface visible at each pixel and the surface normal can be determined in a preprocessing phase and stored in a map. Phong interpolation is more time consuming but the generated image is not only numerically precise, but is also visually pleasing.

The final image is the average of these estimates. The complete algorithm — which requires just one variable for each patch i, the max d-bounce irradiance $\mathbf{I}[i]$ — is summarized in the following:

> **for** $m = 1$ **to** M **do**   // *samples of global walks*
>   Generate the mth random or low-discrepancy point: $(\omega_1^{(m)}, \omega_2^{(m)}, \ldots, \omega_D^{(m)})$
> $\mathbf{I} = 0$
>   **for** $d = 0$ **to** $D - 1$ **do**   // *a random or quasi-random walk*
> $\mathbf{I} = \mathbf{A}(\omega'_{D-d}) \cdot \left(\mathbf{L}^e(\omega'_{D-d}) + 4\pi \cdot \mathbf{F}(\omega'_{D-d+1}, \omega'_{D-d}) \cdot \mathbf{I} \right)$
>   **endfor**
>   Calculate the image estimate from the irradiance $\mathbf{I}$
>   Divide the estimate by M and add to the final image
> **endfor**
> Display Image

3.3 Combined and bi-directional walking techniques

The algorithm that has been derived directly from the quadrature formulae uses direction ω_1 to evaluate the contribution of 1-bounces, directions (ω_1, ω_2) for the 2-bounces, $(\omega_1, \omega_2, \omega_3)$ for the 3-bounces, etc. This is just a little fraction of the information that can be gathered during the complete walk. We could also use the samples of ω_1, ω_2, ω_3, etc. to calculate the 1-bounce contribution, (ω_1, ω_2), (ω_1, ω_3), $\ldots$, (ω_2, ω_3), etc. combinations of directions for 2-bounces, etc. This is obviously possible, since if the samples of $(\omega_1, \omega_2, \ldots \omega_D)$ are taken from a uniform sequence, then all combinations of its elements also form uniform sequences in lower dimensional spaces.

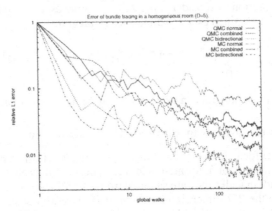

Fig. 3. Combined and bi-directional walking techniques versus normal walk

If all possible combinations are used, then each random walk generates $\binom{D}{d}$ samples for the d-bounces, which can be used to increase the accuracy of the method. Note that the increased accuracy of this "*combined*" method is for free in terms of additional computation. However, due to the dependence of the BRDF functions on two directions and due to the fact that different bounces will be estimated by different numbers of samples, the required storage per patch is increased to $D(D + 1)/2$ variables. Since D is 5 to 8 in practical cases, this storage overhead is affordable. Furthermore, when the radiance is transferred to a direction, the required information to transfer the radiance to the opposite direction is also available. Taking advantage of this, a single *"bi-directional"* walk will generate $\binom{D}{d} \cdot 2^d$ samples for the d-bounce transfer.

The additional samples of the "combined" and particularly the "bi-directional" walking techniques increase the accuracy as shown by figure 3. The test scene was the homogeneous Cornell-box where all surfaces have constant 0.5 diffuse reflectance and emission, which allowed to solve the rendering equation analytically [10] (the solution is $L = 1 - 2^{-(D+1)}$) to find a reference for the error analysis. Note that although quasi-Monte Carlo sampling is generally better, the improvement provided by the combined and bi-directional methods is less for the quasi-Monte Carlo walk than for the Monte-Carlo walk since the low-discrepancy points are so "well-designed" that mixing different sets of them does not improve the quadrature much further.

4 Metropolis solution of the directional integrals

The global illumination method proposed so far is particularly efficient if the lighting distribution in the scene does not exhibit high variations. For difficult lighting conditions importance sampling can help, that prefers those sequences of directions that transport significant radiance towards the eye. Since no a-priori information is available which these important directions are, some kind of adaptive technique must be used. In this section the application of Metropolis sampling is considered.

The Metropolis algorithm is a Monte-Carlo quadrature method that incorporates adaptive importance sampling by exploring the properties of the integrand automatically. Suppose that integral $I = \int_V f(\mathbf{z}) \, d\mathbf{z}$ needs to be evaluated. Let us define importance $\mathcal{I}(\mathbf{z})$ that should be approximately proportional to f. Importance sampling requires the generation of samples $\{\mathbf{z}_1, \mathbf{z}_2, \ldots \mathbf{z}_M\}$ according to a probability density $p(\mathbf{z})$ — which is proportional to $\mathcal{I}(\mathbf{z}) = b \cdot p(\mathbf{z})$ — and using the following formula:

$$I = \int\limits_V \frac{f(\mathbf{z})}{\mathcal{I}(\mathbf{z})} \cdot \mathcal{I}(\mathbf{z}) \, d\mathbf{z} = b \cdot \int\limits_V \frac{f(\mathbf{z})}{\mathcal{I}(\mathbf{z})} \cdot p(\mathbf{z}) \, d\mathbf{z} = b \cdot E[\frac{f(\mathbf{z})}{\mathcal{I}(\mathbf{z})}] \approx \frac{b}{M} \cdot \sum_{i=1}^{M} \frac{f(\mathbf{z}_i)}{\mathcal{I}(\mathbf{z}_i)} \quad (7)$$

In order to generate samples according to $p(\mathbf{z}) = 1/b \cdot \mathcal{I}(\mathbf{z})$ a Markovian process is constructed whose stationary distribution is just $p(\mathbf{z})$.

The definition of this Markovian process $\{\mathbf{z}_1, \mathbf{z}_2, \ldots \mathbf{z}_i \ldots\}$ is as follows:

> **for** $i = 1$ **to** M **do**
> Based on the actual state $\mathbf{z}_i$, choose another random, tentative point $\mathbf{z}_t$
> **if** the tentative point is more important ($\mathcal{I}(\mathbf{z}_t) \geq \mathcal{I}(\mathbf{z}_i)$) **then** accept ($\mathbf{z}_{i+1} = \mathbf{z}_t$)
> **else**   *// accept randomly with probability of the importance degradation*
> Generate random number r in $[0, 1]$.
> **if** $r < \mathcal{I}(\mathbf{z}_t)/\mathcal{I}(\mathbf{z}_i)$ **then** $\mathbf{z}_{i+1} = \mathbf{z}_t$ *// accept*
> **else** $\mathbf{z}_{i+1} = \mathbf{z}_i$ *// keep the previous*
> **endif**
> **endfor**

The generation of the next tentative sample is governed by a *tentative transition function* $T(\mathbf{x} \rightarrow \mathbf{y})$. In the algorithm we use a symmetric tentative transition function, that is $T(\mathbf{x} \rightarrow \mathbf{y}) = T(\mathbf{y} \rightarrow \mathbf{x})$. The transition probability of this Markovian process is:

$$P(\mathbf{x} \rightarrow \mathbf{y}) = \begin{cases} T(\mathbf{x} \rightarrow \mathbf{y}) & \text{if } \mathcal{I}(\mathbf{y}) > \mathcal{I}(\mathbf{x}), \\ T(\mathbf{x} \rightarrow \mathbf{y}) \cdot \mathcal{I}(\mathbf{y})/\mathcal{I}(\mathbf{x}) & \text{otherwise.} \end{cases} \quad (8)$$

In equilibrium state, the transitions between two states $\mathbf{x}$ and $\mathbf{y}$ are balanced, that is $p(\mathbf{x}) \cdot P(\mathbf{x} \to \mathbf{y}) = p(\mathbf{y}) \cdot P(\mathbf{y} \to \mathbf{x})$. Using this and equation (8), and then taking into account that the tentative transition function is symmetric, we can prove that the stationary probability distribution is really proportional to the importance:

$$\frac{p(\mathbf{x})}{p(\mathbf{y})} = \frac{P(\mathbf{y} \to \mathbf{x})}{P(\mathbf{x} \to \mathbf{y})} = \frac{T(\mathbf{y} \to \mathbf{x})}{T(\mathbf{x} \to \mathbf{y})} \cdot \frac{\mathcal{I}(\mathbf{x})}{\mathcal{I}(\mathbf{y})} = \frac{\mathcal{I}(\mathbf{x})}{\mathcal{I}(\mathbf{y})}. \tag{9}$$

If we select initial points according to the stationary distribution — that is proportionally to the importance — then the points visited in the walks originated at these starting points can be readily used in equation (7).

4.1 Definition of the importance function

Let the importance function $\mathcal{I}$ be the sum of luminances of all pixels of the image, resulting from the walk. This importance function really concentrates on those walks that have a significant influence on the image. Using the luminance information is justified by the fact that the human eye is more sensitive to luminance variations than to color variations.

The Metropolis method achieves the sampling according to this probability by establishing a Markovian process on the space of global walks, whose limiting distribution is proportional to the selected importance function.

The Metropolis approximation of the radiance vector is:

$$\mathbf{L}(\omega) = (\frac{1}{4\pi})^D \int_\Omega \ldots \int_\Omega [\mathbf{L}^e(\omega) + 4\pi \cdot \mathbf{F}(\omega_1', \omega) \cdot \mathbf{I}_D(\omega_1', \omega_2', \ldots, \omega_D')] \, d\omega_D' \ldots d\omega_1' \approx$$

$$(\frac{b}{M}) \sum_{m=1}^M \frac{\mathbf{L}^e(\omega) + 4\pi \cdot \mathbf{F}(\omega_1', \omega) \cdot \mathbf{I}_D(\omega_1^{(m)}, \omega_2^{(m)}, \ldots, \omega_D^{(m)})}{\mathcal{I}(\omega_1^{(m)}, \omega_2^{(m)}, \ldots, \omega_D^{(m)})}. \tag{10}$$

where M is the number of mutations and b is the integral of the importance function over the whole space.

4.2 Definition of the tentative transition function

The statespace of the Markovian process consists of D-dimensional vectors of directions that define the sequence of directions in the global walks. Thus the tentative transition function is allowed to modify one or more directions in these sequences.

4.3 Generating an initial distribution

The Metropolis method promises to generate samples with probabilities proportional to their importance in the stationary case. To ensure that the process is already in the stationary case from the beginning, initial samples are also selected according to the stationary distribution, i.e. proportionally to the importance function. Selecting samples with probabilities proportional to the importance can be approximated in the following way. A given number of seed points are found in the set of sequences of global directions. The importances of these seed points are evaluated, then, to simulate the distribution following this importance, the given number of initial points are selected randomly from these seed points using the discrete distribution determined by their importance.

4.4 Automatic exposure

Equation (10) also contains an unknown b constant that expresses the luminance of the image. The initial seed generation can also be used to determine this constant. Then at a given point of the algorithm the total luminance of the current image — that is the sum of the importances of the previous samples — is calculated and an effective scaling factor is found that maps this luminance to the expected one.

5 Calculation of the radiance transport in a single direction

In order to compute the irradiances of different bounces recursively by formula (6), the geometry matrix **A** must be determined for the actual direction and used to weight the radiance vector. Since the geometry matrix contains the relative areas of patches as visible from other patches of the scene, the determination of the geometry matrix requires the solution of a global visibility problem where the eye position visits each patch of the scene.

A possible solution is the application of the painter's algorithm that renders patches having sorted according to the given direction into an image buffer and computes the radiance transfer during rendering for each "pixel" [25].

In this section another method is proposed that traces back the visibility problem to a series of z-buffer steps to allow the utilization of the z-buffer hardware of workstations.

The proposed algorithm solves the discretized visibility problem placing a rasterized window perpendicularly to the given direction.

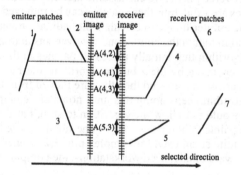

Fig. 4. Calculating the power transfer

The radiance is transferred by dynamically maintaining two groups of patches, an emitter group and a receiver group, in a way that no patch in the receiver group may hide a patch in the emitter group looking from the given direction.

Let the two classes of patches be rendered into two image buffers — called the emitter and receiver images, respectively — setting the color of patch j to j and letting the selected direction be the viewing direction for the receiver set and its inverse for the emitter set. Looking at figure 4, it is obvious that a pair of such images can be used to calculate the radiance transfer of all those patches which are fully visible in the receiver image. The two images must be scanned parallely and when i (that is the index of patch i) is found in the receiver image, the corresponding pixel in the emitter image is read and its value is used as a patch index to find the source radiance which is then added to the irradiance of patch i. Finally, the irradiance is scaled by P/A_i where P is the size

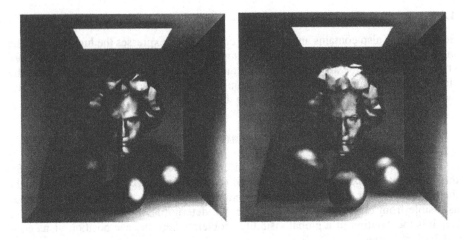

Fig. 5. A scene after the "first-shot"(left) and after 500 Metropolis walks (right)

of a pixel and A_i is the size of the receiver patch.

In order to find out which patches are fully visible in the receiver image, the number of pixels they cover is also computed during scanning and then compared to the size of their projected area. For those patches whose projected area is approximately equal to the total size of the covered pixels, we can assume that they are not hidden and their accumulated irradiances are valid, thus these patches can be removed from the receiver set and rendered into the emitter image to calculate the radiance transfer for other patches (this is the strategy to maintain the emitter and receiver sets automatically). This leads to an incremental algorithm that initially places all patches in the receiver set. Having calculated the receiver image by the z-buffer algorithm, the radiance transfer for the fully visible patches are evaluated, and then they are moved from the receiver set to the emitter set. The algorithm keeps doing this until no patch remains in the receiver set (cyclic overlapping would not allow the algorithm to stop, but this can be handled by a clipping as in the painter's algorithm [15]). The number of z-buffer steps required by the algorithm is quite small even for complex practical scenes [18]. Exploiting the built-in z-buffer hardware of advanced workstations, the computation can be fast.

6 Preprocessing of point lightsources

As other global radiosity methods, this method is efficient for large area lightsources but loses its advantages if the lightsources are small [19]. This problem can be solved by a "*first-shot*" that shoots the power of the point lightsources onto other surfaces, then removes them from the scene [2]. Since the surfaces can also be non-diffuse, the irradiance received by the patches from each point lightsource should be stored (this requires l additional variables per patch, where l is the number of point lightsources). The secondary, non-diffuse emission to a direction is computed from these irradiances.

7 Simulation results

Figure 5 shows a scene as rendered after the first shot and after 500 walks of length 5. The scene contains specular, metallic objects tessellated to 9602 patches, and is illuminated by both area (ceiling) and point (right-bottom corner) lightsources. A global radiance transfer took about 0.7 seconds on a Silicon Graphics O2 computer. Since the radiance information of a single patch is stored in 18 float variables (1 for the emission, 1 for the irradiance generated by the point lightsource, $D(D+1)/2 = 15$ for the irradiances and 1 for the accumulating radiance perceived from the eye), the extra memory used in addition to storing the scene is only 0.7 Mbyte.

The color plate shows a fractal terrain containing 14712 patches after 500 walks which provide an accuracy within 2 percents. A global radiance transfer took approximately 1.1 seconds and the radiance information required 1 Mbyte.

8 Conclusions

This paper presented a combined finite-element and random-walk algorithm to solve the rendering problem of complex scenes including also glossy surfaces. The basic idea of the method is to form bundles of parallel rays that can be traced efficiently, taking advantage of the z-buffer hardware. Unlike other random walk methods using importance sampling [7] [11] [28], this approach cannot emphasize the locally important directions, but handles a large number (1 million) parallel rays simultaneously instead, thus it is more efficient then those methods when the surfaces are not very specular.

The memory requirement is comparable to that of the diffuse radiosity algorithms although the new algorithm is also capable to handle non-diffuse reflections or refractions. Since global ray-bundle walks are computed independently, the algorithm is very well suited for parallelization.

In order to incorporate importance sampling, the Metropolis sampling technique was applied. However, for homogeneous scenes, we could not demonstrate significant noise reduction compared to quasi-Monte Carlo walks. This is due to the fact that the integrand of equation (4) is continuous and is of finite variation unlike the integrand of the original rendering equation, thus if its variation is modest then quasi-Monte quadrature is almost unbeatable. If the radiance distribution has high variation (difficult lighting conditions), then the Metropolis method becomes more and more superior. Future research should concentrate on the tuning of the Metroplis method in this application and on other adaptive importance sampling techniques.

References

1. C. Buckalew and D. Fussell. Illumination networks: Fast realistic rendering with general reflectance functions. *Computer Graphics (SIGGRAPH '89 Proceedings)*, 23(3):89–98, July 1989.
2. F. Castro, R. Martinez, and M. Sbert. Quasi Monte-Carlo and extended first-shot improvements to the multi-path method. In *Proc. of SCCG'98*, pages 91–102, 1998.
3. P. H. Christensen, D. Lischinski, E. J. Stollnitz, and D. H. Salesin. Clustering for glossy global illumination. *ACM Transactions on Graphics*, 16(1):3–33, 1997.
4. P. H. Christensen, E. J. Stollnitz, D. H. Salesin, and T. D. DeRose. Global illumination of glossy environments using wavelets and importance. *ACM Transactions on Graphics*, 15(1):37–71, 1996.

5. R. Cook, T. Porter, and L. Carpenter. Distributed ray tracing. In *Computer Graphics (SIG-GRAPH '84 Proceedings)*, pages 137–145, 1984.

6. I. Deák. *Random Number Generators and Simulation*. Akadémia Kiadó, Budapest, 1989.

7. P. Dutre, E. Lafortune, and Y. D. Willems. Monte Carlo light tracing with direct computation of pixel intensities. In *Compugraphics '93*, pages 128–137, Alvor, 1993.

8. D. S. Immel, M. F. Cohen, and D. P. Greenberg. A radiosity method for non-diffuse environments. In *Computer Graphics (SIGGRAPH '86 Proceedings)*, pages 133–142, 1986.

9. J. T. Kajiya. The rendering equation. In *Computer Graphics (SIGGRAPH '86 Proceedings)*, pages 143–150, 1986.

10. A. Keller. A quasi-Monte Carlo algorithm for the global illumination in the radiosity setting. In H. Niederreiter and P. Shiue, editors, *Monte-Carlo and Quasi-Monte Carlo Methods in Scientific Computing*, pages 239–251. Springer, 1995.

11. E. Lafortune and Y. D. Willems. Bi-directional path-tracing. In *Compugraphics '93*, pages 145–153, Alvor, 1993.

12. E. Lafortune and Y. D. Willems. A 5D tree to reduce the variance of Monte Carlo ray tracing. In *Rendering Techniques '96*, pages 11–19, 1996.

13. N. Metropolis, A. Rosenbluth, M. Rosenbluth, A. Teller, and E. Teller. Equations of state calculations by fast computing machines. *Journal of Chemical Physics*, 21:1087–1091, 1953.

14. L. Neumann. Monte Carlo radiosity. *Computing*, 55:23–42, 1995.

15. M. E. Newell, R. G. Newell, and T. L. Sancha. A new approach to the shaded picture problem. In *Proceedings of the ACM National Conference*, pages 443–450, 1972.

16. H. Niederreiter. *Random number generation and quasi-Monte Carlo methods*. SIAM, Pennsilvania, 1992.

17. S. N. Pattanik and S. P. Mudur. Adjoint equations and random walks for illumination computation. *ACM Transactions on Graphics*, 14(1):77–102, 1995.

18. M. Sbert. *The Use of Global Directions to Compute Radiosity*. PhD thesis, Catalan Technical University, Barcelona, 1996.

19. M. Sbert, X. Pueyo, L. Neumann, and W. Purgathofer. Global multipath Monte Carlo algorithms for radiosity. *Visual Computer*, pages 47–61, 1996.

20. P. Shirley. Discrepancy as a quality measure for sampling distributions. In *Eurographics '91*, pages 183–194. Elsevier Science Publishers, 1991.

21. P. Shirley, C. Wang, and K. Zimmerman. Monte Carlo techniques for direct lighting calculations. *ACM Transactions on Graphics*, 15(1):1–36, 1996.

22. F. Sillion and C. Puech. A general two-pass method integrating specular and diffuse reflection. In *Computer Graphics (SIGGRAPH '89 Proceedings)*, pages 335–344, 1989.

23. F. X. Sillion, J. R. Arvo, S. H. Westin, and D. P. Greenberg. A global illumination solution for general reflectance distributions. *Computer Graphics (SIGGRAPH '89 Proceedings)*, 25(4):187–198, 1991.

24. I. Sobol. *Die Monte-Carlo Methode*. Deutscher Verlag der Wissenschaften, 1991.

25. L. Szirmay-Kalos and T. Fóris. Sub-quadratic radiosity algorithms. In *Winter School of Computer Graphics '97*, pages 562–571, Plzen, Czech Republic, 1997.

26. E. Veach and L. Guibas. Optimally combining sampling techniques for Monte Carlo rendering. In *Rendering Techniques '94*, pages 147–162, 1994.

27. E. Veach and L. Guibas. Bidirectional estimators for light transport. In *Computer Graphics (SIGGRAPH '95 Proceedings)*, pages 419–428, 1995.

28. E. Veach and L. Guibas. Metropolis light transport. *Computer Graphics (SIGGRAPH '97 Proceedings)*, pages 65–76, 1997.

29. J. R. Wallace, M. F. Cohen, and D. P. Greenberg. A two-pass solution to the rendering equation: A synthesis of ray tracing and radiosity methods. In *Computer Graphics (SIGGRAPH '87 Proceedings)*, pages 311–324, 1987.

Editors' Note: see Appendix, p. 334 for colored figure of this paper

Hierarchical Monte Carlo Radiosity

Philippe Bekaert*, László Neumann°, Attila Neumann°,
Mateu Sbert* and Yves D. Willems*

* Department of Computer Science, Katholieke Universiteit Leuven
Celestijnenlaan 200 A, B-3001 Leuven, Belgium
e-mail: Philippe.Bekaert@cs.kuleuven.ac.be

° Maros u. 36, H-1122 Budapest, Hungary
e-mail: neumann@mail.datanet.hu

* Institut d'Informàtica i Aplicacions, Universitat de Girona
Lluis Santaló s/n, E 17071 Girona
e-mail: mateu@ima.udg.es

Abstract. Hierarchical radiosity and Monte Carlo radiosity are branches of radiosity research that focus on complementary problems. The synthesis of both families of radiosity algorithms has however received little attention until now. In this paper, a procedure is presented that bridges the gap. It allows any proposed hierarchical refinement strategy to be investigated in the context of an arbitrary discrete Monte Carlo radiosity algorithm. The synthesis of Monte Carlo radiosity and hierarchical radiosity yields more reliable and easy-to-use radiosity algorithms. Our experiments show that storage requirements for rendering complex models can be reduced to about 20% compared to hierarchical radiosity. At the same time, computation times for images of very reasonable quality can be reduced by one order of magnitude.

Keywords: hierarchical radiosity, Monte Carlo radiosity, quasi-Monte Carlo methods

1 Introduction

Hierarchical radiosity (HR) deals well with the error due to discretisation [1] by automatically constructing a mesh capturing illumination variations accurately. It is however sensitive to computational error, e.g. inexact visibility estimation. HR also suffers from quite high storage requirements.

On the other hand, Monte Carlo radiosity (MCR) provides reliable form factor computation while explicit form factor storage is avoided. MCR algorithms quickly yield "complete" radiosity solutions, showing the effect of higher order interreflections, but with noticeable noise or aliasing effects. These disappear only slowly as the computations proceed. Quasi-Monte Carlo (QMC) sampling [11, 18] in combination with a multiresolution approach like HR will considerably improve the convergence rate.

Recently, an adaptive meshing technique for a continuous MCR algorithm has been proposed [32]. Our research has focussed on discrete rather than continuous MCR algorithms because QMC sampling appears to be significantly more effective in the former [3]. A discrete MCR algorithm solves the radiosity system of linear equations obtained by Galerkin discretisation of the rendering integral equation as opposed to continuous MCR algorithms, which solve this integral equation avoiding discretisation. Although extra discretisation error is introduced in discrete MCR algorithms, the sum of discretisation and computational error often turns out to be considerably lower for the same amount of work (see figure 1).

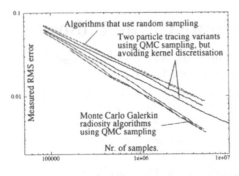

Fig. 1. Quasi-Monte Carlo sampling appears to be significantly more effective when used in the Monte Carlo radiosity algorithms that do not avoid kernel discretisation. The test scene is shown in the right image. The radiance solution is constant and equal to one everywhere.

We first briefly review hierarchical radiosity and discrete Monte Carlo radiosity algorithms in §2 and §3. In §4 we present our new procedure to make a synthesis of both. Implementation issues are discussed in §5. Experimental results and a discussion of our approach are presented in §6. §7 summarises and gives directions for future research.

2 Hierarchical radiosity

Hierarchical refinement techniques [5, 9] yield automatic adaptive meshing and result in lower than quadratic storage complexity and log-quadratic time complexity. Lower complexity is obtained by using a multi-resolution representation of radiosity.

We consider the full multilevel hierarchy introduced in [9] including clustering [25, 24]. In this approach, the single candidate interaction representing all light transport in the complete scene is recursively refined. The result is a tree-structure of cluster elements and surface elements. With each element, a list is maintained of interactions that allow the light transport to the element to be computed with sufficient accuracy. Basically, all interactions are first enumerated. Next, the form factor between the pair of elements involved in each interaction is computed by element-pairwise integration.

Whether or not a candidate interaction needs to be refined is decided by an *oracle* function. The oracle function will also return information about *how* to carry out the proposed refinement. It will typically estimate how accurately a candidate interaction will represent light transport. Hanrahan et al. [9] proposed to use a cheap form factor estimate as error estimate. Others, e.g. [26, 14, 28], proposed to use discretisation error estimates, also called *smoothness*-estimates. Several degrees of freedom exist in refining a deemed inaccurate candidate interaction. In [9], the larger of the two elements is regularly subdivided. Lischinski et al. [14] proposed to subdivide the element contributing the larger of propagated and approximation error. Non-regular element subdivision has been examined as well, e.g. in [16]. The oracle function uses knowledge about the geometry and optical properties of the scene, the intrinsic properties of light transport and/or a current partial radiosity solution.

The resulting number of interactions per element is independent of the number of elements in the hierarchy after refinement. In practice, a few tens of form factors have to be computed and stored per element. Although storage complexity is linear in terms of the number of elements, it is still quite high [33]: usually, a few hundred bytes per element are needed. Storage requirements can be substantially reduced by avoiding form factor storage, e.g. by recomputing form factors whenever needed as in [27].

While HR deals well with discretisation error, computational error is often a major source of reliability problems. Computational error on the form factors, and the oracle functions that drive the hierarchical refinement, will result in incorrect or even missed energy transport. An example will be shown in §6. Computational error may even prevent convergence of the iterative radiosity system of equations solver if energy conservation is not explicitly enforced.

The main sources of computational error on the form factors and the oracle function are due to the singularity in the form factor integral when elements touch as well as various discontinuities due to partial occlusion. Discontinuity meshing [10, 15] and back projection [29, 6, 7] algorithms have been proposed to deal with such discontinuities. These algorithms are quite costly and hard to implement robustly or require considerable amounts of storage. They result in exceptional image quality but are restricted to moderately complex scenes.

A very common approach to deal with computational error is to use simple integration rules with a sufficiently high number (50–100) of samples per form factor. More sophisticated element-to-element integration schemes all induce a higher sampling cost or increased number of samples.

3 Monte Carlo radiosity

Monte Carlo radiosity (MCR) algorithms [23, 8, 20, 17, 19, 22, 18] are based on the following observation: Consider N_i lines crossing a given patch i. The lines are generated such that their intersection points with the patch are uniformly distributed over the patch surface. The directions are distributed according to the cosine of the angle w.r.t. the patch normal. If N_{ij} of these N_i lines have their nearest intersection point with the surfaces in the scene on a second patch j, the ratio N_{ij}/N_i will be an approximation of the form factor F_{ij} between patch i and j. If the lines are generated using random numbers, the variance will be given by $F_{ij}(1 - F_{ij})/N_i$.

MCR algorithms generate a set of such lines. Each line can be viewed as a sample contributing $1/N_i$ to the form factor between the patches that are connected by the line. The lines can be generated using a *local* or *global* technique. In a local technique, uniformly distributed starting points and cosine-distributed directions are explicitly generated w.r.t. each patch (see e.g. [23]). A global technique constructs such lines irrespective of the patches, e.g. using recipes from integral geometry [20, 22, 17, 30].

Because the samples are not generated in patch-pair order, their contribution $1/N_i$ to the form factor is most often immediately translated into a contribution to the received radiosity at the receiver patch. This typically involves a multiplication with total or unshot radiosity of the source patch and the reflectivity of the receiver. The precise way depends on the sampling order and radiosity system solution technique: *breadth-first* (Southwell [23, 8] or Jacobi-like [17, 19, 18]) or *depth-first* (discrete particle tracing [22, 21]). The cost of converting form factor sample contributions into radiosity contributions is negligible while explicit form factor storage can be completely avoided. New samples are always used to improve the implicit form factor estimates.

Under some conditions, the accuracy obtained with a given number of samples is the same regardless whether a local versus global or depth-first versus breadth-first approach is followed.

MCR deals better with computational error on the form factors than with patch-pairwise sampling. First, the singularity in the form factor integral is avoided by using directional sampling. Using directional sampling in a pair-wise sampling procedure requires rejection sampling or expensive transformations that are not needed in MCR. Second, generated samples always connect mutually visible points. There is no need for back-projection algorithms and no samples are wasted between fully occluded patches, solving a problem that can be very bad with HR if no form of visibility preprocessing [31] is used. MCR algorithms generate samples with equal form factor contribution. Fewer samples will be needed for the same accuracy. Finally, energy conservation is easily fulfilled in MCR.

Current MCR algorithms however lack adaptive meshing. The user needs to provide a mesh that will allow illumination variations to be represented accurately enough. HR provides automatic adaptive meshing. A second problem of MCR are noise and aliasing effects such as under-sampling of small patches. Hierarchical refinement with clustering reduces these effects by letting each element interact with a covering of the scene that consists of a fairly small number (tens) of equally important elements.

4 Hierarchical Monte Carlo radiosity

Although HR and MCR are complementary, very little attention has been spent on a synthesis of both approaches (see §6). In this section, we will propose a new, very simple, but effective and general procedure for combining HR and MCR. The procedure can be viewed as a means to extend an arbitrary discrete MCR algorithm to incorporate hierarchical refinement and clustering. Any refinement strategy that has been proposed in the context of HR research can be considered.

MCR algorithms generate lines as described in §3. Each line, connecting mutually visible points x and y, is a sample contributing to the form factor between the patch containing x and the patch containing y. This contribution to the form factor is immediately translated into an energy transfer between the patches. We propose to extend MCR algorithms by deciding, for each sample, which level of the element hierarchies at x and y is appropriate for energy transfer instead of simply taking the patches containing x and y (see figure 2). The element hierarchies are lazily constructed, subdividing elements as needed.

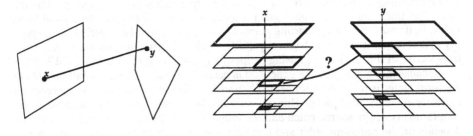

Fig. 2. Hierarchical refinement in Monte Carlo radiosity: for each sample line connecting two points x and y, the algorithm will determine which level of the element hierarchies at x and y is appropriate for computing light transport from x to y. MCR algorithms transfer energy from the patch containing x to the patch containing y. HR algorithms try to determine the interaction levels before form factor sampling.

The decision at what level energy transfer should be computed is made in a top-down recursive manner as in HR. Pseudo-code is shown in figure 3. The main difference is that at each refinement step, only the sub-elements containing x and y are

---

HierarchicalMonteCarloRadiosity:

1. Generate form factor samples as in the chosen MCR algorithm (see §5). For each sample s, connecting a point x_s with a point y_s, call TransferEnergy(x_s, y_s);
2. Make the hierarchical representation of radiosity consistent by calling PushPull(top-level cluster).

TransferEnergy(POINT x, POINT y)

1. initialise element pointers *src* and *rcv* to the top-level cluster element;
2. while the HR oracle deems the *rcv*←*src* candidate interaction not admissible,
   a. if subdivision on the source side is needed, replace *src* by its sub-element containing x,
   b. else, replace *rcv* by its sub-element containing y;
3. compute the actual energy transfer from *src* to *rcv* as in the chosen MCR algorithm.

Fig. 3. Hierarchical Monte Carlo radiosity pseudo-code

considered for further refinement in hierarchical Monte Carlo radiosity (HMC) instead of recursively refining *all* sub-elements in HR. The same refinement criteria and strategies as in HR can be used to check whether a candidate interaction between elements containing x and y is admissible, and, if not admissible, how it should be refined. This is possible because all information that is needed to do so, the geometry and optical properties of the scene as well as a partial radiosity solution, is equally well available in Monte Carlo radiosity.

This approach will result in an adaptively refined mesh and (implicit) link structure that is very similar as in HR. These structures are however lazily constructed as needed during form factor sampling rather than beforehand as in HR. MCR switches the enumeration of form factors and their sampling.

5 Implementation

Breadth-first versus depth-first. In figure 3, a breadth-first MCR approach is assumed. With a breadth-first approach, the multiresolution representation of radiosity needs to be made consistent only once after drawing all samples. In a depth-first algorithm however, a slightly adapted push-pull operation will be needed after *each* sample since the next sample will carry on the transported power, possibly starting from another element level. The adapted push-pull operation will consider only the new deposited radiosity and only the elements in the hierarchy that contain or are part of the receiver element *rcv* at y_s need to be updated. Although the associated overhead is not dramatic, we prefer a breadth-first approach in our implementation for this reason. Another reason is that more suitable partial radiosity solutions for the refinement oracle are available early-on. With a breadth-first approach, the push-pull procedure is identical as in hierarchical radiosity.

Local versus global. MCR with a local line generation technique contains a loop over all patches in order to take samples using importance sampling based on the power or unsent power of the patches. In the hierarchical extension, the loop shall be over the leaf elements of the element hierarchy since the most detailed representation of radiosity is available there. In a global line approach, the sampling is irrespective of patches and needs no changes for hierarchical refinement.

Quasi-Monte Carlo sampling. Quasi-Monte Carlo (QMC) sampling yields considerably improved convergence rates (see figure 1). It was our main motivation for considering discrete MCR algorithms. The hierarchical extension of a local breadth-first MCR algorithm is however slightly less straightforward when QMC sampling is used rather than random sampling. We recommend the use of a QMC sampling series with appropriate properties, e.g. [4] for reasons we will explain next:

When QMC sampling is used in a non-hierarchical local breadth-first Monte Carlo radiosity algorithm [23, 8, 19, 18], a sample index needs to be kept with each patch. This index determines the next sample to be generated from the patch. In the Well Distributed Ray Set (WDRS) algorithm [18] the same index also indicates the number of samples already generated from the patch.

Hierarchical refinement requires sample sequences on sub-elements. As a sample sequence on a sub-element, we will take a subset of all samples generated on the patch containing the sub-element: those with origin within the sub-element. Taking the next sample on a sub-element involves generating subsequent samples on the patch until one is found with origin within the sub-element. In the WDRS algorithm, the number of samples previously generated from a sub-element and the index of the next sample will not be the same anymore. Separate sample counts and indices will need to be kept.

Whenever a leaf element is subdivided, a sample count and index need to be computed for each resulting new leaf element. The next-sample-index can be copied from the parent element. In the WDRS algorithm, the sample count will need to be determined by regenerating all samples from the parent element and counting how many of those have their origin in the freshly created sub-element.

When regular subdivision is used, finding a subsequent sample on a sub-element and the sample counting are considerably facilitated with a so called (t, k) low discrepancy sample sequence. We have used a base-2 31-bits 4-dimensional Niederreiter sequence [4]. The (t, k)-property of this sequence implies that of each four successive samples with index $4i, 4i + 1, 4i + 2, 4i + 3$ on a parent element, one sample will fall within each regular sub-element. The (t, k) property also ensures that directions of rays with origin restricted to a sub-element still will be properly distributed.

An appropriate sub-element numbering scheme and quadrilateral-to-triangle mapping was constructed that allows the same sample generation and counting routines to be used for triangular and quadrilateral elements [2].

6 Results and discussion

Comparison with previous work. HMC can be viewed as a variant of HR in which form factors are computed as in MCR. It inherits the benefits of MCR: more reliable and efficient form factor computation while avoiding form factor storage without double work. Form factor storage was previously avoided in HR by recomputing form factors whenever needed (see e.g. [27]). In HMC algorithms, new form factor samples always *improve* the (implicit) form factor estimates.

Since as in MCR, every HMC form factor sample connects mutually visible points, *no back-projection* algorithm [29, 6, 7] is needed for accurate form factor computation in partial occlusion cases. Discontinuity meshing [10, 15] will however still be needed for high quality penumbrae. Unfortunately, such algorithms are restricted to moderately complex scenes ($\lesssim 10^4$ input patches). HMC without discontinuity meshing will yield reasonably good image quality for *much more complex scenes*. We report on experiments indicating the reliability, efficiency and low storage requirements of HMC further in this section.

Hierarchical meshing and clustering have been applied before in the context of the progressive refinement radiosity algorithm in which Monte Carlo simulation was used for form factor computation instead of a hemicube algorithm [13, 12]. HMC however supports *full hierarchical refinement*, e.g. the source element can be refined as well. These previous approaches also required a large amount of rays to be shot simultaneously from the source. This is most often not the case with more modern MCR algorithms that always display "complete" radiosity solutions and obtain progressiveness by gradually increasing the number of samples in small steps. Finally, unlike these previous approaches, HMC avoids form factor storage without repeated form factor computation.

Recently, an adaptive meshing technique was proposed in the context of a continuous MCR algorithm [32]. By simultaneously keeping track of incident particles on successive hierarchical element levels, smoothness assumption violations can be detected. When an element is subdivided, part of the incident samples had to be forgotten, resulting in about 25% of the samples being "wasted". This technique also doesn't solve the under-sampling of small patches. We have studied discrete rather than continuous MCR algorithms because QMC sampling appears to be much more effective in the latter, leading to lower total error for the same amount of work (see §1). We propose to base refinement on the geometric and optical properties of a scene, without "wasting" samples. With clustering in HMC, the under-sampling problem is solved well.

Reliability. Figure 4 shows an example of a problem due to computational error on the refinement oracle in HR. The horizontal patch was erroneously classified as fully occluded from the light source early on during the refinement process in HR. The problem could be reduced by deferring visibility estimation until other criteria, such as small enough unoccluded form factor, are fulfilled [9]. This would however lead to many form factor samples being "wasted" in fully occluded candidate interactions. Visibility preprocessing [31] has been proposed as a solution. In HMC, all form factor samples always connect mutually visible points. The appropriate level of interaction is determined for each sample, after visibility detection. HMC is considerably less sensitive than HR to missing important interactions such as shown in this example.

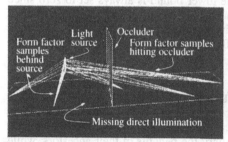

Fig. 4. With HR (left), the horizontal patch was erroneously classified as fully occluded from the light source because all form factor sample lines either hit the receiver behind the source, or intersected the occluder surface. As a result, direct illumination is missing on the horizontal patch. HMC (right) is much less sensitive to such visibility mis-estimation problems.

Computation time. Figure 5 (see colour section) shows some results obtained with the hierarchical extension of WDRS radiosity [18] as described in this paper. Although the models that are shown are quite complex, the images only required a few minutes of computation time[1]. While noisy effects are still visible, the most disturbing artifacts in the radiosity solutions are due to a bad initial mesh quality. These artifacts are however not unique to HMC and also show up with other radiosity algorithms.

Comparing running times of HR and HMC for the cubicle office space, we found that a first image showing only direct illumination was obtained after 14 minutes with HR. Six Jacobi iterations, yielding a more complete illumination solution took 89 minutes. With HMC, a first solution, containing the effect of interreflections, was obtained already after 3.6 minutes with less than 2 visibility samples per interaction. We found that often no more than 5 to 10 samples per form factor suffice in order to obtain images in which noise is not noticeable. Many more samples are needed to ensure sufficiently reliable element-to-element integration.

Storage requirements. In HMC, interactions need not to be stored explicitly. Storage is reduced from (at least) 8 bytes per interaction with HR to 8 (WDRS), 4 (other local MCR algorithms using QMC sampling) or zero (other MCR algorithms) bytes per element with HMC. Experiments within our test-bed rendering system[2] reveal that a reduction of storage requirements to about 20% is feasible.

Smoothness-based oracles. In the experiments above, the popular low-cost refinement oracle of [9] was used. A smoothness-based oracle (see §2) will however result in fewer elements and interactions. With the same number of samples, lower variance will be obtained. We have found that with the smoothness-based oracle of [14], the number of elements is typically reduced by a factor of about two. The analytical form factor and visibility computations in this oracle are however carried out several times for each sample, resulting in an increased computation cost by a factor of 10 to 20. In order to avoid repeated computation of the smoothness estimates, they can be stored for each link (the form factors themselves don't need to be stored with HMC). The observed increase in computation time (factor three) still wasn't compensated by the gain. The need for good low-cost refinement criteria is more urgent in HMC than in HR.

We achieved acceptable smoothness-based refinement similar to [14] by using cheap point-to-element form factor estimates at the vertices and midpoint of an element for estimating an upper form factor bound. The lower bound is always set to zero since it is not determined whether there is full or partial visibility in a link (full occlusion never has to be considered in HMC). Radiosity variation over an element for estimating propagated error is approximated by the difference between maximum and minimum radiosity on descendant leaf elements.

7 Conclusion and directions for future research

We have presented a synthesis of hierarchical radiosity and Monte Carlo radiosity algorithms. The new algorithms that result inherit the benefits of their ancestors, while important drawbacks are annihilated. They are well suited for reliable and fast rendering of complex diffuse models on low-cost platforms.

Several problems with hierarchical radiosity however remain: for high-quality rendering of penumbrae, discontinuity meshing is still needed. Neither have the problems of over-refinement and over-bright regions near abutting surfaces [33] disappeared.

---

[1] All experimental data were obtained on a SGI Octane workstation with 195MHz R10000 processors and 256MB RAM.

[2] RenderPark: http://www.cs.kuleuven.ac.be/cwis/research/graphics/RENDERPARK/

When rendering real-world models with radiosity, a poor initial mesh quality is another permanent source of problems. Even without a solution for these problems, the technique presented in this paper will still serve as a fast view-independent rendering technique yielding image quality that is sufficient for many applications.

Directions for future research include application of the technique for non-diffuse reflection, for rendering dynamic environments, for higher order approximations and as a fast first pass in two-pass methods.

Acknowledgements

Philip Dutré provided invaluable comments on early versions of this paper, contributing significantly to the presentation. The first author also acknowledges financial support by a grant from the Flemish 'Institute for the Promotion of Scientific-Technological Research in Industry' (IWT#941120).

Model credits for figure 5: Candlestick Theatre: Design: Mark Mack Architects, 3D Model: Charles Ehrlich and Greg Ward (work conducted as a research project during the Architecture 239X course taught by Kevin Matthews formerly at UC Berkeley, College of Environmental Design). Conference room and cubicle space models by Anat Grynberg and Greg Ward (Lawrence Berkeley Laboratory, Berkeley, California).

References

1. J. Arvo, K. Torrance, and B. Smits. A framework for the analysis of error in global illumination algorithms. In *SIGGRAPH '94 Proceedings*, pages 75–84, July 1994.
2. Ph. Bekaert. An appropriate sub-element numbering scheme and triangle-to-qaudrilateral mapping for hierarchical monte carlo radiosity. Technical report, K. U. Leuven, Department of Computer Science, 1998.
3. Ph. Bekaert, R. Cools, and Y. D. Willems. The effectiveness of quasi-Monte Carlo sampling for solving the radiosity problem. In *third International Conference on Monte Carlo and quasi-Monte Carlo Methods in Scientific Computing, Claremont, California, USA*, June 1998.
4. P. Bratley, B. Fox, and H. Niederreiter. Algorithm 738: Programs to generate Niederreiter's low-discrepancy sequences. *ACM Transactions on Mathematical Software*, 20(4):494–495, December 1994.
5. M. F. Cohen, D. P. Greenberg, D. S. Immel, and P. J. Brock. An efficient radiosity approach for realistic image synthesis. *IEEE Computer Graphics and Applications*, 6(3):26–35, March 1986.
6. G. Drettakis and E. Fiume. A Fast Shadow Algorithm for Area Light Sources Using Backprojection. In *Computer Graphics Proceedings, Annual Conference Series, 1994 (ACM SIGGRAPH '94 Proceedings)*, pages 223–230, 1994.
7. F. Durand, G. Drettakis, and C. Puech. The visibility skeleton: A powerful and efficient multi-purpose global visibility tool. In *SIGGRAPH 97 Conference Proceedings*, pages 89–100, August 1997.
8. M. Feda and W. Purgathofer. Progressive ray refinement for Monte Carlo radiosity. In *Fourth Eurographics Workshop on Rendering*, pages 15–26, June 1993.
9. P. Hanrahan, D. Salzman, and L. Aupperle. A rapid hierarchical radiosity algorithm. In *Computer Graphics (SIGGRAPH '91 Proceedings)*, volume 25, pages 197–206, July 1991.
10. P. Heckbert. Discontinuity meshing for radiosity. *Third Eurographics Workshop on Rendering*, pages 203–226, May 1992.
11. A. Keller. Quasi-Monte Carlo radiosity. In *Eurographics Rendering Workshop 1996*, pages 101–110, June 1996.

12. A. Kok. *Ray Tracing and Radiosity Algorithms for Photorealistic Images Synthesis.* PhD thesis, Technische Universiteit Delft, The Netherlands, 1994.

13. E. Languenou, K. Bouatouch, and P. Tellier. An adaptive discretization method for radiosity. *Computer Graphics Forum,* 11(3):C205–C216, 1992.

14. D. Lischinski, B. Smits, and D. P. Greenberg. Bounds and error estimates for radiosity. In *Proceedings of SIGGRAPH '94,* pages 67–74, July 1994.

15. D. Lischinski, F. Tampieri, and D. P. Greenberg. Discontinuity meshing for accurate radiosity. *IEEE Computer Graphics and Applications,* 12(6):25–39, November 1992.

16. D. Lischinski, F. Tampieri, and D. P. Greenberg. Combining hierarchical radiosity and discontinuity meshing. In *SIGGRAPH '93 Proceedings,* pages 199–208, 1993.

17. L. Neumann. Monte Carlo radiosity. *Computing,* 55(1):23–42, 1995.

18. L. Neumann, A. Neumann, and Ph. Bekaert. Radiosity with well distributed ray sets. *Computer Graphics Forum,* 16(3), 1997.

19. L. Neumann, W. Purgathofer, R. Tobler, A. Neumann, P. Elias, M. Feda, and X. Pueyo. The stochastic ray method for radiosity. In P. Hanrahan and W. Purgathofer, editors, *Rendering Techniques '95 (Proceedings of the Sixth Eurographics Workshop on Rendering),* July 1995.

20. M. Sbert. An integral geometry based method for fast form-factor computation. *Computer Graphics Forum,* 12(3):C409–C420, 1993.

21. M. Sbert. Error and complexity of random walk Monte Carlo radiosity. *IEEE Transactions on Visualization and Computer Graphics,* 3(1):23–38, March 1997.

22. M. Sbert, X. Pueyo, L. Neumann, and W. Purgathofer. Global multipath Monte Carlo algorithms for radiosity. *The Visual Computer,* 12(2):47–61, 1996.

23. P. Shirley. Radiosity via ray tracing. In James Arvo, editor, *Graphics Gems II,* pages 306–310. Academic Press, San Diego, 1991.

24. F. Sillion. A unified hierarchical algorithm for global illumination with scattering volumes and object clusters. *IEEE Transactions on Visualization and Computer Graphics,* 1(3):240–254, September 1995.

25. B. Smits, J. Arvo, and D. Greenberg. A clustering algorithm for radiosity in complex environments. In *SIGGRAPH '94 Proceedings,* pages 435–442, July 1994.

26. B. Smits, J. Arvo, and D. Salesin. An importance-driven radiosity algorithm. In *Computer Graphics (SIGGRAPH '92 Proceedings),* volume 26, pages 273–282, July 1992.

27. M. Stamminger, H. Schirmacher, and Ph. Slusallek. Getting rid of links in hierarchical radiosity. In *Computer Graphics Forum (Proceedings EUROGRAPHICS '98),* 1998.

28. M. Stamminger, P. Slusallek, and H-P Seidel. Bounded radiosity — illumination on general surfaces and clusters. *Computer Graphics Forum,* 16(3), 1997.

29. A. Stewart and S. Ghali. Fast computation of shadow boundaries using spatial coherence and backprojections. In *Proceedings of SIGGRAPH '94 (Orlando, Florida, July 24–29, 1994),* pages 231–238, July 1994.

30. L. Szirmay-Kalos, T. Foris, L. Neumann, and B. Csebfalvi. An analysis of quasi-Monte Carlo integration applied to the transillumination radiosity method. *Computer Graphics Forum,* 16(3):C271–C281, September 1997.

31. S. Teller and P. Hanrahan. Global visibility algorithms for illumination computations. In *Computer Graphics Proceedings, Annual Conference Series, 1993,* pages 239–246, 1993.

32. R. Tobler, A. Wilkie, M. Feda, and W. Purgathofer. A hierarchical subdivision algorithm for stochastic radiosity methods. In *Eurographics Rendering Workshop 1997,* pages 193–204, June 1997.

33. A. Willmott and P. Heckbert. An Empirical Comparison of Radiosity Algorithms. Technical report, Carnegie Mellon university, department of Computer Science, May 1997.

Editors' Note: see Appendix, p. 336 for colored figure of this paper

Importance Driven Construction
of Photon Maps

Ingmar Peter

Wilhelm-Schickard-Institut
GRIS
Universität Tübingen
Germany
peter@gris.uni-tuebingen.de

Georg Pietrek

Lehrstuhl 7
Fachbereich Informatik
Universität Dortmund
Germany
pietrek@ls7.cs.uni-dortmund.de

Abstract. Particle tracing allows physically correct simulation of all kinds of light interaction in a scene, but can be a computationally expensive task. Use of visual importance is a powerful technique to improve the efficiency of global illumination calculations. We describe a three pass solution for global illumination calculation extending the two pass approach proposed by Jensen. In the first pass particle tracing of importance is performed to create a global data structure, called importance map. Based on this data structure importance driven photon tracing is used in the second pass to construct a photon map containing information about the global illumination in the scene. In the last pass the image is rendered by distributed ray tracing using the photon map.

The photon tracing process, improved by the use of importance information, creates photon maps with an up to 8-times higher photon density in important regions of the scene. This allows a better use of memory and computation time resulting in better image quality.

1 Introduction

The generation of photorealistic images requires the simulation of illumination in a virtual scene. Several approaches were tried to solve this task. Early illumination models took only direct illumination into account and neglected interreflection between objects (therefore called local illumination models) [Gou71, Pho75, Whi79]. The first approaches for physically correct simulation of global illumination were Monte Carlo ray tracing [Kaj86, CPC84] and radiosity [GTGB84] resulting in two families of algorithms derived from the original approaches. Stretching these algorithms beyond their best suited application domains results in inefficiency. Monte Carlo ray tracing is best suited for specular interaction of light with objects because the sampling can be concentrated on small regions of the hemisphere. Correctly simulating diffuse interaction results in tracing prohibitively many rays because the whole hemisphere has to be sampled. The radiosity algorithm was designed for scenes limited to objects with ideal diffuse reflection. Generalization of this method to specular reflection is possible [ICG86, SP89] but leads to algorithms which need much time and memory.

Better approaches to the global illumination problem are two pass methods that start by distributing light from the light sources (in a radiosity or particle tracing manner), and storing the light reaching diffuse surfaces (in surface oriented [Hec90a, PM92] or spatial data structures [War94, Jen96a]). In the second pass the image is rendered using the information stored in the data structures.

After introducing the notion of visual importance in the following section, we give in section 3 a short overview how photon maps can be used in global illumination calculation. Section 5 describes the importance driven construction of photon maps. After that the results are discussed. The last section contains a few remarks on future work. For explanation of the symbols used see Appendix A.

2 Importance Driven Lighting Simulation

Simulation of global illumination needs much computation time and, depending on the algorithm, lots of memory. In fact, most approaches need a lot of both ressources. Therefore we should look for algorithms that make the best use of invested time and memory. Importance driven algorithms are an important step towards this goal.

Light leaving a point in a given direction contributes directly or indirectly to the illumination of other points in the scene. This is formalized by the notion of visual importance. For image generation we are interested in the amount of light reaching the eye through a set of given viewing directions. This defines a light receiver S. The visual potential W is the fraction of light leaving a point x in a given direction Ψ that finally reaches the receiver.

The value of W for a point x and a direction Ψ is defined by the potential equation [PM93b]

$$W(x, \Psi) = W_e(x, \Psi) + \int_{\Omega_x} W(x, \Theta) f_r(y, \Psi, \Theta) |\vec{N}_y \cdot \Theta| d\omega^\Theta$$

where W_e is the self-emitted potential which has the value 1 for all points and directions in S.

The potential equation describes the propagation of importance through the scene. We can think of importance as a theoretical quantity that is emanating from the eye through the set of viewing directions. The mechanisms for interaction of importance with objects are the same as for light transport. Therefore the same algorithms for approximation of light distribution in a scene can be used to get an approximate importance distribution.

Global illumination algorithms calculate approximate solutions of the light distribution. To make best use of the resources, importance can be used to guide the calculation process. In regions with less importance a coarser approximation with a larger error is acceptable, because light from these regions has small influence on the image. The calculation efforts should be concentrated on the regions with high importance. Smits et. al. devise an efficient radiosity algorithm based on this principle [SAS92]. Their algorithm simultaneously calculates approximate solutions for light and importance in a hierarchical interaction scheme. Links are refined, if the product of importance, transferred energy, and the estimated formfactor error exceeds a given threshold. Pattanaik and Mudur introduced this principle to the field of particle tracing for lighting simulations [PM93a]. A first particle tracing using the bidirectional reflectance distribution function (brdf) of surfaces is used to get approximations of the importance function. This information is used to construct modified probability distribution functions (pdf) for the particle tracing so that more particles are guided in regions with high importance. They restrict themselves to diffuse reflection and use a surface subdivision as data structure for accumulation of the particle hits. Dutré and Willems present a particle tracing algorithm with adaptive probability functions for light source sampling and surface reflectance sampling [DW94, DW95]. The adaptive PDFs are based on visual importance and gradually built up during the tracing process. All these approaches showed substantial improvements compared to previous algorithms not using importance information.

In this paper we present an approch for a three pass solution of the global illumination problem. The first pass constructs a spatial data structure, called importance map, that approximates the importance distribution in a scene for a given view. The second pass consists of an importance driven construction of a photon map. In the third pass the image is rendered by Monte Carlo ray tracing using the photon map.

2.1 A Litmus Test for Importance Driven Algorithms

In a normal particle tracing approach the number of photons traced from a light source is determined by the fraction of power emitted from this light source. Let us think of a scene where one of the light sources is positioned in a closed box, so that no light from this source gets out into visible parts of the scene. Then it is obviously unnecessary to emit any photon from this light source. Any effort spent in tracing photons from this light source is wasted. This is a hypothetical example but it can be used as a litmus test for importance driven algorithms. We are looking for algorithms that have the inherent property to identify such situations and avoid tracing photons from hidden light sources.

3 Photon Map

A very promising global illumination approach is Jensen's two pass method using the photon map [JC95b, Jen95, JC95a, Jen96b, Jen96a]. The photon map is constructed in the first pass by tracing particles from the light sources into the scene, and storing particles if they hit diffuse surfaces. The photon map is a spatial data structure, storing position, direction, and flux of a photon. An efficient data structure for the purposes in Jensen's algorithm is the kd-tree [Ben75]. The image is rendered in the second pass by Monte Carlo ray tracing. In this pass the photon map can be used in two ways. If an accurate computation is needed, the photon map is used to construct a more efficient probability density function for the Monte Carlo sampling of indirect illumination [Jen95]. If an approximate solution is sufficient this approximation is directly calculated from the photon map. An additional photon map, called caustics map, can be used to incorporate caustics into the images [Jen96b].

We will give short descriptions of the operations performed with the photon map. Analogous operations are performed with our additional data structure, the importance map.

3.1 Importance Sampling

Estimating an integral with Monte Carlo techniques requires the generation of samples according to a pdf. The speed of convergence of Monte Carlo integration depends on the pdf used. The optimal choice is a pdf that is proportional to the integrand. Choice of this pdf would require exact knowledge of the integrand which is usually not available. But if an approximation of the integrand is known this can be used as a nearly optimal pdf.

The integrand in the rendering equation is the brdf f_r times the incoming radiance L_i

Fig. 1. Hemisphere Ω_x divided into 7×16 patches.

at a hitpoint x. A piecewise constant approximation $\hat{L}$ of $L_i \cdot f_r$ is used as pdf $p \propto \hat{L}$ for a Monte Carlo sampling of the hemisphere. Therefore the hemisphere is subdivided into $n \times m$ regions with equal solid angle with respect to x (figure 1). Now the N photons with the shortest distance to x are used to construct the approximation. For an outgoing direction Ψ_o each photon $\wp$ with flux $\Phi_\wp$ and incoming direction $\Psi_\wp$ contributes with

$$I_\wp = f_r(x, \Psi_\wp, \Psi_o)\, \Phi_\wp \tag{1}$$

to a region of $\hat{L}$. The region, where the value of $I_\wp$ is added, is determined by the incoming direction $\Psi_\wp$ of $\wp$. Regions that receive no flux because no photon enters through their directions get a small overall fraction of intensity added. This avoids that the pdf becomes zero in regions where the integrand is possibly nonzero.

3.2 Radiance Approximation

An approximate radiance value can be obtained directly from the photon map. Since each photon's flux $\Phi_\wp$ and incoming direction $\Psi_\wp$ is stored in the photon map, the contribution to reflected radiance can be calculated for a given brdf f_r and outgoing direction Ψ_o. The N photons $\wp_1, \ldots, \wp_N$ with the shortest distance to a hitpoint x give an estimate of the incoming radiance for this point. Then, the outgoing radiance is estimated as

$$L(x, \Psi_o) \approx \frac{1}{\pi r^2} \sum_{j=1}^{N} f_r(x, \Psi_{\wp_j}, \Psi_o)\, \Phi_{\wp_j} \tag{2}$$

Since the reconstruction of irradiance from particle hits is in fact a density estimation problem [Hec90b, SWH+95] an area estimation is needed. It is calculated as the area of a disc with the maximum distance between x and any from the N photons as radius.

4 Photon Tracing with Arbitrary PDFs

In the orginal approach in photon tracing pdfs $p \propto L_e$ are used to select the starting points and directions and pdfs $p \propto f_r$ are used to select the reflection directions of the photons. By choosing these pdfs, all photons emitted from one light source have the same flux. To direct photons into the parts of the scene with the highest importance, we have to use suitable pdfs for emission and reflection of the photons.

The radiant flux Φ_ℓ that leaves a light source ℓ is described by

$$\Phi_\ell = \int\limits_{A_\ell} \int\limits_{\Omega_x} L_e(x, \Psi)d\omega^\Psi dx \tag{3}$$

where A_ℓ is the surface of ℓ and Ω_x the hemisphere over x. The emission of photons from a light source ℓ during photon tracing is a Monte Carlo integration [Kal86] of the self emitted radiance L_e of ℓ. The selection of the starting points and directions of the photons is a stochastic process, controlled by an arbitrary pdf p. According to the theory of Monte Carlo integration the flux $\Phi_\wp$ of every photon $\wp$ is determined by

$$\Phi_\wp = \frac{1}{N} \frac{L_e(x, \Psi)}{p(x, \Psi)} \tag{4}$$

where N is the number of photons emitted from ℓ.

The total flux of all photons emitted from ℓ is

$$\Phi_\ell \approx \frac{1}{N} \sum_{i=1}^{N} \frac{L_e(x_i, \Psi_i)}{p(x_i, \Psi_i)} \tag{5}$$

where $(x_i)_{i=1,...,N}$ and $(\Psi_i)_{i=1,...,N}$ are the starting points and directions of the photons. (5) is the secondary Monte Carlo estimator of integral (3). The theory of Monte Carlo Integration guarantees, that the right side of (5) converges for $N \to \infty$ against Φ_ℓ.

If a photon $\wp_i$ hits an object at hitpoint x, the distribution of reflected intensity I_o is given by

$$I_o(\Psi_o) = \frac{f_r(x, \Psi_{\wp_i}, \Psi_o)}{\rho(x, \Psi_{\wp_i})} \Phi_{\wp_i}$$

where $\Psi_{\wp_i}$ is the incoming direction of $\wp_i$ and ρ the directional-hemispherical reflectance. To select the reflection direction Ψ_o of $\wp_i$ an arbitrary pdf p can be used. The flux of the reflected photon $\wp_o$ is calculated by the primary Monte Carlo estimator for $\int_{\Omega_x} I_o(\Psi) d\omega^\Psi$:

$$\Phi_{\wp_o} = \frac{f_r(x, \Psi_{\wp_i}, \Psi_o)}{\rho(x, \Psi_{\wp_i})} \frac{\Phi_{\wp_i}}{p(\Psi_o)} \tag{6}$$

The use of arbitrary pdfs in photon tracing makes it possible to direct the photons into particular parts of the scene. The flux of the photons can be calculated using (4) and (6).

5 Algorithm

We have developed an algorithm to construct photon maps using importance information, which can be split up into three main steps:

1. Construction of an importance map.
2. (a) Estimation of importance at the light sources.
   (b) Photon shooting using importance information.
3. Rendering.

In the following, each of the steps is described.

5.1 Construction of the Importance Map

To store a rough estimate of the distribution of the importance in the scene, we use a spatial data structure, which is dual to the photon map: The *importance map* is implemented as kd-tree and stores *importons*. Importons are dual to photons and carrying an amount of importance, simulating the transport of importance through the scene.

The distribution of importance in the scene is estimated by performing a particle tracing similar to photon tracing. But instead of emitting photons from the light sources, importons are emitted from the viewpoint though the viewing plane into the scene. Every importon carries the same amount of importance, because the viewing plane is randomly sampled with uniform pdf. If an importon w hits a scene object, the importon is stored together with its incoming direction Ψ_w in the importance map.

The information about the distribution of importance in the scene, stored in the importance map, can be used to direct the photons in the following steps of the algorithm into the "important" parts of the scene.

5.2 Estimation of Importance at the Light Sources

Before emitting photons from a light source, we have to determine the points and directions on the surface of the light sources with the highest values of visual potential W, to direct the photons into the parts of the scene with a high amount of importance.

To create an approximation $\widehat{W}$ of the potential function for all points and directions on the surface of a light source, we need samples of W. But compared to the total number of elements in a scene, the number of light sources is in most cases very small. Additionally a light source can have a very small geometric extent but may emit a high amount of radiant power. So the probablity, that a sufficient high number of importons hits the surface of a light source during step 1 of the algorithm, providing a good estimate of W directly at the light source, is very low.

So we obtain sample values of W for every light source carrying out a simple path tracing starting at the light source:

1. The starting point x_0 and direction Ψ_0 of the path are selected randomly using a pdf $p \propto L_e$.
2. At every hitpoint x_i of the path the following steps are carried out:
   (a) The path terminates at x_i with probability ρ_d/ρ, where ρ_d is the value of the diffuse reflectance of the material at x_i.
   (b) If the path is extended, its reflection direction is randomly selected using a pdf $p \propto f_s$, where f_s is the specular part of the brdf at x_i.
   (c) If the path terminates at x_i, the value of $W(x_i, -\Psi_{i-1})$ is estimated using a modified form of the density estimation suggested by Jensen. Similar to (2) we use the N importons $w_1, \ldots, w_N$ with the shortest distance to x and adding them weighted with the brdf:

$$\widehat{W}(x_i, -\Psi_{i-1}) = \frac{1}{\pi r^2} \sum_{j=1}^{N} f_d(x, -\Psi_{i-1}, \Psi_{w_j})$$

   where $-\Psi_{i-1}$ is the direction from x_i to the previous hitpoint x_{i-1}, Ψ_{w_j} the incoming direction of importon w_j and r the maximum distance between x and one of the N importons. Because all importons are carrying the same amount of importance, this constant value need not to be mentioned here.

   After that, the value of $\widehat{W}$ is retransferred along the path onto the surface of the light source.

The sample values, obtained by the path tracing algorithm described above, are used to build a piecewise constant approximation $\widehat{W}$ of W for all points and directions on the surface of a light source. Although the algorithm is very simple the coherence in the scene leads to good results.

5.3 Photon Shooting Using Importance Information

The emission of the photons from a light source during photon tracing is a random process, which is controlled by a pdf p (see section 4).

To emit the photons from the light sources, according to the importance information obtained in the previous step of the algorithm, we use the following pdf:

$$p(x, \Psi) \propto L_e(x, \Psi)\widehat{W}(x, \Psi)$$

Using this pdf to select the starting points and directions for the photons, most of the photons are emitted in directions where both, the potential function *and* the emission of light, is high.

During the following photon tracing every time a photon hits a diffuse scene object the photon is stored in the photon map like in Jensen's algorithm. A photon's path is randomly terminated with probality $1 - \rho$. If a photon $\wp$ is reflected at a hitpoint x, a new direction is selected using the pdf

$$p(\Psi) \propto W(x, \Psi) f_r(x, \Psi_\wp, \Psi)$$

where x is the actual hitpoint and $\Psi_\wp$ the incoming direction of $\wp$.

To obtain p, an approximation of $W \cdot f_r$ at hitpoint x must be constructed. We use a technique similar to the construction of $\hat{L}$ (section 3.1), but now using the entries of the importance map. The only difference is, that instead of the contribution $\Phi_{r,\wp}$, the number of importons for every region of the hemisphere over x are determined. Every importon contributes with

$$W_{r,w} = f_r(x, \Psi, \Psi_w)$$

to the approximation, because all importons are carring the same amount of importance (see section 5.1).

After normalizing p, a reflection direction Ψ_o is selected. The flux $\Phi_{\wp_o}$ of a reflected photon $\wp_o$ is calculated according to (6) with

$$\Phi_{\wp_o} = \frac{f_r(x, \Psi_{\wp_i}, \Psi_o)}{\rho(x, \Psi_{\wp_i})} \frac{\Phi_{\wp_i}}{p(\Psi_o)}$$

where $\Psi_{\wp_i}$ is the direction of the incoming photon $\wp_i$.

5.4 Rendering

The rending of the image is carried out using backward ray tracing from the eye. To calculate the illumination of a hitpoint x, we split up the rendering equation [Kaj86] into four parts:

$$
\begin{aligned}
L(x, \Psi) \quad = \quad & L_e(x, \Psi) \\
& + \int_{\Omega_x} f_d(x, \Theta, \Psi) L_{i,l}(x, \Theta) |\Theta \cdot \vec{N_x}| d\omega^\Theta && (7) \\
& + \int_{\Omega_x} f_s(x, \Theta, \Psi) L_i(x, \Theta) |\Theta \cdot \vec{N_x}| d\omega^\Theta && (8) \\
& + \int_{\Omega_x} f_d(x, \Theta, \Psi) L_{i,r}(x, \Theta) |\Theta \cdot \vec{N_x}| d\omega^\Theta && (9)
\end{aligned}
$$

where $L_{i,l}$ is the direct illumination and $L_{i,r}$ the radiance, which is at least one time reflected before arriving at x. The brdf is divided into a specular f_s and a diffuse f_d part.

The direct illumination (7) of x is calculated using well known techniques like distributed ray tracing. The incoming radiance L_i from the specular reflection directions (8) is estimated by a simple Monte Carlo ray tracing using a pdf $p \propto f_s$.

At points visible to the viewer direct or via specular reflection, a Monte Carlo ray tracing is carried out to estimate the indirect illumination (9) at hitpoint x. The used pdf $p \propto f_d \cdot L_{i,r}$ is constructed using the entries of the photon map similar to the method described in section 3.1. But instead of adding contribution (1), every photon is equally weighted. Similar to the construction of $\widehat{W}$, described in the previous section, we only use the occurence of a photon to construct p. Every photon $\wp$ contributes with

$$\Phi_{r,\wp} = f_r(x, \Psi, \Psi_\wp)$$

to the approximation. This is done due to the different flux of the photons.

At the hitpoints of the secondary rays shooted from x according to p, (2) is used to estimate the diffuse illumination reflected towards x.

6 Results

6.1 Visualisation of Photon-Map

To demonstrate the effect of importance driven construction of photon maps, we use the scene shown in figure 2. The scene consists of a 'U'-shaped room that contains a small spherical light source in the left part. The viewer is placed in the right part of the scene in a way, that he can not see any directly illuminated parts of the scene.

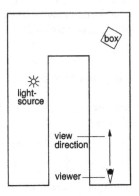

Fig. 2. Top view of the scene used to demonstrate importance-driven particle tracing.

Figure 3 (a) to (c) shows different particle maps. The brighter a point the higher is the density of the shown particle map. Figure 3 (a) shows a photon map generated without the use of importance information. The photons are symmetrically distributed according to the light source. Most of the photons are stored in the left part of the scene. In the importance map shown in figure 3 (b) the parts of the scene, which can be seen directly from the viewer, are marked by the entries of the map. The last figure 3 (c) contains the visualisation of a photon map, which was generated using the importance map shown in 3 (b). The density of the photon map in the right part of the room is strongly increased. Most of the photons are stored in the visible parts of the scene.

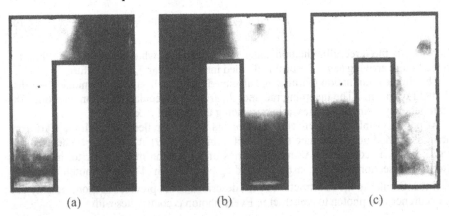

Fig. 3. Visualisation of different particle maps. (a) Photon map constructed without importance information. (b) Importance map. (c) Photon map constructed using the importance map as shown in (b).

6.2 Images

Figure 5 (see Appendix) shows images that were rendered using the photon maps shown in figure 3. The parameters used are listed in table 1.

figure		step 1 # importons	step 2 # photons	step 3 # secondary rays	N
5	(a)	10.000	10.000	50	50
	(b)	-	10.000	50	50
6	(a)	25.000	100.000	50	50
	(b)	-	100.000	50	50

Table 1. Parameters used to render the images shown in figures 5 and 6.

Image 5(a), which was rendered using the new importance driven algorithm, is slightly less noisier than image 5(b). Both images differ mostly in the illumination of the ceiling of the room. Due to the higher density of the photon map in image 5(a) the color bleeding effect at the ceiling near to the upper right corner of the room is much better determined than in image 5(b).

To measure the quality of the created photon maps, we used the following method: At every hitpoint x of a primary ray, the N photons next to x were retrieved from the photon map, where r is the maximum distance between x and one of the N photons. The value of r^2 can be used as measure for the density of the photon map at x. $\bar{r}^2$ is the mean value of r^2 for all pixels of an image. The lower the value of $\bar{r}^2$ the higher is the density of a photon map.

figure		$\bar{r}^2$
5	(a) Importance driven	0.616
	(b) Without importance	1.499
6	(a) Importance driven	0.250
	(b) Without importance	1.993

Table 2. Densities of photon maps constructed in different scenes.

As shown in table 2 the density of the photon map created using importance information is up to 8 times higher than the density of the photon map created without use of importance information.

A complex scene containing 28 light sources has been used to show the improvements through importance driven photon map construction in real world scenes. The test scene consists of 2 large and 16 small rooms as shown in figure 4. Each of the small rooms contains a small spherical light source. The large rooms are indirectly illuminated by 6 lights directed to the ceiling. All images were rendered using the marked viewer position.

Comparing the images in figure 6 (see Appendix) we see that image (a) contains less noise and also shows a better representation of the illumination in the scene than image (b). The density of the importance driven constructed photon map is 8 times higher than the density of the standard photon map (table 2).

7 Conclusion

We have demonstrated the advantages of the use of importance in a global illumination algorithm. It was shown that importance driven construction of photon maps leads to better

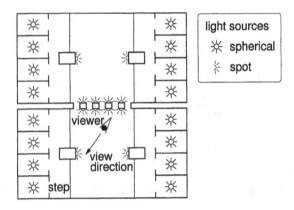

Fig. 4. Top view of the complex scene containing 28 light sources.

results. The proposed algorithm stands the litmus test for importance driven algorithms described in section 2.1. The photon maps constructed with the use of importance information possess a high density at the points where they are used for global illumination calculation.

Further works include:

- The importance information should also be used to enhance other parts of the algorithm: The kd-tree used to store the photon map can be optimized using the knowledge about the regions of high interest while rendering the image.
- Techniques like Ward's irradiance caching [WRC88] have to be incorporated into the algorithm to reduce its computation time to a reasonable extent.
- Our current implementation does not support the rendering of caustics. This can easily be incorporated into the algorithm using a technique like the *caustic map* proposed by Jensen [Jen96a].
- The photon map can be used to store information about the illumination at arbitrary points of the room. For this reason the photon map is well suited to be used in scenes which contains participating media, improving the global illumination calculations.
- Importance driven construction of photon maps could be much more efficient if the proposed three pass algorithm would be extended to a multi pass algorithm, which refines both the solutions of the rendering and the potential equation iteratively until a given error bound for the resulting image is reached.

References

Ben75.   Jon Louis Bentley. Multidimensional binary search trees used for associative searching. *Communications of the ACM*, 18(9):509–517, September 1975.

CPC84.   Robert L. Cook, Thomas Porter, and Loren Carpenter. Distributed ray tracing. *Computer Graphics*, 18(3):137–145, July 1984.

DW94.   Philip Dutre and Yves D. Willems. Importance-Driven Monte Carlo Light Tracing. In *Fifth Eurographics Workshop on Rendering*, pages 185–194, Darmstadt, Germany, June 1994.

DW95.   Philip Dutre and Yves D. Willems. Potential-Driven Monte Carlo Particle Tracing for Diffuse Environments with Adaptive Probability Density Functions. In P. M. Hanrahan and W. Purgathofer, editors, *Rendering Techniques '95 (Proceedings of the Sixth Eurographics Workshop on Rendering)*, pages 306–315, New York, NY, 1995. Springer-Verlag.

Gou71.   H. Gouraud. Continuous shading of curved surfaces. *IEEE Transactions on Computers*, C-20(6):623–629, June 1971.

GTGB84.   Cindy M. Goral, Kenneth K. Torrance, Donald P. Greenberg, and Bennett Battaile. Modelling the interaction of light between diffuse surfaces. *Computer Graphics (SIGGRAPH '84 Proceedings)*, 18(3):213–222, July 1984.

Hec90a.   Paul S. Heckbert. Adaptive radiosity textures for bidirectional ray tracing. *Computer Graphics*, 24(4):145–154, August 1990.

Hec90b.   Paul S. Heckbert. Adaptive radiosity textures for bidirectional ray tracing. *Computer Graphics (SIGGRAPH '90 Proceedings)*, 24(4):145–154, August 1990.

ICG86.   David S. Immel, Michael F. Cohen, and Donald P. Greenburg. A radiosity method for non-diffuse environments. In *Proceedings of Siggraph '86*, volume 20, pages 133–142, Ithaca, New York, 1986. Cornell University.

JC95a.   Henrik Wann Jensen and Niels J. Christensen. Efficiently rendering shadows using the photon map. In Harold P. Santo, editor, *Edugraphics + Compugraphics Proceedings*, pages 285–291, December 1995.

JC95b.   Henrik Wann Jensen and Niels Jorgen Christensen. Photon maps in bidirectional monte carlo ray tracing of complex objects. *Computers & Graphics*, 19(2):215–224, 1995.

Jen95.   Henrik Wann Jensen. Importance driven path tracing using the photon map. In P. M. Hanrahan and W. Purgathofer, editors, *Rendering Techniques '95 (Proceedings of the Sixth Eurographics Workshop on Rendering)*, pages 326–335. Springer-Verlag, New York, NY, 1995.

Jen96a.   Henrik Wann Jensen. Global illumination using photon maps. In Xavier Pueyo and Peter Schröder, editors, *Eurographics Rendering Workshop 1996*, pages 21–30, New York City, NY, June 1996. Eurographics, Springer Wien.

Jen96b.   Henrik Wann Jensen. Rendering caustics on non-lambertian surfaces. In Wayne A. Davis and Richard Bartels, editors, *Graphics Interface '96*, pages 116–121. Canadian Information Processing Society, Canadian Human-Computer Communications Society, May 1996.

Kaj86.   J. T. Kajiya. The rendering equation. *Computer Graphics*, 20(4):143–150, August 1986.

Kal86.   Malvin H. Kalos. *Monte Carlo Methods*. John Wiley & Sons, 1986.

Pho75.   Bui-Tuong Phong. Illumination for computer generated pictures. *Communications of the ACM*, 18(6):311–317, June 1975.

PM92.   S. N. Pattanaik and S. P. Mudur. Computation of global illumination by monte carlo simulation of the particle model of light. *Third Eurographics Workshop on Rendering*, pages 71–83, May 1992.

PM93a.   S. N. Pattanaik and S. P. Mudur. Efficient potential equation solutions for global illumination computation. *Computers and Graphics*, 17(4):387–396, July–August 1993.

PM93b.   S. N. Pattanaik and S. P. Mudur. The potential equation and importance in illumination computations. *Computer Graphics Forum*, 12(2):131–136, 1993.

SAS92.   Brian E. Smits, James R. Arvo, and David H. Salesin. An importance-driven radiosity algorithm. *Computer Graphics*, 26(2):273–282, July 1992.

SP89.   François Sillion and Claude Puech. A general two-pass method integrating specular and diffuse reflection. *Computer Graphics*, 23(3):335–344, July 1989.

SWH+95. Peter Shirley, Bretton Wade, Philip M. Hubbard, David Zareski, Bruce Walter, and Donald P. Greenberg. Global Illumination via Density Estimation. In P. M. Hanrahan and W. Purgathofer, editors, *Rendering Techniques '95 (Proceedings of the Sixth Eurographics Workshop on Rendering)*, pages 219–230. Springer-Verlag, New York, NY, 1995.

War94.   Gregory J. Ward. The RADIANCE lighting simulation and rendering system. In Andrew Glassner, editor, *Proceedings of SIGGRAPH '94*, Computer Graphics Proceedings, Annual Conference Series, pages 459–472. ACM SIGGRAPH, ACM Press, July 1994.

Whi79.   T. Whitted. An improved illumination model for shaded display. *Computer Graphics (Special SIGGRAPH '79 Issue)*, 13(3):1–14, August 1979.

WRC88.   Gregory J. Ward, Francis M. Rubinstein, and Robert D. Clear. A ray tracing solution for diffuse interreflection. *Computer Graphics*, 22(4):85–92, August 1988.

A Symbols

f_r	Bidirectional reflectance-distribution function (brdf)
f_s	Specular part of the brdf
f_d	Diffuse part of the brdf
ρ	Directional-hemispherical reflectance
ρ_d	Diffuse reflectance
ρ_s	Specular reflectance
L	Radiance function
L_e	Self emitted radiance
W	Potential function
W_e	Self emitted potential
ℓ	Light source
$\wp$	A photon
$\Phi_\wp$	Flux of photon $\wp$
$\Psi_\wp$	Incoming direction of photon $\wp$
w	An importon
Ψ_w	Incoming direction of importon w
$\vec{N}_x$	Surface normal at point x
Ω_x	Hemisphere above x with pole $\vec{N}_x$
pdf	probability distribution function

Lazy decompression of surface light fields for precomputed global illumination

Dr. Gavin Miller, Dr. Steven Rubin, Interval Research Corp. and Dr. Dulce Ponceleon, IBM Almaden Research Center

miller@interval.com

Abstract. This paper describes a series of algorithms that allow the unconstrained walkthrough of static scenes shaded with the results of precomputed global illumination. The global illumination includes specular as well as diffuse terms, and intermediate results are cached as surface light fields. The compression of such light fields is examined, and a lazy decompression scheme is presented which allows for high-quality compression by making use of block-coding techniques. This scheme takes advantage of spatial coherence within the light field to aid compression, and also makes use of temporal coherence to accelerate decompression. Finally the techniques are extended to a certain type of dynamic scene.

1 Introduction

1996 saw the introduction to the graphics community of the idea of light fields and lumigraphs. The appearance of a scene or isolated object is captured in terms of a 4-dimensional table, which approximates the plenoptic function sampled on an image plane for light fields [11], and on an envelope surface in the case of lumigraphs [5]. Reconstruction of views for locations not on the light field image plane is then possible using interpolation of table entries, typically using quadralinear interpolation.

Light fields have a number of advantages - the rendering time is independent of scene or object complexity; they may represent arbitrary illumination functions and they may be captured for isolated real objects and scenes. However, light fields also have a number of limitations. The camera is restricted to certain regions of space and the finite angular resolution leads to depth of field effects. Objects become blurry in proportion to their distance from the image plane. To overcome these restrictions the lumigraph paper [5] resampled the light field of an object onto a crude approximation of the surface geometry. This helped to reduce the depth of field effect and allowed the camera to be placed in any region outside the approximating surface. In related work, the idea of view-dependent texture mapping was introduced in [4], where a surface texture is created from a blend of images depending on the direction from which it is viewed. A small number of views were combined using blending, or a stereo algorithm was used to reconstruct intermediate views of surface detail. The method was unable to capture detailed surface reflectance effects because of the sparse sampling of orientation space for each surface point. A fine sampling of orientation space leads to an explosion in the memory requirements. It is the use of compression algorithms to best handle such data that forms the basis of this paper.

1.1 Goals of this paper

In this paper, we explore the use of surface light fields to capture the appearance of global illumination for synthetic scenes. The particular examples were generated with ray-tracing, since the specular terms present the most difficulty in terms of sampling and compression, but effects due to interreflection [2] could be incorporated with little difficulty. This approach enables us to explore compression issues for a scene that contains specular objects.

We have a number of objectives for the resultant algorithms. The camera should be free to move anywhere in the scene. Coherence in the light fields should allow for efficient data compression where applicable. Efficient decompression should be accomplished by making use of temporal and cache coherence. (In this case, temporal means from frame-to-frame as the camera moves. However there is also cache coherence within a single frame.) We wish to match the angular resolution of the light fields to the material reflectance properties of the surface and the resultant radiance distribution. Perfect specular reflectors will have higher angular frequencies than nearly diffuse surfaces. Lack of angular resolution in the light fields should appear as changes in surface properties rather than lack of depth of field in the camera. In a progressive refinement scheme, for instance, the surfaces would start diffuse and then become more and more crisply specular.

Unfortunately to make our scheme efficient, the requirement that rendering cost be independent of scene complexity has to be abandoned. By placing the light field directly on the surface we are no longer using light fields to represent detailed surface geometry. They now only store the variations with angle of surface radiance.

The rendering scheme for this paper works in the following way:

- As a pre-processing step, a surface light field is generated for each parametric surface in the scene. This is a four dimensional table for each surface indexed by two surface parameters and two orientation parameters. The light fields are generated using stochastic ray tracing. Twenty-five rays per sample were used for the examples in this paper. The indices were jittered to prevent aliasing in both surface location and ray orientation.

- During real-time interaction, a texture map is generated for each (non-diffuse) surface every frame. Each texture pixel is computed using the position of the camera relative to the surface. The textures are downloaded to the graphics hardware and the geometry is scan-converted.

1.2 Structure of this paper

Section 2 discusses the parameterization of the light fields, and efficient resampling on playback. Section 3 talks about the internal representation of the light fields. Section 3.1 describes the block compression scheme and its internal representation. Section 3.2 explores the use of cache coherence as part of a block decompression scheme. Section 3.3 extends the block-coding algorithm to make use of angular coherence within the light field. Section 4 explores the use of light fields to render a dynamic scene in which a single surface is bump mapped on the fly using a water ripple simulation.

2 Geometry of surface light fields

The examples for this paper consist of smooth parametric surfaces. In a spirit similar to [5] we sample the light field on each parametric surface directly. (We assume that the surfaces are closed, otherwise a light field would be required for each side.) The location on the parametric surface is defined by $P(U, V)$ whereas the orientation relative to the tangent coordinate frame is defined by S and T. Figure 1 illustrates the surface parametrization and the tangent coordinate frame.

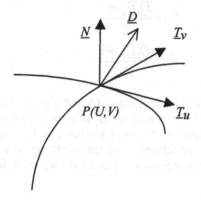

Fig. 1. Parameterization of the surface light field.

Unlike [11], we decided to use the non-linear mapping between (S, T) and orientation given in Equation 1. This is because we wished to capture a hemisphere of orientation in a single 2-D parameterization, where (S, T) span the range from zero to one.

$$s = 2S - 1$$

$$t = 2T - 1$$

$$n = \sqrt{1 - s^2 - t^2}$$

$$\underline{D} = s\underline{T}_U + t\underline{T}_V + n\underline{N} \tag{1}$$

Where $\underline{D}$ is the view direction vector, $\underline{T}_u$ and $\underline{T}_v$ are the surface tangents and $\underline{N}$ is the surface normal. When n is not a real number (because of a negative square root) n is set to zero. This corresponds to values for S and T that are outside the surface's orientation hemisphere. Note that the light field captures the light leaving the surface in the $\underline{D}$ direction. For an idealized purely diffuse surface, the light leaving the surface will be independent of (S, T). A degenerate light field results, which may be expressed as a single texture map. This case will occur frequently in practice.

2.1 Efficient resampling

To reconstruct the surface texture for a given camera location, it is necessary to compute the (S, T) values for each texture pixel at (U, V) and then index into the light field using (U, V, S, T). This could be done at each texture pixel using the surface equation. In practice, we approximate the surface using a finite resolution mesh. At the vertices of this mesh we compute (S, T). We then scan-convert the polygon into the texture map interpolating S and T values in the interior of the polygon.

We can compute (S, T) at polygon vertices using:

$$S = \frac{D.T_u + 1}{2}$$

$$T = \frac{D.T_v + 1}{2}$$

Unfortunately, the use of linear interpolation of S and T over a surface region in U, V parameter space is inexact. To minimize errors, large polygons are diced into smaller ones for the purpose of creating the texture (rasterizing in U, V space), but the larger polygons are used directly for textured polygon rendering in screen-space. For the examples in this paper, the polygons were diced if they were larger than 8 texture pixels along a side.

An advantage of parametric light fields is that they may be reconstructed with bilinear interpolation rather than quadralinear interpolation. This is because the surface texture is already aligned with the surface (U, V) space. A second resampling occurs when the surface is rendered using texture mapping, which is done efficiently by standard hardware.

2.2 Visibility culling and back-face culling

The cost of resampling a light field to create a surface texture is significant, both because of the interpolation cost and also, where used, the decompression cost. For this reason, such resampling is to be avoided whenever possible. When a surface is not visible in the current view, its texture does not need to be computed from the light field. Such surfaces may be culled using a visibility precomputation such as [6].

A special form of visibility culling, worth particular mention, is back-face culling. For a light field on a closed surface, back-face culling may remove up to half of the triangles. These regions will not have a texture computed, and the savings are useful. However, having undefined regions of the texture image can lead to problems when generating a Mip map [15]. The examples for this paper were scan-converted using bilinear interpolation of the full resolution texture pixels without pre-filtering. The texture maps were of sufficiently low resolution that they did not alias. If, on the other hand, the surfaces were being rendered using Mip-mapping, we would need to compute a Mip map for each surface texture, each time that the texture changed. Unfortunately, portions of the texture would be undefined in back-facing regions.

Those regions may contribute to visible portions of the final Mip map because of blurring. This can happen near an outline where the effective filter footprint of a pre-filtered region straddles the border of the visible region in texture space. Resolving this problem is an area for future work.

3 Compression

Given the large data size of high-resolution light fields, there is an obvious need for compression. A 256-squared by 64-squared light field takes 803 Megabytes of uncompressed data. In many ways this is comparable to the compression problem with video sequences. However, there are some key differences. As stated in [11], the most salient difference is the need to only access a sparse sampling of the decompressed light field at any frame.

In [11], vector quantization was used to make use of this property. VQ has the advantage that it allows for local decompression, so a random sample may be decompressed in an efficient manner. Unfortunately, the use of VQ in isolation does not allow for the most memory efficient compression since coherence in the image is not fully exploited. Also, transform-based methods, such as Discrete Cosine Transform (DCT) [3], may allow for compression with less artifacts at comparable compression ratios, and have a more naturally scalable quality control mechanism. In [5], it was suggested that transform-based block coding techniques could be used instead, but few details were given. This section explores issues related to block coding strategies including cache coherence and angular coherence.

3.1 Light field as an array of images
If we decide to store a light field as a 2-dimensional array of 2-D images, then it is important to decide which representation to use. It was pointed out in [11], that we could have images that were adjacent in U, V space or S, T space. We chose to store images that span (U, V) space as contiguous pixels for a given value of (S, T). This choice was based on the characteristics of the decoding scheme rather than the degree of coherence in the data. (This will be addressed further in Section 3.2.)

It is worth noting that for dull-specular surfaces, their appearance may be characterized adequately using surface light fields with angular resolutions of (S, T) which are far less than the spatial resolution of the texture (U, V). This arises from the surface microstructure acting as a low-pass filter on the angular dependence of the surface-bidirectional-reflectance-distribution function [1]. Since we are sampling discrete orientations, we will refer to the discrete directions as S_i and T_i. If the angular resolution is low, i.e. adequate for a dull reflector, then S_i and T_i vary only slowly over the surface in a particular view, being the same for many adjacent pixels in U, V space. (Of course, for proper antialiasing, the S_i and T_i each really refer to intervals of orientation.)

A resolution of 16 by 16 (S, T) values, as compared to 128 by 128 (U, V) values, leads to a plausible looking dull-reflective surface, as is illustrated in Figure 2a and Figure 2b (see Appendix). Figure 2a shows a typical U, V slice of the surface light field for the reflective floor in Figure 2b. Since the floor is planar, this corresponds to an orthographic projection of the scene. The depth of field effect is achieved using distributed ray tracing to jitter the S and T values (and hence the direction) of the rays. The visible shadow is seen because of the Phong highlight component of the surface shader [14]. Computation of this highlight involves casting a ray to the light to see if the region is illuminated. For this particular surface there is no diffuse component of the shader, so the shadow only becomes apparent when the highlight is near the shadowed region.

In contrast to the planar floor example, Figure 3a shows a surface light field for the reflective bottle in Figure 3b (see Appendix). Ripples in the orientation of the bottle surface create the striped appearance of the light field U, V slice. The angular S, T resolution was 32 by 32. Correct reflections are seen, which are more accurate than those from reflection mapping, and have the expected depth of field effect. Regions close to the bottle surface are seen reflected in crisper focus, those further away are seen in softer focus. However, the bottle itself and the surrounding scene remain crisp.

3.2 Block coding and temporal coherence

A goal of this paper is to explore the use of block coding schemes to represent light fields. Image compression schemes such as JPEG typically store an image as a set of 8 by 8 blocks [13], each of which is converted into DCT coefficients and then Huffman coded to remove redundancy [8]. This is especially efficient for whole images, since, during decoding, all of the image pixels are required.

Unfortunately, if each light field U, V slice is stored as an array of blocks, then we may need to decode an entire block to access one of its pixels. Vector quantization was used to overcome this problem in [11], by making use of the ability to efficiently decode pixels within a block. In DCT-based methods the inverse DCT may be used to find single pixels. However, Fast Fourier Transform methods are much more efficient when decoding several pixels in a single block. In addition, the Huffman decoding scheme has to reconstruct all of the coefficients, which is a significant cost. To make use of such block coding, we too dice the U, V slice images into blocks of 8 by 8 pixels. The light field is stored as an array of compressed blocks or C-Blocks, each of which may be decompressed into an uncompressed block or U-Block. Figure 4 (see Appendix) shows examples of compressing Figure 2b at different compression ratios. At 36 to 1, the compression artifacts show up clearly in the reflections. At a compression ratio of 24 to 1 the artifacts are less apparent. We found that at 11 to 1, the images were indistinguishable from Figure 2b.

Given a block-based encoding scheme, we need an algorithm to decompress blocks on the fly during real-time rendering. A compute-hungry algorithm is as follows:

> For a given (U, V, S, T) decompress the corresponding C-Block and then sample the U-Block. (In practice you will need to decompress 4 blocks if bilinear interpolation is being used for fractional values of (S, T).)

This is the most memory efficient scheme, but requires a block (or four) to be decompressed for every sample in the U, V texture map.

As noted in Section 3.1, low angular resolutions lead to discrete orientations that vary only slowly across the surface. This implies that a pixel in U, V space will often need to decode information from blocks accessed by its neighbors. A second observation is that, if the camera moves only slowly, the same blocks will probably be accessed in the subsequent time frame. For infinitesimal camera motions, all of the same blocks will be decompressed from frame to frame. This effect will decrease as the angular motion of the camera relative to the surface region approaches the interval between S, T blocks.

A memory-hungry algorithm, that exploits these ideas, is as follows:

> For each (U, V, S, T) interpolated in the texture map, decompress the corresponding C-Block, and keep it in memory. Subsequent accesses merely use the U-Block directly.

In practice, of course, there is a continuum between the two algorithms. By employing a block manager, we can keep the last N decompressed blocks in a block cache, only decompressing additional blocks when required. The block manager size determines the number of blocks that must be decompressed each frame.

A test was made by repeatedly rendering Figure 3b for a static camera. This has an S, T resolution of 32 by 32 images. Table 1 shows the results of this experiment. Rendering times are for a 180 MHz PowerPC microprocessor with an Apple QuickDraw3D accelerator card. The surface U, V map was computed at 256 by 256 pixels. (Uncompressed, this is a 200 Mbytes of data consisting of 1,048,576 blocks.) For reference, it is worth noting that the number of blocks accessed for that frame is 10,194. When the cache size exceeds this value, all of the blocks are kept in cache from the previous frame and the decompression count goes to zero. This is an expression of the frame-to-frame coherence of the cache.

When the cache is just slightly smaller than the number of accessed blocks, all of the blocks need to be decompressed for each frame. With an even smaller cache, each block has to be decompressed many times each frame, since it has been flushed from the block manager before it needs to be reused. It is worth noting that little performance is lost when the cache size is reduced from 10000 to 100. This is because a triangle of the tessellated surface accesses the images in a small region that can be mostly cached in 100 blocks. It is also worth noting that use of no compression block cache, i.e. storing the images as arrays directly, speeds up the rendering by a factor of 7 in this case. In contrast Table 2 shows similar data for the bottle with an S, T resolution of 16 by 16 (and 4 times less data). For small block cache sizes, the timings are similar with the number of block decompressions determining the frame rate. However, when the cache is large enough to store all of the decompressed blocks, the rendering time is only 2.5 times longer than the uncompressed case. We suspect that this performance improvement relative to the 32 by 32 case arises because the recently visited blocks are stored within the processor cache.

Table 1. Block decompressions versus cache size for 32x32 Silver Bottle.

Cache Size in Blocks	Time in Seconds	Blocks Decompressed	Block Decompressions
Uncompressed	0.24	Uncompressed	Uncompressed
10194	1.75	0	0
10193	2.44	10194	10194
1000	2.44	10194	10194
100	2.99	10194	1,042
10	4.84	10194	47646
4	8.15	10194	97646
1	16.18	10194	226088
0	17.59	10194	226332

Table 2: Block decompressions versus cache size for 16x16 Silver Bottle.

Cache Size in Blocks	Time in Seconds	Blocks Decompressed	Block Decompressions
Uncompressed	0.23	Uncompressed	Uncompressed
6136	0.59	0	0
6135	1.01	6136	6136
1000	1.01	6136	6136
100	1.19	6136	9587
10	2.1	6136	25288
4	4.6	6136	65876
1	12.7	6136	197274
0	13.8	6136	197320

To measure frame-to-frame coherence, a camera was moved relative to an object by a given linear distance. The test object was the glass cube in Figure 5 (see Appendix). Initial camera distance from the cube was 5 units of length. The camera was moved perpendicular to the view vector to the cube. When rendering the resultant new view, some additional blocks needed to be decompressed because the S, T values for the visible surface region depend on the position of the camera relative to the local surface frame.

It is instructive to measure the number of block decompressions as the linear distance is increased. Table 3 shows the number of blocks to decompress (assuming a very large cache size) as a function of the distance moved. This is recorded for the same scene but with different surface light field angular resolutions. In the 4 by 4 case, the frame-to-frame coherence meant that no new blocks needed to be decompressed up to a motion of 0.5 units. As the angular resolution is increased, the number of decompressed blocks increases more rapidly with camera motion. So, as expected, the frame-to-frame coherence is most applicable to fairly low angular resolution light fields. Cache coherence within the same frame, however, is still useful as was shown in Table 2.

3.3 Block coding and angular coherence
An important area for improvement is the use of coherence within the light field images to improve the compression rates. In the future, a general motion

compensation scheme may be applicable in the style of MPEG [12] and low bit-rate coding schemes such as [9]. Unfortunately, such a scheme is harder to implement in a block caching framework, since motion difference blocks may depend on four or so blocks from an S, T adjacent (in MPEG previous) image. The number of decompressed blocks will grow alarmingly with the number of motion-compensated (in this case angle-compensated) frames.

Table 3. Block decompressions versus camera motion.

Offset Distance	64x64	32x32	16x16	4x4
0.01	29	8	0	0
0.1	148	56	0	0
0.2	337	90	6	0
0.5	583	312	237	0
1.0	828	800	539	256
2.0	847	800	796	256
5.0	861	798	770	512

Fortunately, given the unique structure of surface light fields, there are examples where a very simple frame-differencing scheme is helpful. Consider the box in Figure 6a (see Appendix). Some regions of each face are purely diffuse, with a marble texture. Other regions are reflective in a decorative pattern. This is typical of a composite shader [7].

For purely diffuse regions, the image blocks will be identical for changing values of S, T. By comparing each block to the corresponding block with S and T values of zero, we can detect blocks which do not change. These can be coded with a single bit rather than storing the image coefficients. By detecting such blocks it is possible to double the compression ratio in the example of Figure 6a. Figure 6b was rendered with 17 to 1 block compression, but with the additional use of identical-block elimination, the compression ratio increased to 30 to 1.

A second approach to exploiting angular coherence is to compare a block with its corresponding neighbors in S, T space assuming some relative motion. We implemented a scheme in which a block was compared to its left neighbor in S space. MPEG style motion compensation was used to find the optimal offset between neighboring blocks. The key difference was that pixel queries outside of the reference block merely replicated the value of the nearest pixel inside the block (rather than decoding other adjacent blocks). For offsets of a few pixels, the overlap was still most of the block area. Differences between the offset adjacent block and the current block were encoded and stored. Every eighth block was made a key-block and encoded without motion compensation. Our experience with this scheme was disappointing. For the example of the silver bottle in Figure 3b (see Appendix), best results were found with an angular (S, T) resolution of 64 by 64. For comparable image quality the compression ratio was doubled. However, for smaller angular resolutions, the compression improvement was less noticeable. This is because the amount of angular change increases between blocks as the angular resolution goes down. Motion vectors need to be stored with the blocks and the motion is poorly approximated by a uniform offset. A better scheme for encoding angular coherence is an important area for future work.

4 Dynamic scenes

So far we have considered static scenes. An area for future work is how far these ideas may be generalized to dynamic scenes. In general this is very difficult, but there is one simple example where we can add some degree of animation and get accurate results.

Consider the scene in Figure 7 (see Appendix). A shiny floor reflects a room and a ball. We can use the light field for the floor to not only render the appearance of a flat floor, but also that of any bump-mapped reflective floor in the same location. Bump mapping changes the orientation of the surface normal, which may be used to compute the reflected ray direction. The reflected ray then gives an associated viewing direction for that surface point viewed in the flat case. This, in turn, is used to find the reflected ray color by indexing into the surface light field. This will give accurate images of the scene, provided that the other objects do not in turn show reflections of the floor.

The algorithm proceeded as follows:

- Detect hit events from the mouse. If the water polygon is hit, pass the U, V location of the hit to the water simulation, adding a disturbance at that location.
- Compute the water depth values based on a finite difference equation [10].
- Compute the normals of the water depth.
- Compute the reflected ray direction for each texture pixel.
- Index into the surface light field to compute the reflected color and place in the corresponding texture pixel.
- Download the completed texture to the rendering hardware.
- Scan-convert the scene.

Figure 7 illustrates the results of this algorithm. The water simulation was computed on a 64 by 64 grid and then the texture was computed using reflected ray S, T values interpolated up to 128 by 128. Figure 7 was rendered at 13 frames per second when no compression was used, and at (a maximum of) 6 frames per second when using 10 to 1 compression. When the surface was newly disturbed, the frame rate dropped to 2.5 frames per second. This was with a block cache that was double the size required to cache the same view of the smooth water. Figure 8a (see Appendix) shows the cache map for the flat floor. Each small box is a U, V image for a given value of S, T. Each pixel in the image represents an 8 by 8 block of U, V pixels in the light field. Blocks decompressed this frame are shown as red pixels. Those previously cached are shown in green. Because of the smooth surface and relatively distant camera, the blocks are clustered in one region of the S, T map. Since the cache is large, all of the blocks are green. In Figure 8b, on the other hand, the cache map shows a much more scattered structure and the red blocks indicate a large number are being decompressed this frame (about a quarter of the number of blocks in the cache).

Note that unlike the use of reflection maps for water ripples, these ripples have the correct appearance, as if they were being ray-traced on the fly. This is apparent from the reflections of the vertical lines on the walls. The reflections near the base of the wall wiggle less than those further from the wall, a characteristic of ray-traced images of rippling water, as opposed to reflection-mapped images which tend to wiggle different regions of objects all by the same amount. As given, the technique would be best applied to architectural scenes with ornamental ponds. To extend the scheme to more general dynamic scenes is an area for future work.

5 Conclusions

We explored the use of surface light-fields for walk-throughs of pre-ray-traced scenes. A caching scheme for decompressed blocks was introduced to allow the use of transform coding techniques with near real-time playback. Block cache coherence led to order of magnitude improvements in efficiency for on-the-fly decompression. It was found that frame-to-frame coherence was most effective for low-angular resolution light fields (up to about 16 by 16). However, cache coherence within the frame was effective for all angular resolutions, and great performance increases were achieved for relatively small block caches. Finally a simple example was given of a dynamic scene rendered accurately using a precomputed light field.

While the block-based compression schemes proved effective, the compression ratios achieved for good image quality were rarely greater than 20 to 1. To achieve higher compression ratios with comparable quality, a different formulation will probably be required. However, it is worth noting that the choice of angular resolution to match the angular content of the surface radiance was already a form of compression. Other experiments with light field compression tend to concentrate on mostly diffuse surfaces, which yield higher compression ratios. However, with our formulation, no compression of the angular component would be required for such diffuse surfaces. The trade-off is the requirement for more complexity in the geometric model.

6 Acknowledgements

This work was completed at the Graphics Research Group of Apple Research Laboratories at Apple Computer, Inc. in Cupertino, California. Special thanks to Douglass Turner for the use of his ray tracing code.

References

1. Arvo, James, "Application of Irradiance Tensors to the Simulation of Non-Lambertian Phenomena", Computer Graphics Proceedings, Annual Conference Series, 1995, pp. 335-342.

2. Chen, Schenchang Eric, Holly E. Rushmeier, Gavin Miller, Douglass Turner, "A Progressive Multi-pass Method for Global Illumination", Computer Graphics, Vol. 25, No. 4, July 1991.

3. Chen, W. H., C. H. Smith and S. C. Fralick "A fast computational algorithm for the discrete cosine transform," IEEE Trans. Commun., vol. COM-25, pp. 1004-1009, Sept 1977.

292

4. Debevec, Paul E., Camillo J. Taylor, Jitendra Malik, "Modeling and Rendering Architecture from Photographs: A Hybrid Geometry- and Image-Based Approach", SIGGRAPH '96, Computer Graphics Proceedings, Annual Conference Series, 1996. pp. 11-20.

5. Gortler, S. J, Radek Grzeszczuk, Richard Szeliski, Michael F. Cohen, "The Lumigraph", Proc SIGGRAPH '96 (New Orleans, Louisiana, August 4-9, 1996). In Computer Graphics Proceedings, Annual Conference Series, 1996, ACM SIGGRAPH, pp. 43-54.

6. Greene, Ned, "Hierarchical Polygon Tiling with Coverage Masks", Proc SIGGRAPH '96 (New Orleans, Louisiana, August 4-9, 1996). In Computer Graphics Proceedings, Annual Conference Series, 1996, ACM SIGGRAPH, pp. 65-74.

7. Hanrahan, Pat, and Jim Lawson, "A Language for Shading and Lighting Calculations", Computer Graphics, Vol. 24, No. 4, (August 1990), 289-298.

8. Huffman, D. A., "A method for the Construction of Minimum-Redundancy Codes," Proc. IRE, 40(9):1098-101, Sept 1952.

9. ITU-T (CCITT) "Recommendation H.261: Video Codec for Audiovisual Services at p x 64 kbits/s". , Mar 93.

10. Kass, Michael, Gavin S. P. Miller, "Rapid Stable Fluid Dynamics for Computer Graphics", Computer Graphics, Vol. 24, No. 4, Aug. 1990. pp. 49-58.

11. Levoy M., and P. Hanrahan, "Light Field Rendering", Proc SIGGRAPH '96 (New Orleans, Louisiana, August 4-9, 1996). In Computer Graphics Proceedings, Annual Conference Series, 1996, ACM SIGGRAPH, pp. 31-41.

12. Mitchell, J. L. , W. B. Pennebaker, C. E. Fogg, and D. J. LeGall, MPEG Video Compression Standard, Digital Multimedia Standards Series, Chapman & Hall, New York, 1997.

13. Pennebaker, W. and J. L. Mitchell, JPEG Still Image Data Compression Standard. Van Nostrand Reinhold, New York, 1993. ISBN 0-442-01272-1.

14. Phong, B. T., "Illumination for computer generated pictures", Communications of the ACM, Vol. 18, No. 6, (June 1975), pp. 311-317.

15. Williams, Lance, "Pyramidal Parametrics", Computer Graphics (SIGGRAPH '83 Proceedings), Vol. 17, No. 3, July 1983, pp. 1-9.

Editors' Note: see Appendix, p. 338 for colored figures of this paper

Canned Lightsources

Wolfgang Heidrich, Jan Kautz, Philipp Slusallek, Hans-Peter Seidel

University of Erlangen, Computer Graphics Group
Am Weichselgarten 9, D-91058 Erlangen, Germany
Email: {heidrich,jnkautz,slusallek,seidel}@immd9.informatik.uni-erlangen.de

Abstract. Complex luminaries and lamp geometries can greatly increase the realism of synthetic images.
Unfortunately, the correct rendering of illumination from complex lamps requires costly global illumination algorithms to simulate the indirect illumination reflected or refracted by parts of the lamp. Currently, this simulation has to be repeated for every scene in which a lamp is to be used, and even for multiple instances of a lamp within a single scene.
In this paper, we separate the global illumination simulation of the interior lamp geometry from the actual scene rendering. The lightfield produced by a given lamp is computed using any of the known global illumination algorithms. Afterwards, a discretized version of this lightfield is stored away for later use as a lightsource. We describe how this data can be efficiently utilized to illuminate a given scene using a number of different rendering algorithms, such as ray-tracing and hardware-based rendering.

1 Introduction

Complex lightsources can greatly contribute to the realism of computer generated images, and are thus interesting for a variety of applications. However, in order to correctly simulate the indirect light bouncing off internal parts of a lamp, it is necessary to apply expensive global illumination algorithms.

As a consequence, it was so far not possible to use realistic light sources in interactive applications. Also, for offline techniques such as ray-tracing, complex light source geometry unnecessarily slows down the rendering, since reflections and refractions between internal parts of the lightsource have to be recomputed for every instance of a lightsource geometry in every frame.

In this paper we propose a new method for using realistic lightsources for image synthesis. For a given lamp geometry and luminary, the outgoing lightfield is computed using standard global illumination methods, and stored away in a Lumigraph [6, 9] data structure. Later the lightfield can be used to illuminate a given scene while abstracting from the original lamp geometry. We call a light source stored and used in this fashion a "Canned Lightsource".

Instead of simulating the lightfield with global illumination methods, Canned Lightsources could also be measured [2], or even provided by lamp manufacturers in much the same way far field information is provided today. Thus, a database of luminaries and lamps stored as Canned Lightsources becomes possible.

Our method also speeds up the rendering process by factoring out the computation of the internal reflections and refractions of the lightsource into a separate preprocessing step. As a consequence, realistic lightsources can be efficiently used for interactive rendering, and for other applications where a complete global illumination solution would be too expensive. The cost of computing the global illumination within the lamp can be amortized over a large number of lamp instances and frames.

The remainder of this paper is organized as follows: in Section 2, we review the relevant previous work in this area. Section 3 contains a brief discussion of the acquisition of lightfields using global illumination algorithms. Section 4, the main contribution of this paper, describes ways to use Canned Lightsources in combination with several rendering algorithms, including ray-tracing and hardware-based methods. We conclude with some results and a discussion in Sections 5 and 6.

2 Previous Work

Several researchers have used precomputed global illumination solutions to efficiently generate realistic walkthroughs. A common technique is to use Gouraud shading and texture mapping to render radiosity solutions of diffuse environments from any perspective [12, 3]. Other approaches for diffuse scenes include Instant Radiosity [8] and Irradiance Volumes [7]. Interactive radiosity methods such as [4] incrementally update the illumination in scenes with small numbers of moving objects. Recently, algorithms for walkthroughs of scenes containing specular surfaces have been developed [14, 15].

All of the above methods require a *full* global illumination solution for the complete scene. Thus, they cannot deal with large numbers of changing objects or moving lightsources. Furthermore, the methods are not capable of storing away partial solutions for future use in different environments.

Methods more closely related to our approach include virtual walls [1] for speeding up radiosity computations, and the Lucifer algorithm [5, 10], which computes the radiance distribution on a virtual surface in the environment. Radiance [16] allows the user to replace geometry seen through small openings such as windows by dense radiance samples. None of these methods reuses illumination information in different environments.

In this paper, we encapsulate the lightfield emitted by a lightsource and store it away for future use. To this end, the method presented here relies on a dense sampling of the radiance emitted by a lightsource. The Lumigraph data structure[1] used to store this information has been developed by Levoy and Hanrahan [9], and Gortler et al. [6].

3 Simulating and Sampling the Lightfield of a Lightsource

As mentioned above, we store the lightfield in a Lumigraph data structure. From this data structure the radiance values can be extracted efficiently. Moreover, Lumigraphs can be generated easily using existing rendering systems or real world images, and since they are merely a two dimensional array of projective images, they lend themselves to efficient algorithms using computer graphics hardware (see Section 4.3).

On the down side, the light slabs of a Lumigraph do not uniformly sample the lightfield, and multiple light slabs are required to cover all light directions. Despite these disadvantages, we think that the Lumigraph is well suited for our purposes.

The top of Figure 1 shows the geometry of a single light slab. Every grid point corresponds to a sample location. As in [6] the (u, v) plane is the plane close to the object (lamp), while the (s, t) plane is further away, potentially at infinity.

---

[1]In this paper, we use the term "lightfield" for the continuous 5D function that describes the radiance at every point in space in every direction. The "Lumigraph", which is composed of one or more "light slabs", refers to the 4D data structure used to store a finite number of samples from the lightfield in empty space.

Placing the (s, t) plane at infinity offers the advantage of a clear separation between spatial sampling on the (u, v) plane, and directional sampling on the (s, t) plane. In this situation, all rays through a single sample on the (u, v) plane, that is, the projective image through that point, represent the far field of the lightsource. The higher the resolution of the (u, v) plane, the more nearfield information is added to the Lumigraph. For most applications, the spatial resolution is much smaller than the directional resolution.

If, the (s, t) plane is *not* placed at infinity, it is not possible to clearly separate between spatial and directional resolution. Nonetheless, we can still view the light slab as an array of projective images with centers of projection on the (u, v) plane.

Because of this, a light slab representing a Canned Lightsource can be created by first generating a global illumination solution for the lamp geometry, and then rendering it from multiple points of view on the (u, v) plane. For the global illumination step, we can use any kind of algorithm, such as radiosity, Monte-Carlo ray-tracing, Photon Maps, or composite methods [13]. Alternatively, Canned Lightsources could be generated

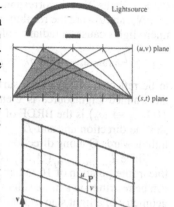

Fig. 1. Geometry of a single light slab within a Lumigraph.

by resampling measured data [2] much in the same way described in [6]. Finally, it is also possible to generate non-physical lightsources for special effects.

One issue is the storage of Canned Lightsources. Due to the large dynamic range required to faithfully represent the lightfield of a lightsource, one of the commonly used 24 bit true color file formats is not sufficient. On the other hand, Lumigraphs can be quite large, and so a memory-efficient representation is mandatory. A good compromise is Larson's LogLuv extension to the TIFF file format [17], which uses 32 bit/sample to store high dynamic range images.

4 Reconstruction of Illumination from Lightfields

Given one or more Canned Lightsources, we can use them to illuminate a scene by reconstructing the radiance emitted by the lightsource at every point and in every direction. Like in [6] and [9], we use quadrilinear interpolation between the samples stored in the light slabs. For a ray-tracer, we could simply obtain a large number of radiance samples from the Lumigraph. However, the challenge is to obtain a good reconstruction with the smallest number of samples possible. It is important not to miss any narrow radiance peaks, because this would lead to artifacts in the reconstructed illumination.

In the following, we discuss three different approaches for reconstructing the illumination at every point in the scene. Section 4.1 describes a high-quality technique which can be used as a reference solution for the other methods. Section 4.2 deals with a generalization of this method for efficient ray-tracing. Finally, in Section 4.3, we describe a way to use the graphics library OpenGL for the reconstruction.

4.1 High Quality Reference Solutions

In order to describe the illumination in a scene caused by a Canned Lightsource, we view a light slab as a collection of $N_u \times N_v \times N_s \times N_t$ independent little area lightsources ("micro lights"), each corresponding to one of the light slab's 4D grid cells. Here, N_u, N_v, N_s, and N_t are the resolutions of the slab in each parametric direction. Each of the micro lights causes a radiance of

$$L^o_{n_u,n_v,n_s,n_t}(\mathbf{Q}, \omega_o) = \int f(\mathbf{Q}, \omega_i \to \omega_o) L^i_{n_u,n_v,n_s,n_t}(\mathbf{Q}, \omega_i) \cos\theta d\omega_i \qquad (1)$$

to be reflected of any given surface point $\mathbf{Q}$. The total reflected radiance caused by the Canned Lightsource is then the sum of the contributions of every micro light. $f(\mathbf{Q}, \omega_i \to \omega_o)$ is the BRDF of the surface, θ is the angle between the surface normal and the direction ω_i, and $L^i_{n_u,n_v,n_s,n_t}(\mathbf{Q}, \omega_i)$ is the radiance emitted from the micro light towards $\mathbf{Q}$ along direction ω_i.

Since $L^i_{n_u,n_v,n_s,n_t}(\mathbf{Q}, \omega_i)$ is obtained by quadri-linear interpolation of 16 radiance values, Equation 1 can be rewritten in a more suitable form. Consider the geometry of a light slab as shown on the right side of Figure 1. Every point $\mathbf{P}$ on the (u, v) plane can be written as $\mathbf{P} = \mathbf{O} + (n_u + u)\vec{u} + (n_v + v)\vec{v}$, where the vectors $\vec{u}$ and $\vec{v}$ determine the dimensions of a 2D grid cell, and $u, v \in [0, 1]$ describe the coordinates of $\mathbf{P}$ within the cell. In analogy we can describe a point $\mathbf{P}'$ on the (s, t) plane, where we assume that the vectors $\vec{s}$ and $\vec{t}$ are parallel to $\vec{u}$ and $\vec{v}$, respectively.

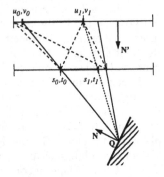

Fig. 2. Clipping a Lumigraph grid cell. $u_{\{0,1\}}$, $v_{\{0,1\}}$, $s_{\{0,1\}}$, and $t_{\{0,1\}}$ describe the portion of the cell that is visible from point $\mathbf{Q}$.

For micro light (n_u, n_v, n_s, n_t) and point $\mathbf{Q}$, we can now determine the region on the (u, v) plane that is *visible* from point $\mathbf{Q}$. This is achieved by projecting the 2D cell (n_s, n_t) onto the (u, v) plane, using $\mathbf{Q}$ as the center of projection (see Figure 2). The resulting rectangle is clipped against the 2D cell (n_u, n_v).

This yields boundaries $u_{\{0,1\}}$, $v_{\{0,1\}}$, $s_{\{0,1\}}$, and $t_{\{0,1\}}$ relative to the 4D cell boundaries of the micro light. In this area, every ray $r(u, v, s, t)$ passing through $\mathbf{Q}$ is characterized by $u = (1 - x)u_0 + xu_1$, $v = (1 - y)v_0 + yv_1$, $s = (1 - x)s_0 + xs_1$, and $t = (1 - y)t_0 + yt_1$ for some $x, y \in [0, 1]$. The values x and y can be used for the quadrilinear interpolation of the radiance emitted towards point $\mathbf{Q}$ by inserting the formulas for u, v, s, and t into the following equation:

$$\begin{aligned}
L^i_{n_u,n_v,n_s,n_t}(x,y) =&(1-u)\cdot(1-v)\cdot(1-s)\cdot(1-t)\cdot L_{0000} + \\
&(1-u)\cdot(1-v)\cdot(1-s)\cdot t \cdot L_{0001} + \\
&+\cdots + u\cdot v\cdot s\cdot t\cdot L_{1111}.
\end{aligned} \qquad (2)$$

$L_{0000}, L_{0001}, \ldots, L_{1111}$ are the 16 radiance values at the corners of the 4D grid cell. These are contained in the Lumigraph. With this result, we can now rewrite Equation 1 as follows:

$$\begin{aligned}
L^o_{n_u,n_v,n_s,n_t}&(\mathbf{Q}, \omega_o) \\
&= \int_A f(\mathbf{Q}, (\mathbf{P}-\mathbf{Q}) \to \omega_o) L^i_{n_u,n_v,n_s,n_t}(\mathbf{Q}, \mathbf{P} \to \mathbf{Q}) \frac{\cos\theta\cos\theta'}{R^2} d\mathbf{P} \\
&= A \cdot \int_0^1 \int_0^1 f(\mathbf{Q}, (\mathbf{P}-\mathbf{Q}) \to \omega_o) L^i_{n_u,n_v,n_s,n_t}(x,y) \frac{\cos\theta\cos\theta'}{R^2} dxdy,
\end{aligned} \qquad (3)$$

where $A = (u_1 - u_0)||\vec{u}|| \cdot (v_1 - v_0)||\vec{v}||$ is the size of the visible region on the (u, v) plane, and $R = ||\mathbf{P} - \mathbf{Q}||$ is the distance of the two points $\mathbf{P}$ and $\mathbf{Q}$. Note that the vectors

$\vec{s}$ and $\vec{t}$ do not occur in this equation. That is, the exact position and parameterization of the (s, t) plane is only required for clipping. The use of an (s, t) plane at infinity is transparent in Equation 3.

While an analytical solution of the integral exists if there is no occlusion, it is too complicated for practical use. However, the integrand in Equation 3 is quite smooth, so that Monte Carlo integration performs well, and only a few samples are required to obtain good results.

In order to compute the complete illumination in $\mathbf{Q}$, we have to sum up the contribution of all micro lights. Although there are $N_u \times N_v \times N_s \times N_t$ micro lights, only $O(\max(N_u, N_s) \cdot \max(N_v, N_t))$ have a non-zero visible area from any point $\mathbf{Q}$. These visible areas are easily determined since 2D grids on the two Lumigraph planes are aligned. Thus, $u_{\{0,1\}}$ and $s_{\{0,1\}}$ only depend on the grid *column* (n_u, n_s), while $v_{\{0,1\}}$ and $t_{\{0,1\}}$ only depend on the grid *row* (n_v, n_t).

The result of this method is that we know how many samples are at least required for faithfully computing the illumination in a given point $\mathbf{Q}$: at least one sample is necessary for each visible micro light. The left side of Figure 3 shows a simple scene illuminated with a Canned Lightsource at full resolution (64×64 on the (u, v) plane). The underlying geometry of the lightsource is described in Section 5.

4.2 Efficient Ray-tracing

While the above method is capable of producing high-quality reference solutions, the number of visible grid cells is prohibitive for real world applications. Independent of the distance from the lightsource, we have to take at least one sample per visible grid cell, or risk missing important features such as thin radiance peaks. Ideally, the number of samples should depend on the solid angle covered by the lightsource.

Fig. 3. A simple test scene illuminated by a Canned Lightsource with full resolution 64×64 (top) and half resolution in every direction (bottom). In the center you see a difference image between the two.

To achieve this kind of adaptive sampling, we have developed a method based on a hierarchy of Lumigraphs at different resolutions, similar to the use of mipmaps. Let us first consider the case where the (s, t) plane is at infinity.

If we ignore occluding geometry, we can say that each point $\mathbf{Q}$ in front of the (u, v) plane sees all sample points there. The distances towards these samples will vary strongly for points $\mathbf{Q}$ close to the (u, v) plane. Thus, the number of visible micro lights is mostly dominated by the directional resolution, which is larger than the spatial resolution for our purposes (see Section 3). Therefore, $O(N_s \cdot N_t)$ micro lights are visible from $\mathbf{Q}$. This number cannot be reduced without compromising quality, since the lightsource covers a large solid angle as seen from $\mathbf{Q}$.

As the distance of $\mathbf{Q}$ from the (u, v) plane increases, the directions of the sample rays become almost parallel. In the end, all rays will pass through a single 2D cell on the (s, t) plane. In this situation, there are exactly $N_u \cdot N_v$ visible micro lights. Although this number is smaller than the number of visible cells for close points $\mathbf{Q}$, we would like to further reduce it for points at very large distances. The number of samples should be reduced linearly with the solid angle covered by the lightsource.

This means that for far away points, a Lumigraph with a lower resolution on the (u, v) plane is sufficient. Similarly to mipmaping, we use a hierarchy of Lumigraphs, where the resolution in the u and v direction is reduced by a factor of two from one level to the next. Since we want to avoid storing the Lumigraph at multiple resolutions, we would like to do the downsampling on the fly during the rendering phase. We use simple averaging of the higher resolution samples. When switching from one level to the next, only the clipping has to be performed at a different resolution, and in Equation 2 averaged values are used for L_{0000}, L_{0001} and so on.

If the (s, t) plane is not located at infinity, a clear separation between spatial and directional resolution is not possible. Reducing the resolution of only one plane is therefore not adequate. Nonetheless, it makes sense to reduce the number of samples quadratically with the distance from the (u, v) plane, in order to maintain a certain number of samples per solid angle. This can be achieved by simultaneously reducing the resolution of both planes from level to level. Due to the lower directional resolution, the results will in general not be quite as good as for (s, t) planes at infinity.

The right side of Figure 3 shows the result of the same image as on the left side, but illuminated with a downsampled Canned Lightsource (1/2 the resolution in each direction). The performance improvement for this scene was roughly a factor of 7.4. Although a minor degradation of quality is noticeable in the difference image, the quality is still very good.

4.3 OpenGL Reconstruction

In addition to ray-tracing, it is also possible to use computer graphics hardware for reconstructing the illumination from a Canned Lightsource. As a hardware abstraction layer, we use the graphics library OpenGL, which is available on a variety of different platforms. Our method can be combined with one of the algorithms for generating shadows in OpenGL, such as shadow maps, or shadow volumes.

For hardware-based rendering, we assume that the BRDF $f(\mathbf{Q}, (\mathbf{P}-\mathbf{Q}) \to \omega_o)$ used in Equation 3 corresponds to the Phong model, which is the OpenGL lighting model. We numerically integrate Equation 3 by evaluating the integrand at the grid points on the (u, v) plane. For these points, the quadrilinear interpolation reduces to a bilinear interpolation on the (s, t) plane, which can be done in hardware.

[11] describes how projective textures can be used to simulate lightsources such as high-quality spotlights and slide projectors. Our method is an extension of this approach, which also accounts for the quadratic falloff and the cosine terms in Equation 3. To see how this works, we rewrite the integrand from this equation in the following form:

$$f(\mathbf{Q}, (\mathbf{P}-\mathbf{Q}) \to \omega_o) \cos\theta \cdot A \frac{\cos\theta'}{R^2} \cdot L^i_{n_u, n_v, n_s, n_t}(x, y) \tag{4}$$

The first factor of this formula is given by the Phong material and the surface normal N (see Figure 1). The second factor is the illumination from a spot light with intensity A, located at $\mathbf{P}$ and pointing towards N' with a cutoff angle of $90°$, a spot exponent of 1, and quadratic falloff. Finally, the third factor is the texture value looked up from the slice of the light slab corresponding to point $\mathbf{P}$ on the (u, v) plane.

In other words, the integrand of Equation 4 can be evaluated by combining projective textures as in [11] with the illumination from a spotlight using the standard

OpenGL lighting model. The results from texturing and lighting are combined using the "modulate" texture environment of OpenGL.

The integral from Equation 3 is approximated by the sum of the contributions from all grid points on the (u, v) plane. We compute this sum by adding multiple images, each rendered with a different spotlight position and projective texture. To this end we use either alpha blending or an accumulation buffer.

5 Results

For the images in Figure 3, we have used a Canned Lightsource that was created by ray-tracing a simple lamp geometry depicted in Figure 4. The lamp itself consists of a diffuse parabolic reflector and a grid of 3×3 long blocker polygons. The shadows cast by these blockers are quite visible in the nearfield of the lightsource, and can thus be used to evaluate the quality of the rendering. A small spherical luminary in the center of the paraboloid was used.

Fig. 4. Lamp geometry used for the images in this paper. A diffuse parabolic reflector with a small spherical luminary is covered with a grid of blockers.

In Figure 6 in the color section, you can see an office scene illuminated by variations of the same Canned Lightsource. The image was generated using ray-tracing. The Canned Lightsource had a resolution of $32 \times 32 \times 16 \times 16$, which occupied about 2MB of space in the LogLuv coding. On the right side the paraboloid is narrower, resulting in a more focussed emission of light. In both images the shadows created by the blockers are noticeable. The blurring of these shadows is a prominent feature of the nearfield of this lightsource.

Finally, we have used a hand-crafted Canned Lightsource that resembles a slide projector. In the top right image of Figure 6, we have pointed this lightsource at a certain angle onto a plane. This results in a focussed area in the center, where the "slide" is in focus, and in blurred outer regions, where the focus of the projector is either in front of or behind the plane. Also note the darkening towards the corners of the projection due to illumination from off-normal directions. Depending on the resolution of the (u, v) plane, the scene can be rendered interactively on current graphics systems. For example, we achieve around 10 frames per second on a Reality Engine 2 with a spatial resolution of 8×8, and roughly the same performance on an $O2$ with a resolution of 4×4. Figure 6 was rendered at a resolution of 8×8.

Fig. 5. Plane illuminated by a hand-crafted Canned Lightsource resembling a slide projector.

6 Conclusions

In this paper we have introduced the notion of a Canned Lightsource, which is a pre-computed, discretely sampled version of the lightfield emitted by a complex lightsource or lamp. This information is stored away in a Lumigraph data structure. Later, the pre-computed information can be used to illuminate a scene while abstracting from the lamp geometry. Thus, Canned Lightsources act as a blackbox for complex lightsources.

Canned Lightsources can be used to create databases of lightsources based on simulation or measurement. Ideally, this information could be directly distributed by the lamp manufacturers in much the same way far field information is provided today.

The precomputed information can be used to speed up the lighting process in standard rendering algorithms such as ray-tracing and hardware-based rendering. Thus, Canned Lightsources are a valuable data structure for efficient image synthesis. Particularly interesting is the possibility to precompute or measure the global illumination of some parts of a scene (the lightsource), and reuse it at a different location or in a completely different scene.

7 Acknowledgments

We would like to thank Marc Stamminger for several interesting discussions regarding this topic. This work was in part supported by the German Research Council through the collaborative research center #603 (Model-based Analysis and Synthesis of Complex Scenes and Sensor Data).

References

1. B. Arnaldi, X. Pueyo, and J. Vilaplana. On the division of environments by virtual walls for radiosity computation. In *Photorealistic Rendering in Computer Graphics*, pages 198–205, May 1991.
2. I. Ashdown. Near-Field Photometry: A New Approach. *Journal of the Illuminating Engineering Society*, 22(1):163–180, Winter 1993.
3. R. Bastos. Efficient radiosity rendering using textures and bicubic reconstruction. In *Symposium on Interactive 3D Graphics*. ACM Siggraph, 1997.
4. G. Drettakis and F. Sillion. Intaractive update of global illumination using a line-space hierarchy. *Computer Graphics (SIGGRAPH '97 Proceedings)*, pages 57–64, August 1997.
5. A. Fournier. From local to global and back. In *Rendering Techniques '95*, pages 127–136, June 1995.
6. S. J. Gortler, R. Grzeszczuk, R. Szelinski, and M. F. Cohen. The Lumigraph. In *Computer Graphics (SIGGRAPH '96 Proceedings)*, pages 43–54, August 1996.
7. G. Greger, P. Shirley, P. Hubbard, and D. P. Greenberg. The irradiance volume. *IEEE Computer Graphics and Applications*, 18(2):32–43, March 1998.
8. A. Keller. Instant radiosity. *Computer Graphics (SIGGRAPH '97 Proceedings)*, pages 49–56, August 1997.
9. M. Levoy and P. Hanrahan. Light field rendering. *Computer Graphics (SIGGRAPH '96 Proceedings)*, pages 31–45, August 1996.
10. R. Lewis and A. Fournier. Light-driven global illumination with a wavelet representation of light transport. In *Rendering Techniques '96*, pages 11–20, June 1996.
11. M. Segal, C. Korobkin, R. van Widenfelt, J. Foran, and P. Haeberli. Fast shadow and lighting effects using texture mapping. *Computer Graphics (SIGGRAPH '92 Proceedings)*, 26(2):249–252, July 1992.
12. F. Sillion and C. Puech. *Radiosity & Global Illumination*. Morgan Kaufmann, 1994.
13. P. Slusallek, M. Stamminger, W. Heidrich, J.-C. Popp, and H.-P. Seidel. Composite lighting simulations with lighting networks. *IEEE Computer Graphics and Applications*, 18(2):22–31, March 1998.
14. W. Stürzlinger and R. Bastos. Interactive rendering of globally illuminated glossy scenes. In *Rendering Techniques '97*, pages 93–102, 1997.
15. B. Walter, G. Alppay, E. Lafortune, S. Fernandez, and D. P. Greenberg. Fitting virtual lights for non-diffuse walkthroughs. *Computer Graphics (SIGGRAPH '97 Proceedings)*, pages 45–48, August 1997.
16. G. Ward. The RADIANCE lighting simulation and rendering system. *Computer Graphics (SIGGRAPH '94 Proceedings)*, pages 459–472, July 1994.
17. G. Ward Larson. LogLuv encoding for TIFF images. Technical report, Silicon Graphics, 1997. www.sgi.com/Technology/pixformat/tiffluv.html.

Editors' Note: see Appendix, p. 337 for colored figure of this paper

Image-Based Rendering for Non-Diffuse Synthetic Scenes

Dani Lischinski   Ari Rappoport

The Hebrew University

Abstract. Most current image-based rendering methods operate under the assumption that all of the visible surfaces in the scene are opaque ideal diffuse (Lambertian) reflectors. This paper is concerned with image-based rendering of *non-diffuse* synthetic scenes. We introduce a new family of image-based scene representations and describe corresponding image-based rendering algorithms that are capable of handling general synthetic scenes containing not only diffuse reflectors, but also specular and glossy objects. Our image-based representation is based on *layered depth images*. It represents simultaneously and separately both view-independent scene information and view-dependent appearance information. The view-dependent information may be either extracted directly from our data-structures, or evaluated procedurally using an image-based analogue of ray tracing. We describe image-based rendering algorithms that recombine the two components together in a manner that produces a good approximation to the correct image from any viewing position. In addition to extending image-based rendering to non-diffuse synthetic scenes, our paper has an important methodological contribution: it places image-based rendering, light field rendering, and volume graphics in a common framework of discrete raster-based scene representations.

1 Introduction

Image-based rendering is a new rendering paradigm that uses pre-rendered or pre-acquired reference images of a 3D scene in order to synthesize novel views of the scene. This paradigm possesses several important advantages compared to traditional rendering: (i) image-based rendering is relatively inexpensive and does not require special purpose graphics hardware; (ii) The rendering time is independent of the geometrical and physical complexity of the scene being rendered; and (iii) The pre-acquired images can be of real scenes, thus obviating the need to create a full geometric 3D model in order to synthesize new images.

While the ability to handle real scenes has been an important motivating factor in the emergence of image-based rendering (particularly in the computer vision community [17]), it soon became clear that image-based rendering is also extremely useful for efficient rendering of complex synthetic 3D scenes [5, 12, 22, 23, 29].

Unfortunately, most current image-based rendering methods operate under the assumption that all of the visible surfaces in the scene are opaque ideal diffuse (Lambertian) reflectors, i.e., it is assumed that every point in the scene appears to have the same color when viewed from different directions. However, in practice, both real and synthetic scenes often contain various non-diffuse surfaces, such as specularly reflecting mirrors, shiny floors, glossy tabletops, etc. If any of these non-diffuse surfaces are visible in the reference images of the

Fig. 1. The result of warping two different reference views of a scene to a common new view. The mirror reflections in the box are incorrect and inconsistent.

scene, new images of the scene produced by most current image-based rendering methods will be incorrect, exhibiting potentially severe inconsistencies, as illustrated in Figure 1. The inability of image-based rendering techniques to handle non-diffuse scenes correctly, is a major limitation of their usefulness, applicability, and power. It should be noted here that one particular variety of image-based rendering, referred to as *light field rendering* [11, 20] is not subject to the above limitation, but at a significant storage penalty.

The main contribution of this paper is a new image-based rendering approach that correctly handles non-diffuse synthetic scenes with general view-dependent shading phenomena, such as specular highlights, ideal specular reflections, and glossy reflections. In our approach, a scene is represented as a collection of parallel projections with multiple depth layers (LDIs) [12, 23]. All view-independent scene information (i.e., geometry and diffuse shading) is represented using three orthogonal high-resolution LDIs. The view-dependent scene information is represented as a separate (larger) collection of low-resolution LDIs. Each of these LDIs captures the appearance of the scene from a particular direction. We describe image-based rendering algorithms that recombine these two components together to quickly produce a good approximation to the correct image of the scene from any viewing position. By separating the representation in the manner outlined above, we are able to obtain results superior in quality to those produced by light field rendering methods, while requiring considerably less storage. Our image-based rendering techniques produce images with view-dependent shading and reflections whose quality is comparable to those generated by a conventional ray tracer, but faster.

Another important contribution of this paper is a methodological one. Our approach places image-based rendering, light field rendering, and volume graphics [13] in a common framework: the framework of discrete raster-based scene representations. This common framework enables better understanding of the relative advantages and disadvantages of these techniques.

2 Related work

An image-based rendering technique can be thought of as a combination of a discrete raster-based scene representation with a rendering algorithm that operates on this representation to produce images from various views. The rendering algorithm typically utilizes the regularity, uniformity, and the bounded size of the raster-based representation to produce the new views rapidly. Four main kinds of raster-based representations have been previously explored: image-based representations, discrete 4D light fields, discrete volumetric representations, and ray representations.

2.1 Image-based representations

Image-based rendering techniques typically represent a scene as a collection of reference images: scene projections sampled on a discrete grid. This projections can be spherical, cylindrical [4, 26], or planar [5, 17, 22, 28]. Each projection pixel can be thought of as a *scene sample*, because it represents the color (and sometimes the depth) of the nearest point in the scene that can be seen along the corresponding projection line. New images can be generated from one or more projections by *warping* these scene samples to the desired view. In order to correctly warp a scene sample, we must know its depth. Depth information is either encoded implicitly in the form of correspondences between pairs of points in different projections [17, 26], or it can be stored explicitly with each scene sample. For example, in the case of synthetic scenes depth information is readily available.

In addition to obtaining depths or correspondences, the key challenge in these approaches is to handle the gaps that appear in the synthesized images. Such gaps appear when areas in the reference images are stretched by the warp to the new image, and when parts that are occluded in all of the reference images come into view. Mark *et al.* [22] handle these problems using multiple reference images, each represented as a mesh of micro-polygons. Gortler *et al.* [12] use a layered depth image (LDI) and "splat" each pixel onto the screen [32].

Layered depth images were also used by Max [23, 24] in order to represent in a hierarchical fashion and render complex photorealistic models of botanical trees. There are similarities between his approach and the one presented in this paper: both approaches use parallel projection LDIs, and store a normal with each scene sample in order to perform shading after the reprojection. However, our approach concentrates on the issue of handling view-dependent shading phenomena, which was not addressed by Max.

As pointed out earlier, creating a new image by warping scene samples is only correct when the samples correspond to ideal diffuse surfaces. There have been a few attempts to extend reprojections and view-interpolation to non-diffuse scenes [1, 2, 6], but these approaches are not completely image-based. They require access to the original scene geometry to handle various cases in which reprojection cannot produce the correct results. Thus, these methods handle diffuse pixels quickly, but still have to do a lot of work for the remaining pixels.

2.2 Light field rendering

While image-based rendering techniques represent both the geometry and the color of each scene sample, light field rendering techniques [11, 20] avoid the representation of geometry altogether. Instead they represent a scene as a 4D subset of the complete 5D light field that describes the flow of light at all positions in all directions. This 4D function, which Gortler *et al.* dub a *Lumigraph*, describes all of the light entering or leaving a bounded region of space. Once the Lumigraph has been constructed, images of the region's contents can be quickly generated from any viewpoint outside the region.

Light field rendering is simpler and more robust than most previous image-based rendering techniques because it does not require depth values, optical flow information, or pixel correspondences, and it is not limited to diffuse scenes. However, it also has several significant limitations: (i) high-fidelity light fields require a very large amount of storage; (ii) computing the light field of a synthetic scene can take a very long time; and (iii) the 4D light field describes the flow of light in a bounded region of space free of objects, and it is not yet clear how to extend light-fields to allow free navigation inside a scene.

2.3 Volume graphics

Image-based rendering techniques were not the first to utilize the advantages of regular and uniform discrete representation of 3D scenes: similar representations have been actively explored in the field of volume visualization [9, 15, 16, 18, 32].

In particular, a subfield of volume visualization, referred to as *volume graphics* [13] is concerned with the synthesis, representation, manipulation, and rendering of volumetric objects stored in a 3D volume buffer. Unlike volume visualization, which focuses primarily on sampled and computed volumetric datasets, volume graphics is concerned primarily with volumetric representation of synthetic 3D scenes. Similarly to image-based rendering, volume graphics is insensitive to the geometric complexity of the scene and of the objects contained in it. Volumes have an advantage over images in that they constitute a more complete representation of the environment: not only the boundary of the objects is represented, but also their interior. However, the price of the extra dimension is much higher storage requirements and longer rendering times.

Several researchers in volume graphics have investigated algorithms for rendering non-diffuse synthetic scenes from their volumetric representation. Volumetric ray tracing [19, 31] can be used for this purpose, handling specular reflections and other non-diffuse phenomena via recursive tracing of secondary rays. Yagel *et al.* [33] describe *discrete ray tracing*, a technique in which discrete (voxelized) rays are tracked through the 3D volume buffer until they encounter a voxel occupied by a voxelized surface. Our proposed approach utilizes a related technique, which traces a ray in a layered depth projection of the scene.

The discrete scene representation and the rendering algorithms that we propose to explore can be viewed as bridging the gaps between the image-based rendering, light field rendering, and volume graphics approaches surveyed above.

2.4 Ray representations

Researchers in solid modeling have explored object representation schemes based on ray sampling. Van Hook [30] extended the ordinary z-buffer by storing a list of objects at each pixel, supporting faster visualization of Boolean operations for simulation of NC machining operations. This method requires special purpose hardware that is not currently available.

Groups from Cornell and Duke universities have devised a hardware architecture called the "ray casting engine", in which a geometric object is represented by a discrete 2D matrix defined by a parallel projection of the surfaces defining the object in a certain direction. Each matrix element contains a list of surfaces pierced by the corresponding ray. The resulting representation was dubbed the "ray representation" or "rayrep" [10]. It was argued that using the rayrep, several important solid modeling operations are easier to solve [27]. One of these operations was ray tracing the solid, but the results were not of high visual quality.

Recently, Benouamer [3] proposed using three sets of rayreps projected in orthogonal directions for computing Boolean operations when converting CSG objects to a boundary representation (Brep). The rayreps are generated by ray casting, and the Brep is obtained by a surface reconstruction algorithm in the spirit of the "marching cubes" algorithm [21]. The emphasis in his work is on the resulting geometry, its tolerance with respect to manufacturing applications, and the simplicity of implementation rather than on rendering and visual quality.

3 A new discrete representation

The main new idea behind our approach is to represent *simultaneously but separately* both view-independent scene information (geometry, diffuse shading, etc.) and view-dependent appearance information. The two components are combined during rendering to efficiently produce correct images of non-diffuse scenes.

Both kinds of scene information are represented using a collection of parallel projections of the scene, where each projection is a layered depth image (LDI) [12, 23][1].

---

[1]The LDIs described by Gortler *et al.* [12] are defined using a general perspective projection of the scene.

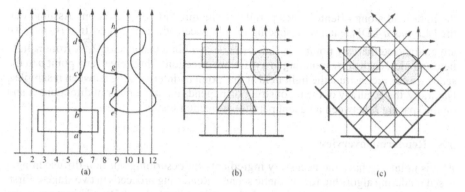

Fig. 2. (a) A parallel LDI of a simple 2D scene. Pixel 6 in the LDI stores scene samples a, b, c, d, and pixel 9 stores scene samples e, f, g, h. (b) and (c) illustrate (in 2D) the view-independent LDIs and the LLF, respectively.

Each LDI pixel contains a list of *scene samples* corresponding to all surface points in the scene intersected by the projection ray through the center of the pixel (see Figure 2a). The scene samples all lie on a 2D raster, when seen from the direction of projection, but their positions along the projection axis can be arbitrary (thus, there is no discretization along the depth axis). The difference between the representation of the view-independent and the view-dependent components is in the type of information stored with each scene sample and in the number of projections.

3.1 The layered depth cube

The view-independent component of our representation consists of three high-resolution parallel LDIs corresponding to three orthogonal directions (see Figure 2b). We refer to this component as the *layered depth cube* (LDC). With each scene sample of the LDC we store its depth value (the orthogonal distance from a reference plane perpendicular to the direction of projection). We also store the surface normal, the diffuse shading, a bitmask describing the visibility of the light sources, and the directional BRDF at the surface point[2]. Note that for diffuse scenes this representation is complete (modulo the finite spatial resolution of the LDIs). Specifically, if h is the spacing between adjacent pixels in the LDIs, we are guaranteed to have at least one scene sample on any surface element of size $\sqrt{3}\,h$, regardless of its orientation.

3.2 The layered light field

The view-dependent component of our representation is referred to as the *layered light field* (LLF). The LLF consists of a potentially large collection of low-resolution parallel LDIs corresponding to various directions (we have experimented with 66, 258, and 1026 directions, uniformly distributed on the unit sphere), as illustrated in Figure 2c. Each scene sample in the LLF contains, in addition to its depth, the total radiance leaving the scene sample in the direction of projection. Thus, each of the LDIs samples

---

Similarly to Max [23], we use parallel projections since they offer computational advantages, as we shall see in section 4.3.

[2]In practice, we store an index into a table containing the material properties of surfaces in the scene.

the light field along oriented lines parallel to the direction of projection. Assume that the LLF has n LDIs, each with spacing of h between adjacent pixels. In this case, we are guaranteed that any point p in the scene lies within a distance of $\frac{\sqrt{2}}{2}h$ from n light field samples, each corresponding to a different direction. Thus, given any point p in the scene we can reconstruct the light field at that point by interpolating between n samples. Naturally, the accuracy of the reconstruction depends on the number of directions n and on the spacing h between adjacent parallel projection lines.

3.3 Rendering overview

At this point, we have the necessary ingredients for describing our non-diffuse image-based rendering algorithm for synthetic scenes. Rendering proceeds in two stages: First, we apply 3D warping to the high-resolution view-independent LDIs. This stage is described in more detail in Section 4.1. The 3D warping results in a *primary* image, which is a correct representation of the geometry of the scene, as seen from the new viewpoint, along with the diffuse component of the shading. In each pixel of the primary image we store the scene sample visible through that pixel.

The second stage utilizes the normal, the light source visibility mask, and the directional BRDF at each visible scene sample to compute the view-dependent component of the shading. First, specular highlights from the light sources in the scene are added, by simply evaluating the local specular shading model (e.g., Phong) for each of the visible light sources (visibility is stored in the light source bitmask). Then, to add reflections of other objects in the scene we use two techniques:

1. *Light-field gather:* this technique integrates the light field at the point of interest using the directional BRDF to weigh each incoming direction. The light field is reconstructed from the LLF as described in Section 4.2. Light-field gather works very well with glossy objects with fuzzy reflections, but in order to generate sharp reflections (e.g., on mirrors) it is more effective to use image-based ray tracing.
2. *Image-based ray tracing:* this technique recursively traces reflected rays from the point of interest. These rays are traced using the view-independent LDIs, as described in Section 4.3. After k reflections recursion is terminated and we perform a light-field gather. Experiments have shown that deeper recursion increases the rendering time, but allows us to use a sparser LLF. In most scenes, tracing just one generation of reflected rays yields very good results.

3.4 Discussion

Our scene representation can be thought of as a parametric family of image-based scene representations. Different members of this family are obtained by different choices of the number of the directions in the LLF, the spatial resolution of the different types of LDIs, and the depth of recursive ray tracing. In particular, it is shown below that within this family we can obtain an image-based representation that is essentially a slightly generalized version of the discrete 4D light field [11, 20]. Another member of the family is an efficient alternative to the 3D volume buffer [13].

Comparison with light field rendering. If the light-field LDIs are of sufficiently high resolution, and if their corresponding directions sample the unit sphere densely enough, we obtain a discrete representation of the light field, similar to the 4D discrete light fields used in light field rendering [11, 20]. Recall that these 4D light fields require the viewpoint to lie in a free region of space, outside the region represented by the light field. The LLF removes this restriction by storing multiple layers per pixel, and by storing a depth value along with each scene sample.

The 4D light field stores only information describing the appearance of the scene from different directions, but it does not contain any explicit geometric information beyond that. This allows images of the scene to be rendered very simply and rapidly, but requires high directional and spatial resolutions to achieve high image quality, resulting in potentially enormous storage requirements. Gortler *et al.* [11] observe that by using approximate geometric information the quality of the reconstruction can be significantly improved. Our representation takes this idea even further: it relies on the LDC to accurately represent the scene geometry, and uses the LLF only to supplement the view-dependent shading where needed. Thus, we obtain high quality images using low directional and spatial resolution of the light field. The price we must pay for these advantages is somewhat slower rendering: a classical case of space-time trade-off.

Comparison with volume graphics. Our view-independent representation component resembles volumetric representations of synthetic scenes [13]. Indeed, the similarity between a projection-based representation and a volumetric one is not at all surprising, since it is easy to see how a volumetric representation can be constructed from the projection-based one (for example, such techniques are used in medical imaging for constructing volume data from CT scans). However, with the aid of 3D image warping, we can produce images of the scene directly from our projection-based representation, without ever constructing the full 3D volume buffer. The main differences between our representation and a full 3D volume buffer are:

1. Three $n \times n$ LDIs require less storage than a volume buffer of the same resolution, as long as the average number of depth layers per pixel is less than $n/3$. It seems reasonable to assume that for high-resolution discrete representations, the spatial resolution of the scene projections is considerably higher than the average depth complexity (i.e., a 512^2 LDI is likely to have average depth complexity much lower than $512/3$). Thus, the LDIs should be much more economical in terms of storage than a full 3D volume buffer, particularly at high resolutions.

2. The LDIs have infinite resolution along the projection axis of each LDI, whereas a volume buffer is not capable of resolving cases where two or more surfaces occupy the same voxel.

3. Similarly to image-based ray tracing, it is possible to render images of non-diffuse scenes by recursively tracing discrete rays in the 3D volume buffer [33]. In our representation, the LLF lets us eliminate the recursion altogether, or terminate it after a small number of reflections.

4 Computational elements for rendering

4.1 3D warping

Rendering begins by 3D warping the view-independent LDIs in order to generate a primary image of the scene. This stage determines which scene sample is visible at each pixel of the new image. We have implemented the fast LDI warping algorithm described by Gortler *et al.* [12]. There are, however, a few important differences between our implementation and theirs. Gortler *et al.* warp a single LDI to produce a new image. Using McMillan's ordering algorithm [25] they are guaranteed to project scene samples onto the new image plane in back-to-front order. Thus, there is no need to perform depth comparisons, and splatting can be used to fix holes due to stretching of the scene samples. In our case, all three LDIs must be warped to ensure that no surfaces are missed (since there might be surfaces in the scene that are only sampled by one of the three LDIs). Therefore, we do perform a depth comparison before writing each scene sample onto the new image. It should be noted that the cost of performing these comparisons is negligible. Also, since scene samples are not guaranteed to arrive in back-to-front order we can't use Gortler's simple splatting approach. Currently, in order to handle holes we first warp the LDIs, and then scan the resulting image looking for holes (pixels not covered by any scene samples and pixels whose depth is significantly larger than that of most of their neighbors). For each pixel suspected as a hole, we follow a ray from the eye to find the visible scene sample, using the image-based ray tracing algorithm of Section 4.3. In the examples shown in Section 5 there were less than two percent of such pixels. However, this problem clearly deserves a more robust solution, and we hope to address it in the future.

4.2 Light field gather

Next, we need to solve the following problem: given a point p in the scene, the directional BRDF (without the diffuse component) ρ_{bd}, and an outgoing direction $\vec{\omega}_o$, estimate the view-dependent component of the radiance reflected from p in direction $\vec{\omega}_o$. We solve this problem by using the LLF to reconstruct the light field at p and "pushing" it through the BRDF:

$$L_{out} = 0$$
$$\textbf{foreach direction } \vec{\omega}_i \text{ in the LLF}$$
$$\quad \textbf{if } \rho_{bd}(\vec{\omega}_i, \vec{\omega}_o) > 0$$
$$\qquad L_{in} = \text{radiance arriving at } p \text{ from } \vec{\omega}_i$$
$$\qquad L_{out} = L_{out} + \rho_{bd}(\vec{\omega}_i, \vec{\omega}_o) L_{in} \cos\theta_i \Delta\vec{\omega}_i$$
$$\quad \textbf{endif}$$
$$\textbf{endfor}$$

To reconstruct the radiance arriving at p from a direction in the LLF we project p onto the reference plane of the corresponding LDI, and establish the coordinates of the four nearest LDI pixels. At each of these pixels, we scan the list of scene samples until we find one visible from p (i.e., with depth greater than that of p), and retrieve the radiance stored in this sample. The requested radiance value is then obtained by bilinear interpolation of the four retrieved values.

The above routine works well for glossy surfaces, but is not well-defined for ideal specular reflectors, because their BRDF is only non-zero in the mirror-reflected direction. In this case, we find all of the directions in the LLF that fall within a small cone around $\vec{\omega}_i$ (the mirror-reflected direction). The opening angle of the cone depends on the directional density of the LLF: it is chosen large enough so that at least three directions would be found. Radiance values are retrieved from each of these directions as above and the radiance arriving at p from $\vec{\omega}_i$ is estimated by a normalized weighted average of these values. The weight for each LLF direction is $\cos^n \alpha$, where α is the angle between the LLF direction and $\vec{\omega}_i$, and n is a user-specified parameter.

4.3 Image-based ray tracing

For the sake of exposition, we begin by describing a rather naive algorithm, which will then be improved. Given a 3D ray and a parallel LDI, we first project the ray onto the reference plane of the LDI, and clip the resulting 2D ray to the square window containing the projection of the scene. We then visit *all* of the pixels in the LDI that are crossed by the 2D ray, starting from the origin. As we step from one pixel to the next, we keep track of the depth interval $[Z_{in}, Z_{out}]$ spanned by the 3D ray while inside each pixel. The depths of the scene samples in each visited pixel are compared against this interval. If a single sample is found whose depth lies inside $[Z_{in} - \epsilon, Z_{out} + \epsilon]$, we report this sample as a ray-surface intersection. If more than one depth value falls inside the depth interval, we report the value nearest to the ray's origin. The magnitude of ϵ depends on the resolution of the LDI and on the angle between the surface normal of the scene sample and the direction of projection. This operation must be performed for each of the three LDIs, because there might be surfaces that are only sampled by one of the LDIs, being parallel to the projection directions of the other two.

The problem with the simple routine described above is that it may have to visit up to $6n$ LDI pixels (at most $2n$ pixels in each of the three $n \times n$ LDIs). At each LDI pixel, it does $O(d)$ work, where d is the depth complexity there. As a result, this algorithm is too slow to be used with high-resolution LDIs. Fortunately, the algorithm can be sped up drastically by using hierarchical data structures. Consider the case of an LDI with a single depth layer — a regular range image. To speed up image-based ray tracing we construct a quadtree for this range image. An internal node in the quadtree stores the minimum and the maximum depth of all the pixels in the corresponding sub-region of the range image. A leaf node stores the depth of a single pixel. Given a 2D ray we can establish whether an intersection exists by a simple depth-first traversal of the quadtree. Whenever we encounter an internal node whose depth range is disjoint from the depth interval of the ray, the entire subtree can be safely skipped. If the depth intervals are not disjoint, we recursively traverse the node's children, visiting them in the same order as the ray. This algorithm is similar to one described by Cohen and Shaked [7] for speeding up ray tracing of terrain. This similarity is not surprising, since digital terrain data is a height field, just like a range image.

To extend the above algorithm to LDIs with multiple layers, think of an LDI as a stack of range images: The topmost layer is the "normal" image of the scene, after parallel projection and hidden surface removal. The second layer is an image showing only those parts of the scene occluded by exactly one surface, and so forth. We could simply construct a quadtree for each range image in the stack, and have each ray traverse each of these quadtrees independently. However, we can perform the traversal in a more efficient manner by taking advantage of the special structure possessed by successive depth layers of the same LDI. Note that a pixel in depth layer $i + 1$ is non-empty (contains a scene sample) only if the same pixel was non-empty in layers 1 through i. This means that if we encounter an empty node in the quadtree of layer i, there is no need to traverse the corresponding nodes in layers $i + 1$ and deeper. Thus, we begin by constructing a separate quadtree for each depth layer of the LDI, but then we create *cross links* between the quadtrees: each non-empty node in the quadtree of layer i points to its twin in the quadtree of layer $i + 1$. Now, given a ray, we start a recursive traversal from the root node of the first layer's quadtree. The traversal routine is given in the pseudocode on the right. As

```
Traverse(node, ray, Z_in, Z_out):
if Z_out < node.Z_min return
if Z_in > node.Z_max then
        Traverse(node.crosslink, ray, Z_in, Z_out)
        return
endif
if Leaf(node) then
        scan layers for intersections
else foreach non-empty child of node
        Update Z_in and Z_out
        Traverse(child, ray, Z_in, Z_out)
    endfor
endif
```

Table 1. Scene statistics and rendering times. All sizes are in megabytes, all timings are in seconds on a 200 MHz Pentium-Pro running Linux.

	sphere & cylinder	teapot	teapot & spheres
# primitives	8	9,121	9,179
LDC size	42.5	28.8	50.6
avg (max) # layers	0.97 (3)	0.53 (4)	1.23 (10)
LLF resolution	258×64^2	1026×64^2	66×64^2
LLF size	11.7	19.3	3.1
construction time	582	726	2420
warping time	2.9	1.3	3.5
view-dependent pass 0 / 1 / 7 reflections	2.7 / 4.3 / 4.5	3.2 / – / –	2.7 / 11.1 / –
total rendering time	5.6 / 7.2 / 7.5	4.5 / – / –	6.2 / 14.6 / –
Rayshade time	8.2	126.1	22.2

before, Z_{in} and Z_{out} is the depth interval spanned by the ray inside the node. Note that the non-empty children of a node should be visited in the order in which they are encountered by the ray.

5 Results

In this section we demonstrate our approach on several non-diffuse test scenes. We compare the resulting images to those generated by conventional ray tracing, which is a very accurate method for computing view-dependent reflections. In all our test cases we have outperformed ray tracing in terms of computing time. The visual quality of the reflections is comparable to that achieved by ray tracing, even on ideal specular reflectors. Furthermore, for glossy surfaces with fuzzy reflections we achieve superior quality along with a dramatic speedup.

We have extended Rayshade [14], a public domain ray-tracer, to generate parallel LDIs of non-diffuse synthetic 3D scenes. One version of the resulting program is used to generate the LDC, while another version computes the LLF. Our current implementation inserts into a particular LDI only scene samples that are front facing with respect to the parallel projection directions. Thus, the LDC actually consists of six LDIs (two for each of the three orthogonal axes), and not three LDIs as described earlier. We expect that using only three LDIs, each containing both front- and back-facing scene samples, would prove more economical both in terms of storage and in terms of processing times.

We have experimented with three test scenes. Various scene statistics and rendering times are summarized in Table 1.

Our first test scene contains a sphere and a cylinder with a mirror-like surface inside a box with diffuse walls and glossy (but not reflective) floor. The scene is illuminated from above by four point light sources. The average depth complexity for this scene is 0.97 (it is less than 1, because the box containing the scene is missing one wall and has no ceiling). The storage requirements for both components of our representation are 54.2 megabytes. Note that this statistic is the raw size of our non-optimized data structures. For example, we currently store the normal as three floating point numbers, whereas a more efficient implementation could have used two fixed point numbers, in addition to some form of geometric compression [8]. Still, note that these storage requirements are considerably lower than those of a 512^3 3D volume buffer, which would have required 128 megabytes, even if only one byte per voxel was used. A $32 \times 32 \times 256^2$ Lumigraph data structure requires 192 megabytes of raw storage, and would have taken about 8,400 seconds to compute for this scene, while our representation was constructed in less 582 seconds. Furthermore, recall that six such Lumigraph slabs are required to cover all viewing directions.

Figure 3 shows four 256×256 images of the same scene from an arbitrarily chosen viewpoint. Image (d) was rendered using Rayshade, while the other three were generated using our (precomputed) image-based representation. In all three images the six view-independent LDIs were first warped to produce a primary image without reflections, which were then added using three different techniques. In (a) view-dependent reflections were computed by performing a light-field gather (Section 4.2) at each reflective pixel, using a specular exponent $n = 40$. In (b), reflections were added by tracing one generation of reflected rays using the image-based ray tracing algorithm described in Section 4.3. In (c), reflections were added by recursive image-based ray tracing with up to seven generations of reflected rays. The LLF in all three images consists of 258 LDIs corresponding to uniformly distributed directions on the sphere. Since the directional resolution of the LLF is relatively low, when we use the light-field gather on ideal specular reflectors, the resulting reflections appear blurry (3a). Note, however, that the geometry of the scene is still sharp and accurate. This is in contrast to light field rendering techniques, where lowering the resolution of the discrete 4D light field results in overall blurring of the image. Tracing one generation of reflected rays drastically improves the accuracy of the reflections at a modest additional cost. Allowing deeper recursion further improves the recursive reflections visible in the image (the reflection of the sphere in the cylinder), and the resulting image (3c) is almost indistinguishable from the one generated by Rayshade (3d).

Note that for a scene as simple as this one (eight simple primitives), ordinary ray tracing is extremely fast: in fact, it took Rayshade only 8.2 seconds to produce the image in Figure 3(d). Still, even for this simple scene, our image-based rendering algorithm outperforms Rayshade.

It is interesting to examine the effect of the number of projections used in the LLF on the quality of the reflections. Figure 4 shows images that were generated using 66, 258, and 1026 directional samples in the LLF. As the directional resolution of the LLF increases, there is noticeable improvement in the accuracy of the reflections in the top row, which uses light-field gather to compute the view-dependent reflections. In the bottom row, which uses one level of image-based ray tracing, the effect is not nearly as drastic. These results illustrate that recursive image-based ray tracing effectively compensates for low directional resolution in the LLF.

The case of near-ideal mirror reflectors is particularly demanding, and, as just shown, such surfaces are better handled by using at least one level of image-based ray tracing. However, for rougher glossy surfaces, an LLF with reasonable directional resolution suffices to produce a good approximation to the exact image. Figure 5 shows a teapot on a glossy floor. The left image was produced by our image-based approach, using a light field gather at each reflective pixel. The right image was produces by Rayshade. Rayshade can only generate glossy reflections by supersampling. As a result, Rayshade takes twenty-eight times longer[3] than our method to generate the image, and the fuzzy reflection is still somewhat noisy. (Although, as a side-effect of the supersampling the Rayshade image is anti-aliased, while our image is not).

---

[3] Rayshade timing were measured using a uniform grid to accelerate ray-scene intersections.

Note that for this scene, the warping stage is roughly twice as fast as in the previous scene (see Table 1). The reason is that the LDIs in the teapot scene have roughly twice lower average depth complexity (see Table 1). The time of the view-dependent pass on the other hand is somewhat higher for this scene. The main reason is that there are more directions in the LLF, and our current implementation of the light-field gather simply loops over all of these directions, rather than employing a more sophisticated search scheme.

Finally, we tested our method on a scene with higher depth complexity, shown in Figure 6. The LDIs in our image-based representation of this scene have up to 10 depth layers per pixel. As expected, the total rendering time for our method is larger than for the simple scene in Figure 3: warping takes slightly longer because of increased depth complexity, and for the same reason image-based ray tracing takes longer in the view-dependent pass. Again, our approach has generated the image faster than Rayshade.

The results in this section demonstrate that our technique can be considered output-sensitive with respect to view-dependent shading: For diffuse scenes, the technique is, in principle, as efficient as previous reprojection methods. We pay an extra price only in non-diffuse scenes, and the price is proportional to the number of visible non-diffuse points. Furthermore, ideal mirror reflections take longer to produce than fuzzier glossy reflections.

6 Conclusions

We have introduced a new image-based representation for non-diffuse synthetic scenes, and have described efficient rendering algorithms that correctly handle view-dependent shading and reflections. Our scene representation is more space-efficient than previous discrete scene representations that support non-diffuse scenes, namely the 4D light field and the 3D volume buffer. Our results demonstrate that our approach is capable of producing view-dependent shading and reflections whose quality is comparable to that of conventional ray tracing, while outperforming ray tracing even on simple scenes. We believe that the performance gap between our technique and ray tracing will continue to grow as we further optimize and tune our implementation, and apply our techniques to more complex scenes.

The main disadvantage of our approach is that the size of our data structures and consequently the rendering time depend on the depth complexity of the scene. We should investigate ways for reducing the number of layers that must be represented in our data structure without affecting the accuracy of the resulting images.

There are many ways to further optimize and improve the techniques that we have described:

- In order to avoid warping all of the LDIs for each frame, we should investigate the possibility of pre-warping the LDIs to a single reference LDI once in several frames, similarly to Gortler et al.[12].[4]
- Geometric as well as entropy compression techniques for our representation should be developed. There is currently considerable redundancy in our representation, and we should explore ways of eliminating it.
- It would be useful to have an automatic way of determining a good combination of representation parameters (LDI resolutions, number and distribution of directions in the LLF) for a particular scene. Furthermore, we should explore an adaptive variant of our representation, rather than using uniform directional and spatial resolutions.

---

[4]Thanks to an anonymous reviewer for suggesting this idea.

- In order to ensure high visual accuracy of the rendered images we should investigate efficient and robust anti-aliasing techniques. This issue has not been resolved yet in the context of image-based rendering, and progress in this area could significantly contribute to other image-based rendering techniques as well.
- Similarly to most previous image-based representations, our representation is only effective for static scenes. A change in the scene (including motion of objects, changes in material properties, and in the positions and intensities of light sources) currently requires the entire representation to be recomputed. Efficiently updating the representation after a change in the scene is an interesting direction for future research.

Acknowledgements:. This research was supported in part by the Israel Science Foundation founded by the Israel Academy of Sciences and Humanities, and by an academic research grant from Intel Israel.

References

1. Stephen J. Adelson and Larry F. Hodges. Generating exact ray-traced animation frames by reprojection. *IEEE Computer Graphics and Applications*, 15(3):43–52, May 1995.
2. Sig Badt, Jr. Two algorithms taking advantage of temporal coherence in ray tracing. *The Visual Computer*, 4(3):123–132, September 1988.
3. M.O. Benouamer and D. Michelucci. Bridging the gap between CSG and Brep via a triple ray representation. In *Proc. Fourth ACM/Siggraph Symposium on Solid Modeling and Applications*, pages 68–79. ACM Press, 1997.
4. Shenchang Eric Chen. QuickTime VR — an image-based approach to virtual environment navigation. In *Computer Graphics Proceedings, Annual Conference Series (Proc. SIGGRAPH '95)*, pages 29–38, 1995.
5. Shenchang Eric Chen and Lance Williams. View interpolation for image synthesis. In *Computer Graphics Proceedings, Annual Conference Series (Proc. SIGGRAPH '93)*, pages 279–288, 1993.
6. Christine Chevrier. A view interpolation technique taking into account diffuse and specular inter-reflections. *The Visual Computer*, 13:330–341, 1997.
7. Daniel Cohen and Amit Shaked. Photo-realistic imaging of digital terrains. *Computer Graphics Forum*, 12(3):363–373, 1993. Proceedings of Eurographics '93.
8. Michael F. Deering. Geometry compression. In *Computer Graphics Proceedings, Annual Conference Series (Proc. SIGGRAPH '95)*, pages 13–20. Addison Wesley, August 1995.
9. Robert A. Drebin, Loren Carpenter, and Pat Hanrahan. Volume rendering. In John Dill, editor, *Computer Graphics (SIGGRAPH '88 Proceedings)*, volume 22, pages 65–74, August 1988.
10. J. L. Ellis, G. Kedem, T. C. Lyerly, D. G. Thielman, R. J. Marisa, J. P. Menon, and H. B. Voelcker. The ray casting engine and ray representations: a technical summary. *Internat. J. Comput. Geom. Appl.*, 1(4):347–380, 1991.
11. Steven J. Gortler, Radek Grzeszczuk, Richard Szeliski, and Michael F. Cohen. The Lumigraph. In *Computer Graphics Proceedings, Annual Conference Series (Proc. SIGGRAPH '96)*, pages 43–54, 1996.
12. Steven J. Gortler, Li-Wei He, and Michael F. Cohen. Rendering layered depth images. Technical Report MSTR-TR-97-09, Microsoft Research, Redmond, WA, March 1997.
13. Arie Kaufman, Daniel Cohen, and Roni Yagel. Volume graphics. *IEEE Computer*, 26(7):51–64, July 1993.
14. Craig E. Kolb. *Rayshade User's Guide and Reference Manual*, 1992. Available at http://www-graphics.stanford.edu/~cek/rayshade/doc/guide/guide.html.

15. Philippe Lacroute and Marc Levoy. Fast volume rendering using a shear–warp factorization of the viewing transformation. In *Computer Graphics Proceedings, Annual Conference Series (Proc. SIGGRAPH '94)*, pages 451–458, July 1994.

16. David Laur and Pat Hanrahan. Hierarchical splatting: A progressive refinement algorithm for volume rendering. In Thomas W. Sederberg, editor, *Computer Graphics (SIGGRAPH '91 Proceedings)*, volume 25, pages 285–288, July 1991.

17. Stephane Laveau and Olivier Faugeras. 3-D scene representation as a collection of images and fundamental matrices. In *Proceedings of the Twelfth International Conference on Pattern Recognition*, pages 689–691, Jerusalem, Israel, October 1994.

18. Marc Levoy. Display of surfaces from volume data. *IEEE Computer Graphics and Applications*, 8(3):29–37, May 1988.

19. Marc Levoy. Efficient ray tracing of volume data. *ACM Transactions on Graphics*, 9(3):245–261, July 1990.

20. Marc Levoy and Pat Hanrahan. Light field rendering. In *Computer Graphics Proceedings, Annual Conference Series (Proc. SIGGRAPH '96)*, pages 31–42, 1996.

21. William E. Lorensen and Harvey E. Cline. Marching cubes: A high resolution 3D surface construction algorithm. In Maureen C. Stone, editor, *Computer Graphics (SIGGRAPH '87 Proceedings)*, volume 21, pages 163–169, July 1987.

22. William R. Mark, Leonard McMillan, and Gary Bishop. Post-rendering 3D warping. In *Proceedings of the 1997 Symposium on Interactive 3D Graphics*. ACM SIGGRAPH, April 1997.

23. Nelson Max. Hierarchical rendering of trees from precomputed multi-layer Z-buffers. In Xavier Pueyo and Peter Schröder, editors, *Rendering Techniques '96*, pages 165–174. Eurographics, Springer-Verlag Wien New York, 1996. ISBN 3-211-82883-4.

24. Nelson Max, Curtis Mobley, Brett Keating, and En-Hua Wu. Plane-parallel radiance transport for global illumination in vegetation. In Julie Dorsey and Philipp Slusallek, editors, *Rendering Techniques '97*, pages 239–250. Eurographics, Springer-Verlag Wien New York, 1997. ISBN 3-211-83001-4.

25. Leonard McMillan. A list-priority rendering algorithm for redisplaying projected surfaces. UNC Technical Report 95-005, University of North Carolina, Chapel Hill, 1995.

26. Leonard McMillan and Gary Bishop. Plenoptic modeling: An image-based rendering system. In *Computer Graphics Proceedings, Annual Conference Series (Proc. SIGGRAPH '95)*, pages 39–46, 1995.

27. J. Menon, R.J. Marisa, and J. Zagajac. More powerful solid modeling through ray representations. *IEEE Computer Graphics and Applications*, 14(3):22–35, 1994.

28. Steven M. Seitz and C. R. Dyer. Physically-valid view synthesis by image interpolation. In *IEEE Computer Society Workshop: Representation of Visual Scenes*, pages 18–27, Los Alamitos, CA, June 1995. IEEE Computer Society Press.

29. Jonathan Shade, Dani Lischinski, David H. Salesin, Tone DeRose, and John Snyder. Hierarchical image caching for accelerated walkthroughs of complex environments. In *Computer Graphics Proceedings, Annual Conference Series (Proc. SIGGRAPH '96)*, pages 75–82, August 1996.

30. Tim Van Hook. Real-time shaded NC milling display. In David C. Evans and Russell J. Athay, editors, *Computer Graphics (SIGGRAPH '86 Proceedings)*, volume 20, pages 15–20, August 1986.

31. Sidney Wang and Arie E. Kaufman. Volume-sampled 3D modeling. *IEEE Computer Graphics and Applications*, 14(5):26–32, September 1994.

32. Lee Westover. Footprint evaluation for volume rendering. In Forest Baskett, editor, *Computer Graphics (SIGGRAPH '90 Proceedings)*, volume 24, pages 367–376, August 1990.

33. Roni Yagel, Daniel Cohen, and Arie Kaufman. Discrete ray tracing. *IEEE Computer Graphics and Applications*, 12(5):19–28, September 1992.

Editors' Note: see Appendix, p. 339 for colored figures of this paper

Photosurrealism

Anthony A. Apodaca

Director of Graphics R & D

Pixar Animation Studios

Abstract. The art of making movies has everything to do with creating the illusion of a world which doesn't actually exist. It is extremely common for various aspects of this world to be inconsistent, illogical and non-physical. This is called being *bigger than life*. When computer graphics imagery is used in movies, it must conform to these rules. This paper will examine how this requirement influences the design and use of programs for realistic image synthesis.

1 History

The first use of computer graphics in the production of a feature film was in 1976, when *Futureworld* included a computer-animated simulation of a robot's hand and face. These animations, created by then graduate-students Ed Catmull and Fred Parke at the University of Utah, started the computer graphics industry down the road which eventually led to the nearly ubiquitous use of computer graphics special effects in feature films.

The road was long, however. The first two films to make significant investments in computer generated imagery (CGI), Disney's *Tron* and Universal's *The Last Starfighter*, were commercial flops. These failures made directors wary. Generally, CGI was limited to imagery that was clearly supposed to be from a computer (fancy displays on computer monitors, for example). It wasn't until 1990 that CGI scored its first real success. *The Abyss* won the Academy Award for Best Visual Effects, partially on the strength of the photorealistic CGI effects produced by Industrial Light and Magic.

Since 1990, computer graphics has been used to create significant special effects for every film which has won the Academy Award for Best Visual Effects. Many Scientific and Technical Academy Awards have been awarded to the authors of computer graphics software, recognizing their contributions to the film industry. The first full-length completely computer-generated feature film, Pixar's *Toy Story*, was released in 1995, and was a huge commercial success. Presently, nearly all films which have any visual effects at all will have at least some effects which are computer generated. Most movie special effects houses in the U.S. have largely switched over from "practical" effects to CGI-based ones. At least 12 more full-length computer-generated feature films are in progress at this time (the summer of 1998). Computer graphics has carried the day.

2 Making Movies

For more than 20 years, one of the stated goals of the computer graphics research community has been to solve the problem of making truly *photorealistic* images. We have strived to make the CG image, as much as possible, look exactly like a photograph of the scene, were that virtual scene to actually exist. Much of this work was done very specifically so that CGI could be used in films. Problems such as accurate light reflection models, motion blur, depth of field, and handling of massive visual complexity, were all motivated by the very demanding requirements of the film industry.

Movies, however, are illusions. They show us a world which does not exist, and use that world to tell a story. The goal of the filmmaker is to draw the audience into this world, and to get the audience to *suspend their disbelief* and watch the story unfold. Every element of the movie — the dialog, costumes and makeup, sets, lighting, sound and visual effects, music, etc. — are all there to support the story, and must help lead the viewer from one story point to the next. In order for the audience to understand and believe the movie's world, it clearly must be realistic. It cannot have arbitrary, nonsensical rules, or else it will jar and confuse the viewers and make them drop out of the experience. However, movie realism and physical realism are two different things. In a movie, it is realistic for a 300' tall radioactive hermaphroditic lizard to destroy New York City, as long as its skin looks like lizard skin we are familiar with. Some things we are willing to suspend our disbelief for, and some things we are not!

There are two people who are primarily responsible for determining what the audience sees when they watch a movie. The director is responsible for telling the story. He determines what message (or "story point") the audience should receive from every scene and every moment of the film. The cinematographer (cinema photographer) is responsible for ensuring that the photography of the film clearly portrays that message.

Over time, filmmakers have developed a visual language which allows them to express their stories unambiguously. This language helps the filmmaker to manipulate the imagery of the film in subtle but important ways, to focus the audience's attention on the story points. Distracting or confusing details, no matter how realistic, are summarily removed. Accenting and focusing details, even if physically unrealistic, are added in order to subtly but powerfully keep the audience's attention focused on what the director wants them to watch. The job and the art of the cinematographer is to arrange that this is true in every shot of the film. The result is that live-action footage does not look like real-life photography. It is distilled, focused, accented, bolder and *bigger than life*. In short, film images manipulate reality so that it better serves the story.

Our computer graphics images must, therefore, also do so. Perhaps even more than other parts of the filmmaking process, CGI special effects are there specifically to make a story point. The action is staged, lit and timed so that the audience is guaranteed to see exactly what the director wants them to see. When a CG special effect is added to a shot, the perceived realism of the effect is more influenced by how well it blends with the existing live-action footage than by the photorealism of the element itself. What the director really wants are images which match as closely as possible the other not-quite-real images in the film. They are based in realism, but they bend and shape reality to the will of the director. Computer graphics imagery must be *photosurrealistic*.

3 Altered Reality

In a CG production studio, the job of the cinematographer is often divided among several CG artists whose jobs include geometric modeling, camera placement, material modeling, lighting and renderer control. These digital cinematographers are collectively known as technical directors. They are the users of image synthesis software.

Computer graphics gives the technical directors a whole host of new tools, removed restrictions and an interesting and growing bag of tricks with which to manipulate reality. Jim Blinn once called these tricks "The Ancient Chinese Art of Chi-Ting"[3]. Technical directors now call them "getting work done".

3.1 Nonphysical lighting

One of the best cheats is to alter the physics of light. Barzel[1] discussed a light model used in computer graphics cinematography which was specifically designed for this type of flexibility. This work is based on the observation that the movie viewer generally does not know where the lights are located in a scene, and even if he does, his limited 2-dimensional view does not allow him to reason accurately about light paths. Therefore, there is no requirement that the light paths behave realistically in order to ensure that an image is believable.

This situation is exploited by the live-action cinematographer in his placement of special-purpose lights to illuminate objects unevenly, kill shadows, create additional specular highlights and otherwise fudge the lighting for artistic purposes. Everyone who has ever seen an image of a movie set knows that even in the middle of the day, every shoot has a large array of artificial lights which are used for a variety of such purposes.

However, the live-action cinematographer, despite his best tricks, cannot alter the physics of the light sources he is using. Computer graphics can do better. Creating objects which cast no shadow, having lights illuminate some objects and not others, and putting invisible light sources into the scene near objects are all de rigueur parlor tricks. Things get more interesting when even more liberties are taken. Consider, for example:

- a light which has negative intensity, and thus dims the scene;
- a light which casts light brightly into its illuminated regions, but into its "shadowed" regions casts a dim, blue-tinted light;
- a spotlight which has not simply two-dimensional barndoors at its source to control its shape, but instead has full three-dimensional volumetric control over the areas which it illuminates and does not, and at what intensities;
- a light which contributes only to the diffuse component of a surface's BRDF, or only to the specular component;
- a light which has independent control over the direction of the outgoing beam, the direction toward which the specular highlight occurs, and the direction that shadows cast from the light should fall.

In fact, such modifications to light physics in CGI are so common that the production community does not think of them as particularly special or noteworthy tricks. They are standard, required features of their production renderers. Other, even more nonphysical cheats, such as curved light rays, photons which change color based on what they hit, lights which alter the physical characteristics of the objects they illuminate, or even lights which alter the propagation of other lights' photons, are all plausible requirements in the production setting. Their use is only limited by their predictability. If the light model is so nonphysical that the technical directors cannot reason accurately about the lighting situation, cannot form a mental model with which to predict what their final rendered images will look like, and cannot get the renderer to emit the image they desire, only then will they not desire a feature.

This is exactly the situation with purely physical photon simulation as well. Remember, the goal of lighting a scene is primarily to accent the action, and secondarily to make a beautiful artistic setting within which to stage the action. Energy conservation, truly accurate diffuse and specular interreflection and other such goals are useful only in so far as they give the technical director a firmer, more intuitive mental model with which to manipulate the photons. Even if perfect, they will only leave him with the same situation that the live-action cinematographer has, with lights that he must finesse and sometimes fight to give him the image he desires.

3.2 Nonphysical optics

Another interesting cheat is to alter the physics of optics, or more generally all reflection and refraction, to create images with the appropriate emphasis and impact.

Much of the trickery involving manipulating optical paths is descended, at least in part, from limitations of early rendering systems, and the techniques that were invented to overcome these limitations. For example, for over 20 years, most CGI was created without the benefit of ray tracing to determine the specular interreflection of mirrored and transparent objects. Even today, ray tracing is generally considered by the production community to be computationally too expensive to use except in particular extreme circumstances. However, mirrored and transparent surfaces still exist, and ways must be found to render them.

As before, the limited knowledge of the viewer comes to our rescue. Since the viewer cannot estimate accurately the true paths of interreflection in the scene, approximations generally suffice. In place of exact reflection calculations, texture maps are used. Environment maps are accurate only from a single point in space (the center of their projection), but work on all surface types and in all directions; planar reflection maps are accurate everywhere on a planar surface, but only when viewed from a particular camera location. Use of environment maps "near" the projection point, and use of planar reflection maps on "almost" flat objects lead to slight imperfections. Refraction is handled similarly, using an environment map indexed by the refracted ray direction, and here the fudging of object thickness causes the largest imperfections.

Experience has shown that viewers have so little intuition about the reflection situation, that almost any approximation will do. Only glaring artifacts, such as constant colored reflections, or the failure of mirrors to reflect objects that touch them, will generally be noticed. In fact, experience has shown that viewers have so very, very little intuition about the reflection situation that wildly and even comicly inaccurate reflections are perfectly acceptable. For example, it is extremely common for production studios to have a small portfolio of "standard reflection images" (such as someone's house, or a distorted photo of someone's face) which they apply to all objects that are not perfect mirrors.

This insensitivity to nonphysical optics gives rise to possibilities for story telling. The objects that are actually visible in a reflection, the intensity and blurriness of the reflection of each object, and the size and location of the reflection of each object, are all generally controlled independently. In *Toy Story*[6], for example, Buzz Lightyear's plastic helmet reflected only three objects of note:

- a prerendered image of the room he was in, quite blurry;
- Woody's face when he was near, sharply;
- the TV set during the commercial, very brightly and sharply.

Just enough blurry room reflection was present to give Buzz the appearance of being in the room, and to remind the viewer that Buzz' helmet was down, but it is only seen near the periphery, so that it never obscured Buzz' face and lip movements. A reflection of Woody was used when the director wanted to emphasize that they were close together, and their fates were intertwined. The key reflection of the TV commercial in the helmet was carefully crafted to be large, bright and sharp, so that the audience could easily watch it, while not obscuring Buzz' crucial facial expression at discovering the truth of his origins (which, of course, marked the turning point of the film).

Even more nonphysically, viewers are generally unaware of the difference between a convex and a concave mirror, to the director's delight. Convex mirrors are generally more common in the world, such as automobile windshields, computer monitors, metal cans, pens, etc. However, mirrors which magnify are generally more interesting objects cinematographically. They help the director focus attention on some small object, or small detail, onto which it would otherwise be difficult to direct the viewers attention. Unfortunately for the live-action cinematographer, magnifying mirrors are concave, and are extremely difficult to focus. But this is no problem in CGI. Because the reflection is generally a texture map anyway (and even if it is not, the reflected rays are under the control of the technical director), it is quite simple (and even quite common) for convex mirrors to have magnifications that are not merely arbitrarily independent of their curvature, but actually magnifying — completely inverted from reality.

Similarly, refractions are extremely difficult for viewers to intuit, and they generally accept almost anything as long as it is vaguely similar to the background, distorted, and more so near the silhouette of the refracting object. They typically are blissfully unaware of when a real refraction would generate an inverted image, and are certainly unable to discern if the refraction direction is precisely in the right direction. For example, in Pixar's short film *Geri's Game*[5], there is a shot where the director asked for Geri to appear innocent and angelic, after having just tricked his opponent into losing. This is accomplished by having Geri's eyes magnified large by his glasses, with pupils widely dilated. Even careful inspection of the frame by trained professionals fails to expose the fact that the image that we would see through Geri's glasses from this camera angle would not magnify nor frame his eyes this way[1]. The refraction is actually zoomed, reoriented and slightly rotated in order to give the director the image he desired.

4 Production Requirements

Once a production studio has made the choice to use CGI, for an effect or an entire film, it starts to make very specific demands on the rendering software developers. Films are very expensive to make, and a lot is riding on the studio's success in using the renderer. Failure is not an option when tens of millions of dollars are on the line.

4.1 Feature Set

The most obvious production requirement is feature set. The studio requires that the renderer do everything they could ever want or need. The features that are required to do a particular film are obviously different from one film to the next, and one year to the next, but there are certain constants.

For example, a production renderer must handle motion-blur. One of the easiest ways for a viewer to detect that an object was filmed independently and added to the scene (*i.e.*, an effect) is for the object to strobe when the background blurs. Audiences simply will not accept a return to the days of stop motion.

Production renderers must accept a wide variety of input geometry. It is not enough for the rendering algorithm to have just polygons. NURBs and other parametric curved surfaces are standard today, and various fads such as primitives for rendering hair, implicit surfaces and other special-purpose models must be handled in a timely fashion.

Production renderers must have a wide variety of tricks available to the technical director. Everything from shadows, texture mapping of all varieties, atmospheric and volumetric participating media effects, depth-of-field, lens flare, and control information for post-rendering compositing and manipulation of images must be provided for.

---

[1]Never mind the fact that he actually has flat lenses.

4.2 Flexibility

A related, yet perhaps even more important requirement is flexibility and controllability. The studio requires that the renderer be user-programmable so that it can be made to do everything that no one ever knew they'd need.

For example, every production renderer now contains some sort of shading language [4], with which the users can redefine the shading model in almost arbitrary ways. It is hard to overstate how critical this one feature of a renderer is to a production environment. Technical directors will not accept the standard Phong shading model on all objects, no matter how many knobs and dials it has. As we have seen, they must be able, on a detailed object-by-object basis, to change surface properties and material characteristics, reactivity to light, and even the physics of light and optics itself, to generate the compelling photosurrealistic image for each shot. The director of a film generally has very little sympathy for limitations of the software environment. He wants his image to look a certain way, and the technical directors must find a way to make it happen.

The best proof of flexibility is when a renderer can be used for purposes not envisioned by the software developer. For example, "photorealistic" renderers are now commonly used in traditional cel-animation studios to create distinctly non-photorealistic imagery which blends with artwork. Cartoon-realism is very distinct, but easily accommodated if the rendering architecture is flexible enough.

Similarly, programmability in other parts of the rendering pipeline is becoming more important. Procedurally-defined geometric primitives are coming back into style, as data amplification is exploited to increase visual complexity of images with minimal human work. Reprogrammable hidden surface algorithms, and easy access to other renderer "internal state" is also valuable.

A production renderer also has a powerful set of speed/quality and speed/memory tradeoffs, which are intuitive and easily controllable. Production deadlines often require that compromises be made, and the variety of hardware to which the studios have access (often some of several different generations of hardware) make it difficult to effectively utilize all available resources when only the "factory settings" available.

4.3 Robustness

A key requirement for any large software package is robustness and reliability. A full-length computer-animated feature film has over 100,000 frames to render (and probably many times that number after counting test frames, rerenders for artistic changes, separate elements for compositing, and rendered texture maps), so the technical directors cannot slave over each one individually. A production renderer absolutely cannot crash, must have vanishingly few bugs, and must reliably generate the correct image each and every time it is executed. If the renderer could guess the intentions of the user even when he gave it the wrong input, that would be a requirement, too.

The renderer should, for all practical purposes, have no limits. Arbitrary limits on number of lights, number of texture maps, number of primitives, image resolution, etc. will all be exceeded by production studios in short order.

Moreover, the feature set of the renderer should be orthogonal. If the renderer can do motion-blur on spheres, and a new primitive type is implemented, production will require motion-blur on the new primitive on day 2. Every feature must interoperate with every other feature, unless it is clearly nonsensical to do so, and perhaps even then.

4.4 Performance

Another obvious production requirement is performance. Those 100,000 frames have to be computed in a matter of months on machines that the production studio can afford to buy. Movies cost a lot to make, but in fact, very little of that money is actually allocated to the computer hardware necessary to make the effects. This means that the time to render a frame must be reasonable, the memory used to render the frame must be limited, and the database which holds the scene description must be compact. For this reason alone, global illumination is rarely used in production rendering, because it cannot generally fulfill these requirements.

And the software developers must not count on Moore's Law to solve this problem. While it is true that computers get faster and memory gets cheaper, seemingly without end, there is a countermanding observation that I call Blinn's Law, which states that regardless of hardware, all renderings take the same amount of time. As computers get faster, all that happens is that the appetites of the users gets greater, and their requirements get more demanding. In some studios, the appetite even grows faster than Moore allows!

Another aspect of performance which is often overlooked is predictability. A technical director should be able to guess, with some accuracy, how long a frame will take to render based on information that he knows (such as the number of objects in the scene, resolution, etc.). Frames which take non-linearly more time because a small thing was added are unacceptable when the studio is trying to make budgets, allocate resources and complete work by totally inflexible delivery dates.

4.5 Image Quality

Perhaps the most subtle, and yet in many ways the most important production requirement is image quality. Production studios require images of the highest possible quality. Those images are going to be projected onto the *big screen*, and every bad pixel will be huge and ugly in front of millions of viewers. Artifacts of any and all varieties are to be eliminated with extreme prejudice. High quality antialiasing in all dimensions, texture and shader filtering, pixel filtering, and accurate geometric approximations are all required features. Polygonal approximation artifacts, patch cracks, intersection mistakes, shadow acne, visible dithering patterns, Mach bands and any other image generation errors are unacceptable.

Since these images are going to be used in animations, frame-to-frame coherence is an important but subtle point. Objects which don't move shouldn't shimmer, and objects which move slowly shouldn't pop. Antialiasing of shading, lighting and geometry must work consistently at all scales and all distances from the camera.

5 Conclusion

In the movie making community, the users of the rendering software are not trying to simulate reality. Rather, they are trying to create an illusionary world, where the eye of the viewer is directed towards the details of the world which advance the story. Just as in live-action cinematography, distracting or confusing details, no matter how physically realistic, are removed. Accenting and focusing details, even if surrealistic, are added. The CG images must subtly but powerfully keep the audience's attention focused on what the director wants them to see. The job and the art of the technical directors is to arrange that this is true in every CGI shot in the film.

The tools that these digital cinematographers use are nontheless photorealistic image synthesis programs, but ones that are carefully designed to solve their problems. Computer graphics is now a mature industry, and advanced rendering techniques have become decreasingly interesting to the research community. This is a shame because there are many important problems yet to be solved. For example,

- **Natural lighting**: The current techniques for cinematographic lighting are powerful, but are ad hoc and difficult for non-experts to control. Meanwhile, what we really want is to be able to quickly reach "natural lighting" as a starting point for our excursions into cinematographic lighting. We want the power of radiosity, but without the box it places us in (pun intended).
- **Vast complexity**: Even the best of the current production renderers fail when geometric complexity reaches tens of millions of objects, and even with intelligent caching, global illumination renderers are limited to orders of magnitude less complexity. Yet look out the window. The real world is a lot more complex than that. We need to render that.
- **Scale independence**: The film camera sees objects in extreme closeups and from miles away. For decades researchers have noticed that there should be smooth scale transitions in geometric levels of detail, as well as in representations of surface details as geometry, texture, and BRDFs[2]. Promising work in these areas have not been followed up.
- **Unrenderable objects**: There are many objects for which truly realistic CG renditions just simply have never been made. Fire and pyrotechnics are one. There has not yet been a totally convincing CG explosion, bonfire or even a match. Realistic humans and their clothing has been an important goal of computer graphics almost from the beginning, but no one has yet passed that Turing Test.

Rather than focusing on more and more esoteric and compute-intensive ways to generate ever more subtle and unusual details in rendered images, I encourage the community to become energized by the use of their technology in the filmmaking industry, and focus on solving the continuously varied problems that come from practical application of our ideas.

- Accommodate the needs of the high-end users, by providing new techniques that are applicable to the production environment;
- Eschew algorithmic solutions which are not general;
- Benchmark with scenes that contain massive visual complexity;
- Never show any image which is low-resolution or is not antialiased;
- Embrace photosurrealism.

References

1. Barzel, Ronen: Lighting Controls for Computer Cinematography. *Journal of Graphics Tools*, 2(1), pp. 1-20, 1997.
2. Becker, Barry, Nelson Max: Smooth Transitions between Bump Rendering Algorithms. *SIGGRAPH 1993 Conference Proceedings*, pp. 183-198, 1993.
3. Blinn, Jim: The Ancient Chinese Art of Chi-Ting. *SIGGRAPH 1985 Course 12: Image Rendering Tricks*, 1985.
4. Lawson, Jim, Pat Hanrahan: A Language for Shading and Lighting Calculations. *Computer Graphics*, 24(4), pp. 289-298, 1990.
5. Pixar Animation Studios: *Geri's Game*, 1997.
6. Pixar Animation Studios: *Toy Story*, 1995.

Appendix: Colour images

Nishita

Fig. 2 (top row, left). Example of sky color taking into account multiple scattering. **Fig. 3** (top row, middle). Example of the earth taking into account light scattering. **Fig. 5** (top row, right). Example of clouds taking into account multiple scattering. **Fig. 6** (bottom row, left). Example of snow scene taking into account multiple scattering. **Fig. 8** (bottom row, right). Example of the optical effects within water.

Gargan and Neelamkavil

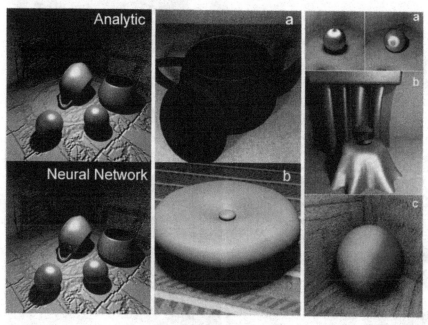

Fig. 11. Scenes rendered with both analytic and neural network models

Fig. 12. Scenes rendered from measured data. **a** Velvet teapot, **b** leather cushion

Fig. 13. BRDFs generated from virtual goniometric simulations of **a** brushed metal, **b** rough fibre, **c** red and green flakes

Tobler et al.

Fig. 9. Examples using the power function as *pdf.*

Zhukov et al.

Actual screenshots from the game.

Kobbelt et al.

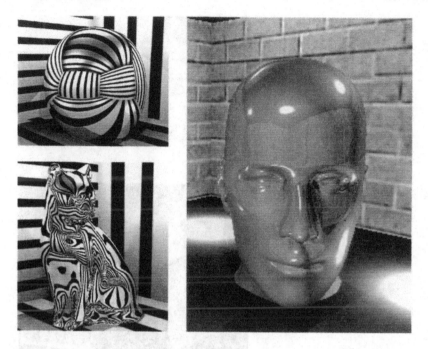

Rushmeier et al.

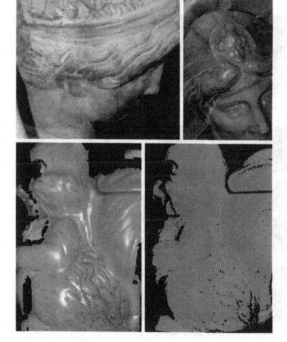

Fig. 2

Fig. 11

Poulin et al.

Fig. 6. Two scenes with extracted geometry and textures, displayed from synthetic cameras in arbitrary positions.

Camahort et al.

Plate 1. Uniformly sampled light fields

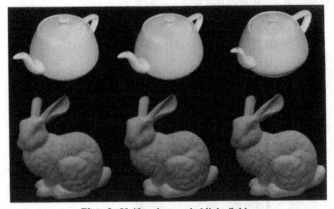

Plate 2. Uniformly sampled light fields

Debevec et al.

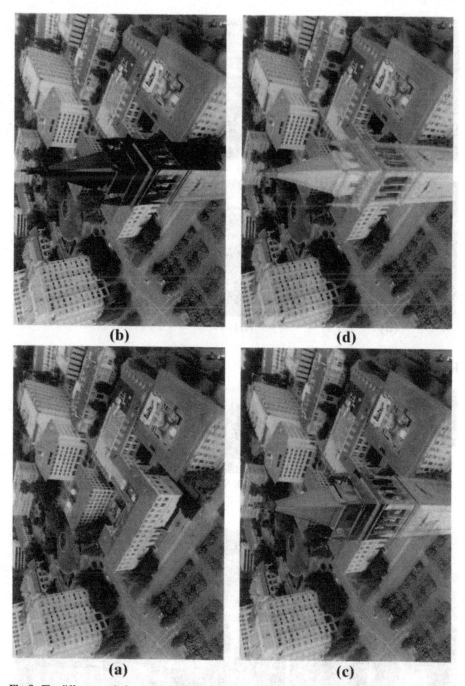

Fig. 8. The different rendering passes in producing a frame from the photorealistic renderings of the Berkeley campus virtual fly-by. The complete model contains approximately 100,000 triangles. **a** The campus buildings and terrain; these areas were seen from only one viewpoint and are thus rendered before the VDTM passes. **b** The Berkeley tower after the first pass of view-dependent texture mapping. **c** The Berkeley tower after the second pass of view-dependent texture mapping. **d** The complete rendering of the scene.

Zhang

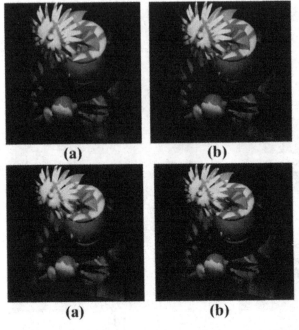

(a) **(b)**

Fig. 3. Shadows for these 512 ×512 final images are produced by forward shadow mapping using a 512×512 shadow map. The projective texture is the same as in Fig. 2. The modulation image is blurred for **a** and not blurred for **b**.

(a) **(b)**

Fig. 4. Shadows for these 512 ×512 final images are produced by forward shadow mapping using a 256×256 shadow map. The projective texture is the same as in Fig. 2. The modulation image is blurred for **a** and not blurred for **b**.

Raskar et al.

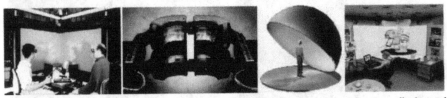

Fig. 1. UNC Protein Interactive Theater – section 5.1, Kaiser tiled HMD – section 5.2, a dome display, and the office of the future – section 5.3.

Fig. 3. Composite view for one eye (left) and derived views for 6 tiles (right).

Meyer and Neyret

Fig. 15. Mapping of 96 texels on a sphere.

Fig. 16. Mapping of 100 texels on a jittered plane made of 200 triangles.

Grossman and Dally

Point Sample Rendering. Four images generated at 256×256 pixels with no shadows and a single directional light source.

Dischler

Plate 1. Limits of the "usual" fast texturing techniques. **a** Usual and view dependent 2D texture mapping. **b** Bump mapping cannot render efficiently complex geometric surface structures. **c** shows the result that we obtained using a BTF.

Plate 2. Mapping a complex geometry onto spheres using the "virtual" ray tracing technique: left: the map's world, right: "virtual" ray tracing.

Plate 3. Some examples of textures rendered using BTFs.

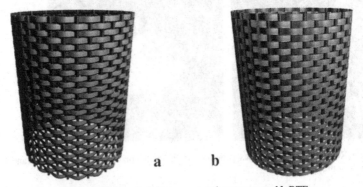

a   b

Plate 4. Comparison between **a** real geometry and **b** BTF.

Miller and Mondesir

Fig. 3. Metal and diffuse sphere with **a** 1 sample per pixel. **b** 4 samples per pixel.

Fig. 4. Embossed text in metal and glass.

Fig. 5. Hyper-sprites of **a** a dropper, **b** a dropper and magnifying glass.

Soler and Sillion

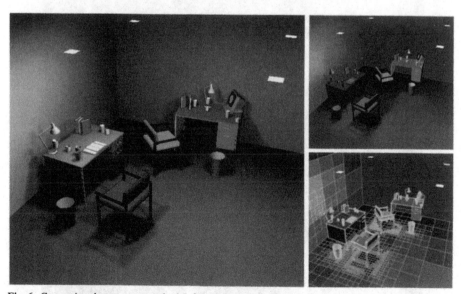

Fig. 6. Comparison between our method (left) and a standard HR algorithm (right) for equivalent computation times (20 sec, 5380 input polygons).

Stamminger et al.

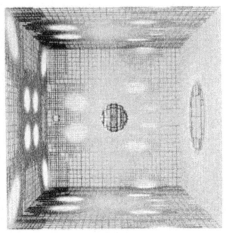

A "disco" sphere decorated with 72 small mirrors is illuminated by a small area light source from the left. The small mirrors produce light spots at the walls. The figure shows the solution obtained with three-point clustering after 800s (left) and the mesh produced (right).

Szirmay-Kalos and Purgathofer

Fig. 1. A fractal terrain illuminated by sky-light and spherical light sources.

Myszkowski

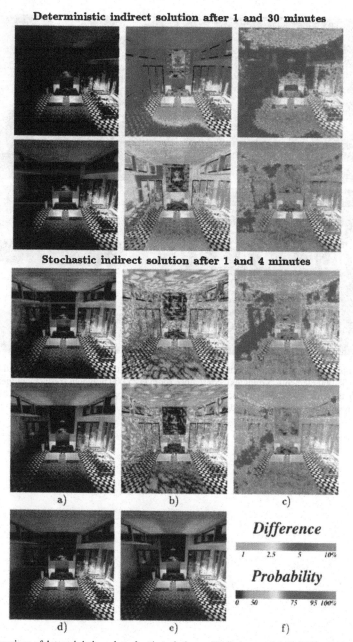

Fig. 4. Comparison of deterministic and stochastic techniques: While images labeled (**d**) and (**e**) are provided for reference, for the HPR and MCPT algorithms, respectively, columns (**a**), (**b**), and (**c**) provide for comparison of different stages in "perceptual" convergence of the indirect lighting simulation: **a** current solution, **b** absolute, and **c** predicted by the VDP differences in respect to the fully converged solutions (**d**) and (**e**). Color scales for encoding the differences in column (**b**) and probabilities in column (**c**) are shown in (**f**).

336

Fig. 5. Complex scenes rendered with hierarchical Monte Carlo radiosity: theatre (39,000 initial polygons, refined into 88,000 elements, 5 minutes), conference room (125,000 polygons, refinement yields 178,000 elements, 9 minutes) and cubicle office space (128,000 polygons, refined into 506,000 elements, 10 minutes). Model credits: see acknowledgements in paper.

Peter and Pietrek

Fig. 5. Images rendered using the photon maps shown in Fig. 3.
a Importance driven, **b** without importance.

Fig. 6. Complex scene rendered using both types of photon maps.
a Importance driven, **b** without importance.

Heidrich et al.

Fig. 6. Office scene rendered with two different Canned Lightsources based on different lamp sizes. The blurring due to nearfield effects is very prominent in both images.

Miller et al.

Fig. 2. A reflective floor as **a** U, V slice of a light field, **b** uncompressed model view.

Fig. 3a. Silver bottle light field⁻ U, V slice.

Fig. 3b. A sliver bottle with reflections.

Fig. 4. Reflective floor compressed **a** 36 to 1, **b** 24 to 1.

Fig. 5. A glass cube compressed 16 to 1.

Fig. 6. A fancy cube **a** uncompressed, **b** compressed 17 to 1.

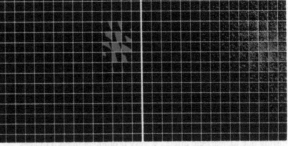

Fig. 7. Water ripples as a bump map.

Fig. 8. Block cache map – green means already in cache, red is not in cache for **a** smooth floor, **b** bump-mapped floor.

Lischinski and Rappoport

Fig. 3. From left to right: **a** warp + light-field gather; **b** warp + one level of reflection + light-field gather; **c** warp + seven levels of reflections; **d** Rayshade.

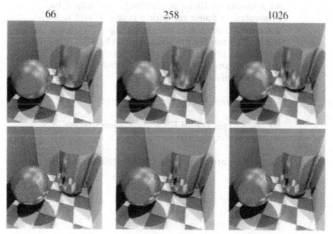

Fig. 4. The effect of the LLF directional resolution on reflection quality. Top row: warp + light-field gather; bottom row: warp + one level of reflection + light-field gather.

Fig. 5. Glossy reflections: left: our method; right: Rayshade (using supersampling).

Fig. 6. Higher depth complexity: right: our method; left: Rayshade.

SpringerEurographics

Dirk Bartz (ed.)

Visualization in Scientific Computing '98

Proceedings of the Eurographics Workshop in Blaubeuren, Germany, April 20–22, 1998

1998. 82 figures. VII, 151 pages.
Soft cover DM 85,–, öS 595,–
ISBN 3-211-83209-2
Eurographics

In twelve selected papers common problems in scientific visualization are discussed: adaptive and multi-resolution methods, feature extraction, flow visualization, and visualization quality.
Four papers focus on aspects of mesh reduction, mesh compression, and increasing the quality of the resulting mesh. Two extentions on particle tracing are presented as well as a paper on the simulation of material transport. Two papers are on feature extraction in dynamics systems and on the accuracy of algorithmic extracted features. Three papers focus on stereoscopic volume rendering, on the visualization of atomic collision cascades and of quality of visualization systems in general.

SpringerWienNewYork

Sachsenplatz 4-6, P.O.Box 89, A-1201 Wien, Fax +43-1-330 24 26
e-mail: order@springer.at, Internet: http://www.springer.at
New York, NY 10010, 175 Fifth Avenue • D-14197 Berlin, Heidelberger Platz 3
Tokyo 113, 3-13, Hongo 3-chome, Bunkyo-ku

SpringerEurographics

Panos Markopoulos,
Peter Johnson (eds.)

Design, Specification and Verification
of Interactive Systems '98

Proceedings of the Eurographics Workshop in Abingdon,
U.K., June 3–5, 1998

1998. Numerous figures. Approx. 310 pages.
Soft cover approx. DM 118,–, öS 826,–
ISBN 3-211-83212-2
Eurographics

Does modelling, formal or otherwise, have a role to play in designing interactive systems? A proliferation of interactive devices and technologies are used in an ever increasing diversity of contexts and combinations in professional and every-day life. This development poses a significant challenge to modelling approaches used for the design of interactive systems.

The papers in this volume discuss a range of modelling approaches, the representations they use, the strengths and weaknesses of their associated specification and analysis techniques and their role in supporting the design of interactive systems.

 SpringerWienNewYork

Sachsenplatz 4-6, P.O.Box 89, A-1201 Wien, Fax +43-1-330 24 26
e-mail: order@springer.at, Internet: http://www.springer.at
New York, NY 10010, 175 Fifth Avenue • D-14197 Berlin, Heidelberger Platz 3
Tokyo 113, 3-13, Hongo 3-chome, Bunkyo-ku